T0295259

Nanoelectronic Devices for Hardware and Software Security

Security, Privacy, and Trust in Mobile Communications

Series Editor: Brij B. Gupta

This series will present emerging aspects of the mobile communication landscape, and focuses on the security, privacy, and trust issues in mobile communication-based applications. It brings state-of-the-art subject matter for dealing with the issues associated with mobile and wireless networks. This series is targeted for researchers, students, academicians, and business professions in the field.

Nanoelectronic Devices for Hardware and Software Security
Balwinder Raj & Arun Kumar Singh

Cross-Site Scripting Attacks: Classification, Attack, and Countermeasures
B. B. Gupta & Pooja Chaudhary

Smart Card Security: Applications, Attacks, and Countermeasures
B. B. Gupta & Megha Quamara

Computer and Cyber Security: Principles, Algorithm, Applications, and Perspectives
Brij B. Gupta

For more information about this series, please visit: https://www.routledge.com/Security-Privacy-and-Trust-in-Mobile-Communications/book-series/SPTMOBILE

Nanoelectronic Devices for Hardware and Software Security

Edited by
Balwinder Raj and
Arun Kumar Singh

CRC Press
Taylor & Francis Group
Boca Raton London New York

CRC Press is an imprint of the
Taylor & Francis Group, an **Informa** business

First edition published 2022
by CRC Press
6000 Broken Sound Parkway NW, Suite 300, Boca Raton, FL 33487–2742

and by CRC Press
2 Park Square, Milton Park, Abingdon, Oxon, OX14 4RN

CRC Press is an imprint of Taylor & Francis Group, LLC

Trademark notice: Product or corporate names may be trademarks or registered trademarks and are used only for identification and explanation without intent to infringe.

ISBN: 978-0-367-64542-7 (hbk)
ISBN: 978-1-032-11695-2 (pbk)
ISBN: 978-1-003-12664-5 (ebk)

DOI: 10.1201/9781003126645

Typeset in Times
by Apex CoVantage, LLC

Contents

Editors

Dr. Balwinder Raj (MIEEE'2006) has been an Associate Professor at National Institute of Technical Teachers Training and Research, Chandigarh, India, since December 2019. Earlier, he worked at National Institute of Technology (NIT, Jalandhar), Punjab, India from May 2012 to December 2019. Dr. Raj also worked as Assistant Professor at ABV-IIITM Gwalior (an autonomous institute established by Ministry of HRD, Govt. of India) July 2011 to April 2012. He received Best Teacher Award from Indian Society for Technical Education (ISTE) New Delhi on 26th July 2013. His areas of research interest are classical/ non-classical nanoscale semiconductor device modeling; nanoelectronics and their applications in hardware security, sensors and circuit design, FinFET-based memory design, low power VLSI design, digital/ analog VLSI design and FPGA implementation.

Dr. Arun Kumar Singh is an Associate Professor in the Department of Electronics and Communication Engineering, Punjab Engineering College (Deemed to be University), Chandigarh. Earlier to PEC, he had worked as Lecturer at Global Institute of Technology from July, 2004 to April 2006. His core research interests focus on novel, high-speed semiconductor nanodevices operating at room temperature, terahertz frequencies, and their application in energy harvesting.

Contributors

Naushad Alam
ZHCET
AMU
Aligarh, India

Shonak Bansal
Electronics and Communication Engineering
    Department
Punjab Engineering College
Chandigarh, India

S. K. Bansal
Department of Electrical and Instrumentation
    Engineering
SLIET
Longowal, India

Brinda Bhowmick
Department of Electronics and Communication
    Engineering
NIT Silchar
Assam, India

B. Biswas
CSIO-CSIR
Chandigarh, India

B. Elizabeth Caroline
Department of ECE
IFET College of Engineering
Villupuram, India

Devendra Chack
Department of Electronics Engineering
Indian Institute of Technology (ISM)
Dhanbad, India

Ankur Garg
Department of Electronics and Communication
    Engineering
Punjab Engineering College
Chandigarh, India

Neena Gupta
Electronics and Communication Engineering
    Department
Punjab Engineering College
Chandigarh, India

Vivek Harshey
Department of Electronics and Communication
    Engineering
SLIET
Longowal, India

M. K. Hooda
Semi-Conductor Laboratory
Department of Space
Government of India

Neelu Jain
Department of Electronics and Communication
    Engineering
Punjab Engineering College
Chandigarh, India

Prince Jain
Electronics and Communication Engineering
    Department
Punjab Engineering College
Chandigarh, India

Birinderjit Singh Kalyan
Chandigarh University
Chandigarh, India

Nehru Kandasamy
Department of Electrical and Computer
    Engineering
National University of Singapore
Singapore

A. Karmakar
Semi-Conductor Laboratory
Department of Space
Government of India

Sathish Kumar Danasegaran
Department of ECE
IFET College of Engineering
Villupuram, India

M. Arun Kumar
Chandigarh University
Chandigarh, India

Sanjeev Kumar
Department of Applied Physics
Punjab Engineering College
Chandigarh, India

M. Margarat
Department of Electronics and Communication
    Engineering
IFET College of Engineering
Villupuram, India

Suman Kumar Mitra
Department of Electronics and Communication
    Engineering
Harcourt Butler Technical University
Kanpur, India

Deepak Kumar Panda
School of Electronics Engineering
VIT AP University
Amaravati, India

Krishna Parkash
Electronics and Communication Engineering
    Department
Punjab Engineering College
Chandigarh, India

Hitendra Singh Pawar
TBRL
Chandigarh, India

S. Poonguzhali
School of Electrical and Electronics Engineering
Sathyabama Institute of Science and Technology
Chennai, India

Balwant Raj
Department of Electronics and Communication
    Engineering
PUSSGRC
Hoshiarpur, India

Balwinder Raj
Department of Electronics and Communication
    Engineering
NITTTR
Chandigarh, India

Mohd Adil Raushan
ZHCET
AMU
Aligarh, India

Rajesh Saha
Department of Electronics and Communication
    Engineering
Malaviya National Institute of Technology
Jaipur, India

Varun Sharma
EEE Department
BITS Pilani
Goa, India

Mohd Jawaid Siddiqui
ZHCET
AMU
Aligarh, India

Arun K. Singh
Electronics and Communication Engineering
    Department
Punjab Engineering College
Chandigarh, India

Balwinder Singh
Chandigarh University
Chandigarh, India

Jeetendra Singh
National Institute of Technology Sikkim
Ravangla, India

Shailendra Singh
ECE
PSIT Kanpur
Kanpur, India

S. Soumiya
Department of Electronics and Communication
    Engineering
IFET College of Engineering
Villupuram, India

A. Sridevi
Department of ECE
M. Kumarasamy College of Engineering
Karur, India

Pankaj Srivastava
Nanomaterials Research Group
ABV-Indian Institute of Information Technology
    and Management
Gwalior, India

Nagarjuna Telagam
Department of Electrical Electronics and
    Communication Engineering
GITAM University
Bangalore, India

K. Vanlalawmpuia
Department of Electronics and Communication
    Engineering
NIT Silchar
Assam, India

Chhaya Verma
National Institute of Technology Sikkim
Ravangla, India

Emerging Nanoelectronic Devices

1

Mohd Adil Raushan, Naushad Alam,
and Mohd Jawaid Siddiqui

Contents

DOI: 10.1201/9781003126645-1

1.1 INTRODUCTION

The era of electronic gadgets has set upon us, giving rise to many smart devices. These smart devices have vastly improved the standard of living of mankind. The ease of life is facilitated by these electronic gadgets, which have brought a revolutionary effect on human race. These electronic gadgets have affected the mankind in all spheres of life, such as engineering, health, education, warfare, entertainment, etc. The basic building block of these smart devices is the transistor, which has led to these innovations.

1.1.1 Transistor Action

The basic building block of these smart electronic gadgets is the transistor. These transistors were invented by Shockley, Bardeen, and Brattain in 1947. The transistor is essentially a three-terminal device where the third terminal (voltage) is used to control the current between the other two terminals. The discovery of the transistor was actually an accident that occurred in the labs of the previously mentioned scientists. Who would have thought this discovery would govern the lives of millions of people in years to come? In the beginning, till the 1970s, the bipolar junction transistors (BJT) served the purpose for the semiconductor industry. But now they are mostly used in radio frequency circuits. The other type of transistor, MOSFETs (metal-oxide semiconductor field effect transistors), has efficiently served the microelectronics market till now. The gate terminal in MOSFETs controls the current in the channel between the source and drain terminals through an insulating dielectric between the gate and the channel. There is also a fourth terminal in a MOSFET called the body, which is generally grounded except in certain situations. The mode of conduction in the channel region in a MOSFET is drift and diffusion. The drain and source terminals are doped heavily, whereas the channel is doped lightly, with opposite impurities. For a MOSFET to behave as a switch, it must have two states, that is, OFF-state and ON-state. For a perfect switch, the current in the OFF-state must be zero, and substantial current must be present in the ON-state. To understand these states, we need a quick review of basic MOSFET physics. Basically, the charge dynamics in the channel region are influenced with the aid of an electric field applied through the gate voltage in these field effect devices. For a negative gate voltage in n-MOSFET (n-type), the gate electrode attracts the majority holes in the lightly doped p-type channel, which increases the hole

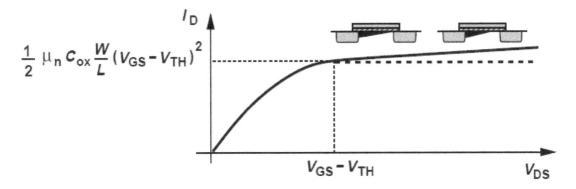

FIGURE 1.1 I_D–V_{DS} curves for MOSFET [1].

concentration at the Si–SiO$_2$ interface. This is called the accumulation mode of the device. For a positive voltage, the holes are pushed from the SiO$_2$–Si interface, thereby depleting the region at the interface. The negative acceptor ions are left behind, which constitutes the depletion region. This is called the depletion mode. For a further increase in positive voltage, the minority carriers' 'electrons' are pulled at the interface, thereby creating an inversion layer at the interface. The gate voltage at the onset of inversion layer at the interface is termed as the threshold voltage. At gate to source voltage (V_{GS}) = threshold voltage (V_{TH}), the carrier concentration of minority electrons at the interface is equal to the hole concentration in the bulk region. Thus, for MOSFET to behave as a switch, the gate voltage can act as the stimulating voltage to trigger an ON-state from the OFF-state. Initially, for a drain to source voltage (V_{DS}), MOSFET remains in OFF-state for zero gate voltage V_G or $V_G < V_{TH}$. For $V_G > V_{TH}$, conduction of electrons take place between the source and drain. The current is not completely zero even in the OFF-state (when $V_G = 0$). In the subthreshold state, a finite leakage current is still present due to various factors. Nevertheless, the two states ON and OFF can be switched by the polarity of gate voltage. As expected, for a given $V_G > V_{TH}$, the drain current increases with increase in V_{DS}. But after some point, the drain current saturates for any further increase in V_{DS}, due to pinch-off of the channel region on the drain side. The drain current v/s drain voltage characteristics are called the output characteristics of the MOSFET shown in Figure 1.1. For a given V_{DS}, the drain current increases with increase in V_G, which is called the inversion mode.

1.1.2 Advent of CMOS

To this point we have discussed the n-MOSFET and its working in different modes, that is, accumulation, depletion, inversion. But it alone cannot perform the complete logic function of the NOT inverter, which is logic '0' to '1', and vice versa when the output load capacitance is not charged initially. It needs a complementary p-MOS to convert the logic '0' to '1'. The complementary metal oxide semiconductor (CMOS) shown in the Figure 1.2 is capable of performing the complete logic function of a basic inverter (NOT gate). This complementary CMOS has been the backbone of the microelectronics industry for the past four decades. However, there is power dissipation in this CMOS inverter, which is static and dynamic in nature. The static power dissipation is caused by the finite leakage current in subthreshold state for both the p-MOS and n-MOS in the CMOS inverter. In the OFF-state, there is a leakage current I_{OFF}, even when switching does not take place. The static power dissipation is given by equation (1.1). The dynamic power dissipation occurs when switching takes place. For every cycle of switching, half of the energy stored or discharged in the load capacitor is dissipated as heat, which is the source for this dynamic power

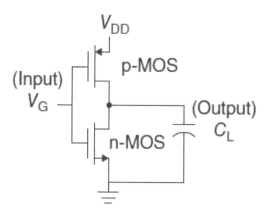

FIGURE 1.2   Complementary CMOS inverter having n-MOS and p-MOS.

dissipation. The dynamic power dissipation is given by equation (1.2). The load capacitance is given by C_L, and f and α represent the frequency of operation and switching probability.

$$P_{S = }V_{DS*}I_{OFF} \tag{1.1}$$

$$P_D = V_{DS\,*}^{\ \ 2}C_L f\alpha \tag{1.2}$$

1.1.3 Scaling of CMOS

In order to make the device more feasible the power consumption of these transistors should be reduced. The static power dissipation could be lowered by reducing the OFF-state current and the supply voltage. The reduction in supply voltage reduces the ON-state current; thus, it should be lowered very efficiently. The reduction in OFF-state current is another option, which can be achieved by controlling certain factors discussed in Section 1.7. For reduction in dynamic power dissipation, we need to reduce the parameters load capacitance C_L, frequency of operation f, and switching probability α. The frequency of operation determines the computational speed, so it needs to be increased. The parameter α determines the switching probability of the digital circuit so it cannot be changed. The load capacitance C_L is basically the input capacitance of other digital circuits. This C_L can be reduced by minimizing the area of MOSFET. This reduction in area of MOSFET calls for scaling of MOSFET. A reduced gate length also increases the ON-state current, as I_{ON} depends inversely on gate length. A reduced capacitance also increases the frequency of operation. The scaling also reduces the numbers of transistors on a single chip, which is essential in this era of miniaturization. Thus, scaling is the way forward for reduced power consumption. Constant field scaling and fixed voltage scaling are two methods of scaling done in the industry. In the constant field scaling, which is called Dennard's scaling rule, the supply voltage is scaled in the same ratio as dimensions of MOSFET. For the fixed voltage scaling, the supply voltage is unchanged. Table 1.1 shows the scaling factor of various device structure parameters.

1.1.4 Moore's Law

One of the famous laws which had predicted the scaling of transistors quite accurately is Moore's law. G. Moore had forecasted that the quantity of transistors on a single chip will double after every 18 months. This law worked pretty accurately for around 40 years through CMOS scaling

TABLE 1.1 Dennard Scaling Rules

DEVICE PARAMETER	FACTOR OF SCALING
Device dimensions T_{ox}, W, L	$1/k$
Doping concentration N_a	k
Delay time per circuit CV/I	$1/k$
Capacitance eA/t	$1/k$
Current I	$1/k$
Voltage V	$1/k$
Power dissipation VI	$1/k^2$
Power density VI/A	1

till 2010 [2]. However, when the MOSFET's gate length became almost equivalent to the depletion region at drain-channel and source-channel interfaces, this law became inaccurate. Due to several effects called short channel effects, the device electrostatics of MOSFET have slightly changed. These short channel effects have a degrading effect on the MOSFET that will be discussed in the next section (section 1.2)

1.1.5 Koomey's Law

A new law that had predicted scaling in a slightly different perspective was Koomey's law. Jonathan G. Koomey had predicted that the number of computations performed by a microprocessor for a unit power consumption would double after 18 months. The number of computations at peak frequency doubled due to reduced power dissipation and increased operating speed as a result of scaling after every 18 months. This was reported for the first time by J. G. Koomey. This law has held better than Moore's law, even after 2010, when the short channel effects began to dominate the MOSFET electrostatics. Thus, in this era of aggressive scaling, the short channel effects have played a vital role, which is discussed in the subsequent section.

1.2 SHORT CHANNEL EFFECTS

When the device dimensions (channel) of the MOSFET are similar to depletion lengths at source-channel interfaces and drain channel interfaces, it is called short channel MOSFET. There is an alarming increase in the OFF-state current attributable to these effects at sub-100 nm gate lengths. The reduction of threshold voltage with device dimensions is also present. The decrease in slope of I–V curves and the effect of drain bias on threshold voltage are some common problems in these devices. These phenomenon are collectively known as short channel effects, which are discussed here.

1.2.1 Subthreshold Slope 'SS'

Subthreshold swing is termed as the quantity of gate voltage required a decade change in drain current. The equation of SS for a MOSFET is given as equation (1.3).

$$SS = (kT/q) * \log \left(1 + (C_D/C_{OX})\right)$$ (1.3)

In this equation, k is the Boltzmann's constant, T is the temperature, q is the charge and C_D, C_{OX} are the capacitances of depletion region and oxide, respectively. The subthreshold slope determines the amount of change of current for a change in gate voltage. As $C_{OX} >> C_D$, the SS for a MOSFET is restricted by the Boltzmann's limit 'kT/q' to 60 mV/dec. For a short channel device, it is severely hampered, as the OFF-state current is larger for a given threshold voltage.

1.2.2 Drain Induced Barrier Lowering 'DIBL'

As the channel length reduces, the depletion regions at drain-channel interface and source-channel interface begin to interact with each other. The electric field coupling takes place, and it affects the barrier height at source-channel interface. This lowering of barrier height at source-channel interface rises the leakage current in MOSFET. Thus, the gate control on the channel is loosened. The channel is also modulated by the application of drain bias. The drain acts as a second gate in the transistor. This is the 2D behavior of the transistor. It also results in poor saturation behavior of the transistor. This behavior is called drain induced barrier lowering (DIBL), as shown in Figure 1.3. In modern devices, as the channel length decreases, this is a serious challenge for device engineers.

1.2.3 Channel Length Modulation (CLM)

When the drain voltage V_{DS} becomes greater than gate overdrive voltage ($V_{GS}-V_T$), pinch-off of the channel occurs at the drain side. This reduction in channel length ΔL_G introduces a change in I_D given by the equation (1.4).

$$I_D = I_{Dsat} / \left(1 - \Delta L_G / L_G\right) \tag{1.4}$$

This ΔL_G is significant with respect to L_G in short channel devices as opposed to long channel devices. ΔL_G has a strong dependence on V_{DS}. Thus, for channel devices. this channel length modulation introduces a non-saturation behavior in the output characteristics of the transistor shown in Figure 1.4. The threshold voltage V_T is also reduced for short channel lengths L_G. This effect is called as the threshold voltage roll-off. This also affects the I_{ON}/I_{OFF} ratio of the device.

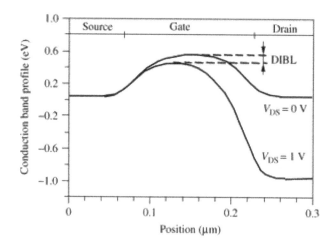

FIGURE 1.3   DIBL in MOSFET [3].

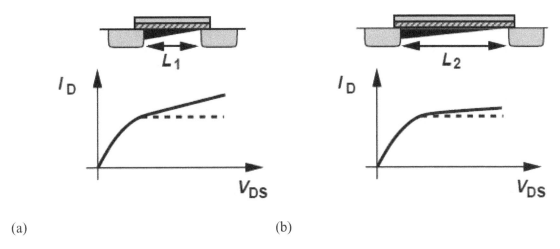

FIGURE 1.4   CLM in MOSFET [1]. (a) Lower saturation in I_D–V_{DS} curves for smaller L_G. (b) Higher saturation for higher L_G.

1.2.4 Gate-Induced Drain Leakage (GIDL)

Before explaining the physics of GIDL, we need a brief review about the tunneling phenomenon. Tunnelling is a quantum-mechanical phenomenon in which the wave particles are able to penetrate through a barrier greater than their energies. Schrodinger had given the wave equation, which stated that there is a transmission probability of finding a wave particle (electron) outside a potential barrier of greater height (energy) than the electron. This was famously called the potential well problem. Similarly, in an ultra-scaled MOSFET or other devices (L_G < 100 nm) in which the drain side conduction band is in close proximity of the channel valence band, the electrons are able to tunnel through the potential barrier shown in Figure 1.5a. The electrons in the filled valence band on the channel side find empty states in the drain side conduction band. The electrons tunnel through the potential barrier instead of going over it, as they have energies lesser than the barrier height. This tunneling of electrons is called as band-to-band tunneling (BTBT). This BTBT phenomenon is responsible for parasitic BJT action in short channel transistors. BTBT increases the leakage current considerably at lower gate voltages. This BTBT is of two types – lateral and transverse. Lateral BTBT takes place when electrons tunnel laterally from the channel to the drain region in the direction of electron flow. Transverse BTBT takes place when electrons tunnel perpendicular to the direction of electron flow. It usually happens under the drain region when a gate/drain overlap is formed due to diffusion of dopant atoms in the channel from either side (Figure 1.5b). This BTBT-induced GIDL is responsible for conduction of electrons even at negative gate voltages, thereby degrading the transistor performance (Figure 1.5c).

1.2.5 Doping Gradient

Another very common problem associated with a short channel MOSFET is the realization of ultra-steep doping gradients or junctions at such short channel lengths. A MOSFET contains a junction on either side of the channel, that is, drain-channel and source-channel junctions. The doping concentration of drain and source regions are normally 10^{20} cm^{-3}, while the channel is complimentarily doped at 10^{17} cm^{-3}. The doping gradients are usually 2–3 nm/decade in ion-implantation processes. This consumes a large chunk of channel region on either side to bring the doping concentration down to 10^{17} cm^{-3} (channel) from 10^{20} cm^{-3} (source/drain). This leads to a large area of channel with high doping concentration in a lower gate lengths MOSFET. This diffusion of dopant atoms is unavoidable, as high temperature annealing is used, which

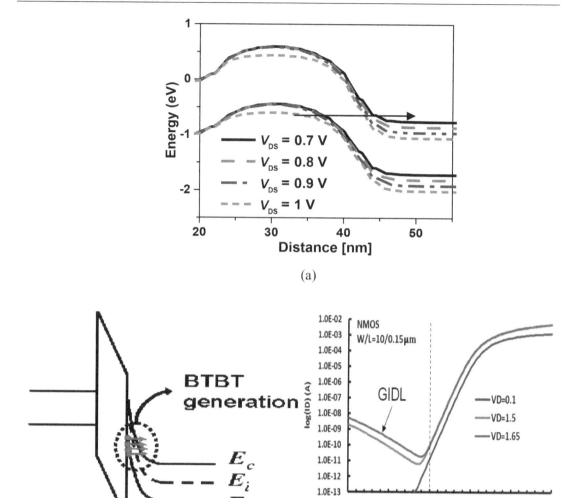

(a)

(b)(c)

FIGURE 1.5 GIDL in MOSFETs. (a) Lateral BTBT in JAMFET [4]. (b) Transverse BTBT generation under an overlap drain region having a high electric field [5]. (c) Conduction for negative gate voltages due to GIDL [6].

assists diffusion. Therefore, alternative device architectures or ultrafast annealing schemes are needed to counter this problem associated with ultra-steep doping gradients.

1.3 JUNCTIONLESS TRANSISTORS

One of the breakthroughs in the device architectures is the realization of junctionless transistors (JLTs). Before the discovery of transistor by Shockley et al., J. Lillienfield had proposed the idea of transistors way back in 1930 [7]. He had proposed the idea of modulating conductivity between two electrodes on a glass apparatus by a third electrode on the channel. The channel was made of copper and sulfur. It behaved as a gated resistor whose conductivity could be controlled by a third electrode. This same idea was implemented in JLTs in 2010 by Colinge et al. [8]. In JLTs, the device layer is composed of a single type of material (n-type or p-type) without any junctions. For an n-type JLT, the depletion of channel is realized by a gate electrode of high work function (> 5 eV). The difference in work function between channel and gate electrodes is responsible for the full depletion of channel. Thus, it also behaves as a gated resistor.

1.3.1 Theory

As explained earlier the difference in work function is responsible for the depletion of channel in uniformly doped JLT. For an n-JLT, the work function difference ϕ_{ms} should be positive, that is, $\phi_m > \phi_s$. There are certain requirements that need to be fulfilled for efficient realization of a JLT. First, the device thickness should be less, such that the channel should be completely depleted for $V_{GS} = 0$ V; otherwise, there will be a large OFF-state leakage current. Second, the gate should be of high work function for an n-type JLT. The structure of a junctionless transistor is shown in Figure 1.6a. The drain, channel and source are all composed of the same material concentration and polarity. These JLTs are devoid of any junctions, making it junctionless in device architecture. The barrier present between source and channel, similar to a MOSFET, inhibits the current flow in OFF-state (Figure 1.6b).

1.3.2 JLFET

The physics underlying the working of junctionless field effect transistor (JLFET) give rise to different conduction mechanism. The volume depletion carried out in a nanowire JLFET can be obtained by adjusting the gate work function so that full volume depletion occurs at $V_{GS} = 0.0$ V, shown in Figure 1.7a. For a p-type JLT, the gate work function should be small (< 4.1 eV). For efficient depletion, the gate control should be strong. Therefore, nanowire structures are best suited for JLTs. For an n-JLT, in the full

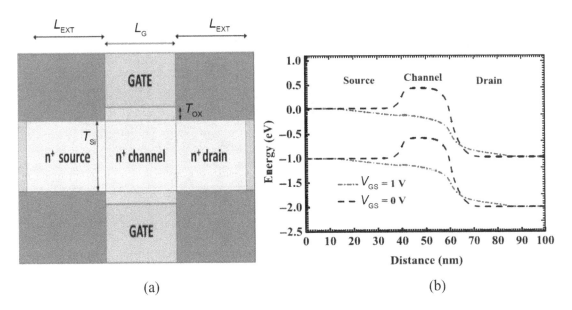

(a)   (b)

FIGURE 1.6 (a) Structure of a JLT, showing the source, channel, and drain regions of the same n+ type concentrations. (b) Band diagram of a JLT in ON- and OFF-states [9].

depletion mode, the channel is completely depleted of majority carriers. The depletion region composed of positive ions engulfs the whole device thickness. On the application of positive gate voltage in n-JLT, the neutral region (bulk n-type) is uncovered in the middle of nanowire. The depletion regions begin to shrink with the application of gate voltage. Some part of the channel remains depleted, and a neutral n-type region begins to appear in the center, facilitating the current flow. This mode is called the partial depletion mode. The threshold voltage for JLT is defined as the voltage where the neutral n-type region disappears due to the fusion of the depletion region realized by the high work function gate electrodes, as shown in Figure 1.7b. On further application of gate voltage ($V_{GS} \gg V_{TH}$), the depletion regions contributed by the gate electrodes disappear completely, and the whole channel thickness becomes conducting, shown in Figure 1.7c. This condition is called the flat band condition, shown in Figure 1.7d. The energy bands of the channel become flat, and the whole channel comes out of depletion. On further increase in gate voltage ($V_{GS} > V_{FB}$), the electrons accumulate at the surface, forming a layer of electrons at the semiconductor surface. This mode of operation is called the accumulation mode. JLFETs are biased to function in flat

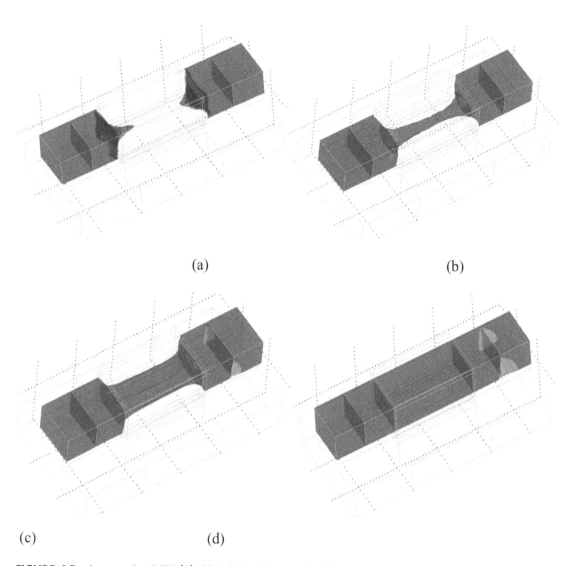

(a)   (b)

(c)   (d)

FIGURE 1.7   A nanowire JLFET [8]. (a) Full depletion mode. (b) At threshold voltage $V_{GS} = V_{TH}$. (c) Above threshold voltage $V_{GS} > V_{TH}$. (d) Flat band mode.

band condition rather than in accumulation mode. In flat band condition, there is bulk conduction, as the whole channel is able to conduct, while in accumulation mode there is surface conduction. In surface conduction, scattering dominates carrier transport, as lattice scattering is prominent, which affects the carrier transport.

1.3.3 JAMFET

The mobility in JLT is affected due to high doping (10^{19} cm^{-3}) in the channel of uniformly doped JLT. The mobility is degraded due to larger impurity scattering in the channel region. The series resistances of source/drain regions are also higher as the source/drain regions are moderately doped (10^{19} cm^{-3}). This affects the ON-current of JLT. Thus the doping of source/drain region were increased to decrease the series resistances and the channel doping was decreased to increase the mobility. These type of transistors were called junctionless accumulation mode field effect transistors (JAMFET) [10]. Although they have high doping and low doping junctions at the source-channel and drain-channel interface, they still were free from p−n junctions. They can also be termed as pseudo-junctionless transistors. The device architecture of JAMFET is shown in Figure 1.8a. The doping of the drain/source region is 10^{20} cm^{-3} and that

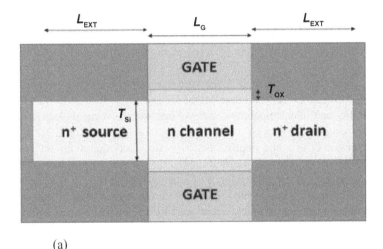

(a)

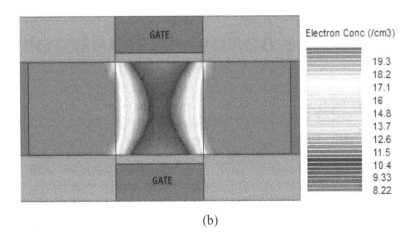

(b)

FIGURE 1.8 (a) Structure of JAMFET with n$^+$ source/drain region and n channel region. (b) Carrier concentrations in JAMFET for thermal equilibrium [4].

of the channel region is 10^{18} cm^{-3}. The working of JAMFET is similar to JLT. The carrier concentration in thermal equilibrium ($V_{GS} = V_{DS} = 0$ V) are shown in Figure 1.8b. The channel is depleted of carriers in the OFF-state by the high work function gate electrode. But the gate work function required in JAMFET (4.5–5 eV) is generally lower than JLT (5–5.5 eV), as the channel doping is lower in comparison to conventional junctionless transistor. The ON-state current is larger in JAMFET because the series resistances are low and mobility is enhanced in the channel region. Although the ON-state performance of JAMFET is superior, there are certain drawbacks in JAMFET, such as higher GIDL, higher short channel effects, and greater fabrication complexity, which will be discussed in section 1.4.4.

1.3.4 Features

Some of the salient features of these junctionless transistors that highlight their importance in this era of aggressive scaling are listed here.

- These JLTs are free from doping gradients or junctions, which makes them a very attractive architecture because the gate lengths are fast approaching the sub-10 nm scale. Thus, they have great potential at lower technology nodes.
- The junctionless structures also relax the thermal budget and ultrafast annealing techniques required for the formation of junctions. They can ease the fabrication complexities.
- The conduction in JLTs is bulk conduction, whereas MOSFETs have surface conduction [8]. Lattice scattering or trap effect is negligible in bulk conduction. Mobility is also unaffected in bulk conduction.
- Although there are short channel effects in JLT, they are lower than a MOSFET device. JLTs show lower DIBL and subthreshold slope SS than MOSFET [8].
- JLTs show better reliability and bias temperature stability because the vertical electric field for the ON-state is absent, whereas maximum electric field is present in ON-state for MOSFETs [11].
- JLTs show lower hot electron effects and improved carrier ballisticity due to absence of junctions at drain-channel and source-channel interfaces [12][13]. The zero or diminished electric field in the ON-state also enhances the performance of junctionless FET.
- Due to the bulk conduction mechanism in junctionless FET, the degradation in transconductance for an increase in gate voltage is much lower in JLFET. This enables a higher speed performance for JLT [8].

1.4 DOPINGLESS TRANSISTORS

JLTs are free from junctions, but they are heavily doped (10^{19} cm^{-3}), which leads to higher random dopant fluctuations (RDFs). These RDFs lead to variabilities in threshold voltage (V_{TH}) of a device. The heavy doping in the channel region leads to greater impurity scattering, which affects mobility of carriers. The high gate work function (> 5.1 eV) requirement to turn off JLT is another strict requirement that needs to be addressed. The moderate doping in source/drain regions increases the series resistances, which reduce the ON-state current of JLT. At higher drain bias, there exists an unavoidable latch-up problem in JLT [14]. This is due to the dominant impact ionization problem, which reduces the turning off capability of JLT. Therefore, another type of transistors that can prove to be a suitable alternative for MOSFETs, called dopingless transistors, was proposed [15]. These transistors are immune to problems associated with doping because they are free from chemical or impurity doping itself. These transistors are electrostatically doped to perform the required transistor action. The concept of electrostatic doping is explained in section 1.4.1.

1.4.1 Electrostatic Doping

The concept of electrostatic doping can be a very suitable alternative to overcome the challenges associated with chemical doping. In electrostatic doping, the concentration of carriers in the body of the
semiconductor bar is influenced by the electrode attached to it. This is mainly doping of an intrinsic semiconductor material by electrodes. When a semiconducting material is brought near a conducting material
or metal electrode, band alignment near the interface occurs, which induces charge carriers. The charge
carriers can be electrons or holes depending upon the type, polarity, and work function of the electrode.
The relative difference amid the Fermi level and energy bands of semiconductor determine the doping
concentration in the semiconducting material. This interaction on the metal-semiconductor interface can
be controlled by metal's work function and semiconductor properties such as work function, electron
affinity, and energy band gap, shown in Figure 1.9 [16]. Many devices employing electrostatic doping
have been proposed such as diodes [17], tunnel field effect transistors (TFETs) [18], impact-ionization
MOSFET (IMOS) [19] and dopingless transistors [15].

1.4.2 Theory of Dopingless Transistors

In dopingless transistors (DLT), the n^+–p–n^+ device structure of n-type DLT required for transistor action
is realized using electrostatic doping. The initial device layer is intrinsic in nature. The n^+ regions required
for source/drain regions are electrostatically doped by placing a low work function (< 4.1 eV) electrode,
for example, hafnium. The same electrode (hafnium) can be used for drain/source contacts, as shown in
Figure 1.10a. The intrinsic region is converted into p-type region by using a high work function (> 5 eV)
gate electrode. The required high electron concentration (~ 10^{20} cm^{-3}) for source/drain regions is achieved

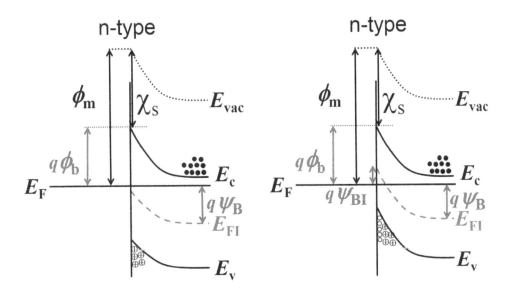

(a)(b)

FIGURE 1.9 (a) Band alignment at metal-semiconductor interface in Schottky contact showing depletion
layer at interface. (b) Excessive band bending at metal-semiconductor interface showing an inversion of n-type
(bulk) material into p-type at the interface [16].

by this technique of electrostatic doping, as shown in Figure 1.10b. The DLT is similar to JLT, as it is also free from impurity doping gradients or junctions. The working of DLT is also similar to JLT. In the OFF-state ($V_{GS} = 0$ V, $V_{DS} = 1$ V), the channel is depleted of carriers because of the high work function of the gate electrode. The channel is devoid of charge carries in the OFF-state, shown in Figure 1.11a. On the application of gate voltage, the channel emanates out of depletion mode similarly to JLT. On further increase in gate voltage ($V_{GS} = 1$ V), the whole channel is able to conduct, as shown in Figure 1.11b. Therefore, these DLTs can effectively perform the transistor action without being chemically doped and are also free from junctions.

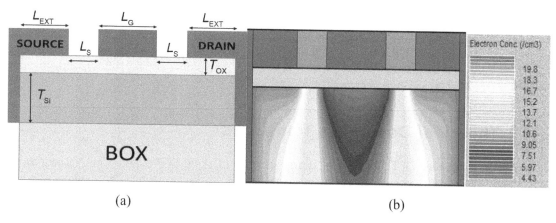

(a)   (b)

FIGURE 1.10   (a) Structure of DLT. (b) Carrier concentration in equilibrium state ($V_{GS} = V_{DS} = 0$ V).

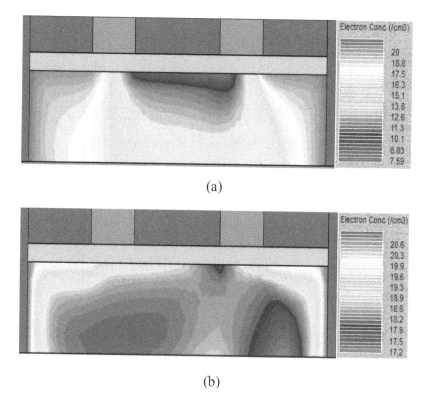

(a)

(b)

FIGURE 1.11   (a) Carrier concentration in ($V_{GS} = 0$ V, $V_{DS} = 1$ V) OFF-state. (b) Carrier concentration in ($V_{GS} = V_{DS} = 1$ V) ON-state.

1.4.3 Features

- Dopingless devices are free from impurity doping and impurity doping gradients, thereby making them dopingless and junctionless in their architecture. Therefore, they are free from complex fabrication processes. They can also relax costly thermal budgets required in annealing processes [20].
- Being dopingless, they are also more immune to random dopant fluctuations, making them a suitable alternative at lower technology nodes [14].
- DLTs are more resistant to short channel effects. This immunity is due to enlargement of depletion region outside the gate length, as the substrate is undoped. This increases the effective or actual gate length (L_{eff}) of the DLT. It also highlights the better scalability of the device [21].
- Due to the dopingless nature of the device, a DLT is less sensitive to parameter variations such as gate oxide thickness, gate length, device thickness, and doping variations. This makes it more suitable to low-power logic technology applications [14].

1.4.4 Challenges in JLT and DLT

There are various issues in JLTs that need to be addressed. Some problems in JLTs, such as random dopant fluctuations, low mobility, and high source/drain resistances, can be curtailed by the use of dopingless transistors. But the problem of gate-induced drain leakage (GIDL) is a common issue that exists for both the nanoscaled JLT and DLT [22][23]. As the gate lengths are fast approaching sub-10 nm scale, the problem of GIDL is a serious issue. All types of junctionless devices – JLTs, JAMFETs, and DLTs – suffer from this problem (Figure 1.12 (a,b,c)). The issue of GIDL is pronounced in JAMFET, owing to the existence of high-low junctions that reduce the effective gate length in JAMFET [24]. This GIDL affects the turn-off capability of the devices, thereby hampering the current ratio (I_{ON}/I_{OFF}) of the devices. It affects various analog performance metrics that have been addressed in this work.

1.4.5 Gate-Induced Drain Leakage in JLT, JAMFET, and DLT

As explained in section 1.2.4, GIDL is a dominant phenomenon in short channel devices. The narrowing of tunnel barriers at low gate voltages ($V_{GS} \leq 0.2$ V) between the channel and drain regions aids the tunneling of electrons from the valence band of channel to the conduction band of the drain region. This BTBT leakage current for short channel lengths is the predominant mechanism for GIDL in JLT [22]. The tunneling of electrons from the channel region to the drain region generates holes in the valence band of the channel region. The accumulation of holes in the channel region serves as the p region of the parasitic bipolar junction transistor, while the drain and source regions function as the collector and emitter regions. This leads to forward biasing of the emitter-base (source-channel) junction of the parasitic BJT, which triggers the BJT (Figure 1.12d). The forward biased emitter-base junction of the parasitic BJT leads to substantial collector current or the leakage current in JLTs [22]. The collector current ($I_C = \beta I_B$) is the product of the parasitic BJT gain and I_B of BJT. The tunneling current by holes in JLT serves as the parasitic BJT's base current. This parasitic BJT formation severely hampers the OFF-state performance in JLT and DLT. For a further decrease in gate voltage, the concentration of holes increases in the valence band of the channel region, which increases the potential of the base region or the floating body potential in parasitic BJT. This leads to saturation of the collector current in parasitic BJT. Thus, for any further increase in negative gate voltage in junctionless transistors, the drain current increases marginally, which signifies low dependence of drain current (I_D) on negative gate voltages (V_G) (Figure 1.12e).

    There have been many solutions proposed for the problem of GIDL in JLT, JAMFET, and DLT. This BTBT-induced GIDL in JLT was first reported by Gundapaneni et al. [22]. The bulk planar JLT (BPJLT),

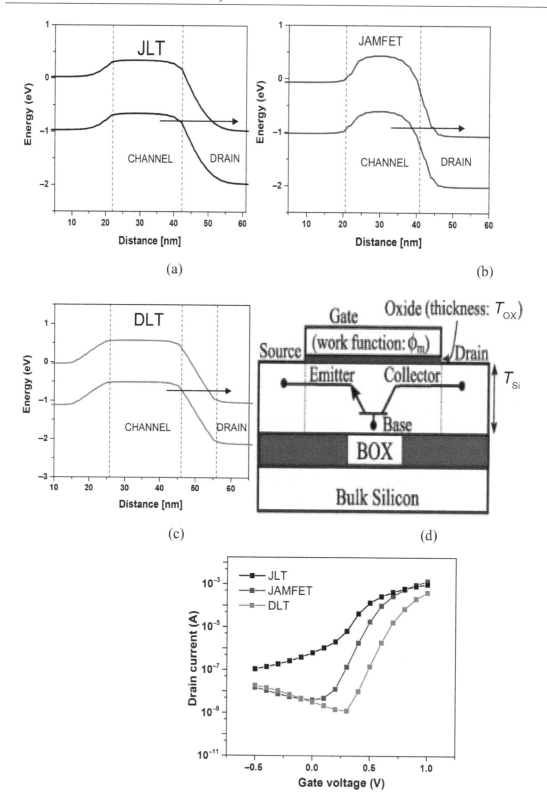

FIGURE 1.12   (a) BTBT in JLT. (b) BTBT in JAMFET. (c) BTBT in DLT. (d) Parasitic BJT action in JLT [22]. (e) I_D–V_{GS} curves for JLT, JAMFET and DLT.

which had a junction in vertical direction due to the presence of p-well, was proposed, which effectively swept the BTBT generated holes [12]. But this BPJLT had reduced ON-state current and increased capacitances owing to the presence of another junction. The extended back gate in double gate JLT was proposed, which widened the tunnel barrier at drain-channel interface and increased the barrier height at source-channel interface in the OFF-state [25]. For nanowire JLTs, many device architectures, such as core shell, hybrid channel, low drain doping, and nanotube JLT, were proposed to mitigate this GIDL by Sahay et al. [26]–[29]. For nanowire JAMFET, the dual material (work function) gate was proposed to widen the tunnel barrier at the drain channel interface in the OFF-state [24]. A dual metal stack gate was also proposed to counter the effect of GIDL in sub-10 nm regime nanowire JLT [30]. All these techniques either (a) widened the tunnel barrier, (b) lowered the electric field at the drain-channel interface, or (c) increased the source-channel barrier height in the OFF-state. This led to reduced BTBT, which in turn suppressed GIDL. The gate sidewall spacer design guidelines and diameter dependence in nanowire JLT and nanowire JAMFET for reduced GIDL were also successfully explained in [31][32]. The proposal for raised source/drain extensions to reduce the effect of GIDL by inhibiting parasitic BJT action in DLT on buried oxide (BOX) was reported by Sindhu et al. [23].

1.5 MOTIVATION BEHIND THE WORK PRESENTED ON JLT AND DLT

Therefore, the objectives presented in this chapter are listed here.

- One of the objectives behind this chapter is to highlight the problem of GIDL in these nanoscaled, junctionless, and dopingless devices.
- The suppression of short channel effects for nanoscaled JLT and DLT is another important problem that has been brought forward in this chapter.
- There have been many techniques proposed for nanowire JLFETs, but for a more robust technological perspective, we need a less complex fabrication process. For these novel JLTs and DLTs, single gate and double gate architectures have been fabricated till now. So we have tried to propose ideas for single gate and double gate architectures. These ideas could be further extended to nanowire architectures by suitable fabrication steps.

1.6 SIMULATION ENVIRONMENT OF JLT AND DLT

The 2D device simulations were implemented by Silvaco Atlas [33]. The quantum confinement model proposed by Hansch [34] is activated, as the device thickness is less than 10 nm. To model the phenomenon of tunneling in OFF-state or low gate voltages, a nonlocal band-to-band tunneling (BTBT) model based on the Wentzel–Kramers–Brillouin method is activated. The nonlocal BTBT used is sufficiently calibrated and is already used in the literature [35]. The tunneling masses 0.10 m_0 for electrons and 0.17 m_0 for holes were used according to [35] for JLT. Fermi–Dirac statistics were involved in calculation of the intrinsic carrier concentration needed for Shockley–Read–Hall (SRH) expressions. SRH, the Lombardi model of mobility, and Auger recombination models were incorporated to model mobility and lifetime phenomena. As the doping is higher in a JLT (~10^{19}), a band gap narrowing (BGN) model is also invoked to model high doping effects. The simulation setup calibration was done by replicating the transfer curves of [36] with DG-JLT for $V_{DS} = 0.02$ V, $T_{Si} = 10$ nm, width $W = 12.5$ nm and work function difference (φ_{MS}) of 1.12 eV. For tunneling in OFF-state, the parameters were used according to [35]. The grid points in the simulation of device is taken at 0.1 nm in the vertical or y-axis across the silicon thickness and at 0.5 nm in the horizontal or x-axis from the source side edge of the gate to the drain. This mesh is used to capture the phenomenon of tunneling for low gate voltages or OFF-state. The cutlines are at 0.5 nm from the insulator-semiconductor interface, as used in [37]. The surface state density value at the interface boundaries has the default concentration of the Silvaco simulator, that is, ~10^{10}. The tunneling masses of 0.60 m_o for holes and 0.40 m_o for electrons are used for JAMFET, as taken by Sahay et al. [24]. For DLT, the simulation setup was calibrated according to [15]. The tunneling masses of $m_e = m_h = 0.65\ m_o$ were taken as reported by Sindhu et al. [23].

1.7 STEEP SLOPE DEVICES

1.7.1 Limitations in CMOS

R. Dennard had proposed a scaling law in 1974 that explained how to scale the device dimensions by keeping the electric fields nearly constant in a device. Although this law worked correctly until the 1.4 μm node, afterwards it became inaccurate. The reason behind this was the inability of threshold voltage to scale down with supply voltage scaling. The required scaling of threshold voltage did not happen, because when the scaling laws are correctly applied, they keep the electric fields constant, which consequently keeps the threshold voltage constant. This led to reduction in gate overdrive, as the supply voltage was continuously being scaled down, as shown in Figure 1.13. The scaling of supply voltage had to be carried on; otherwise, the power consumption would have increased severely, as both the static and dynamic power dissipation depended on supply voltage (V_{DD}). The other option of scaling threshold voltage depended on the device behavior of the MOSFET. As the least subthreshold swing of the conventional MOSFET was fixed at 60 mV/dec, the only option to decrease V_T was to move the I_D–V_{GS} curves horizontally on gate voltage x-axis. For a shift of I_D–V_{GS}, the penalty is one decade of OFF-state current. This also led to serious issues of leakage current in scaling devices, shown in Figure 1.14. Many circuit level solutions such as sleep transistors, dual-V_T circuits, stacked transistors, and multicore processors were proposed to solve this problem. But all of them were not highly effective, as they had their consequences.

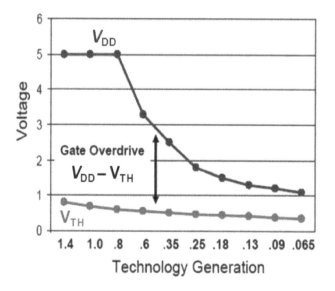

FIGURE 1.13 Decrease in the gate overdrive voltage (V_{GS}–V_{TH}) with scaling [38].

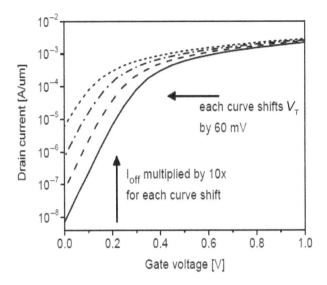

FIGURE 1.14   Increase in leakage current with each step of threshold voltage scaling [39].

Therefore, device-level solutions such as small swing devices were needed. In the subthreshold region, the rate of increase of carriers for an increase in gate voltage depends on the Fermi-Dirac function, given by $f(E) = 1/(1+e^{\wedge}((E - E_F)/kT))$. This function limits the MOSFET subthreshold current by a factor of kT/q. The equation of current for MOSFET in the subthreshold region is given by equation (1.5).

$$I \sim e^{\wedge}((V_{GS} - V_T)/(m*(kT/q))) \qquad (1.5)$$

The subthreshold swing (SS) $d(V_G)/d(\log I_D)$ derived from this equation comes out to be in the form $\ln(10)$ (mkT/q). Whereas 'm' is the body coefficient, which has the value of $1 + (C_{dm}/C_{OX})$. C_{OX} is the oxide capacitance, and C_{dm} is the bulk depletion capacitance at threshold. For $C_{dm} \ll C_{OX}$, m is marginally higher than 1, which makes least value of SS = 60 mV/dec for MOSFET.

1.7.2 Impact Ionization MOS

As explained earlier, the subthreshold swing in the MOSFET is knotted down by the physical phenomenon that produces current in the MOSFET in subthreshold mode. Therefore, for smaller SS, we need to change the physical mechanism for current generation. The impact ionization MOS (IMOS) is single gated p–i–n device where the gate is underlapped from a junction and overlapped from the other side [40]. Avalanche breakdown occurs in the intrinsic region at the non-gated portion, as a very high electric field is present in that region, shown in Figure 1.15. This leads to very low subthreshold swing and larger I_{ON} of the device. But there are many issues pertaining to IMOS such as scalability, low-voltage operation, and hot carrier degradation that need to be addressed.

1.7.3 MEMS and NEMS

The other type of small swing device is an electromechanical relay on a micron scale (MEMS) or nanometer scale (NEMS). There are some 2-terminal MEMS that retain exceptionally low swing value but have restricted applications for circuits. The 3-terminal MEMS have the capability to replace MOSFETs.

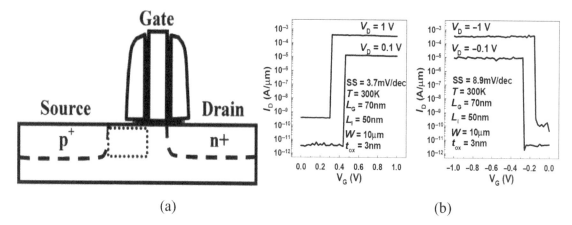

FIGURE 1.15 (a) Structure of IMOS [41]. (b) I_D–V_{GS} curves of n-IMOS and p-IMOS showing very low subthreshold slopes [42].

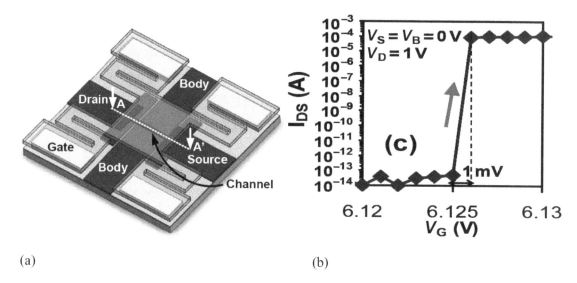

FIGURE 1.16 (a) Structure of MEMS [44]. (b) I_D–V_{GS} curves of MEMS [45].

There have been many device architectures proposed that could be fabricated (Figure 1.16a). In one of the architectures, a bendable cantilever beam is electrically coupled to the source terminal [43]. This beam is actuated by a gate electrode and lowered to touch the conducting drain electrode. Another configuration has the drain and source regions on the substrate and gate/channel, which can be lowered to form the connection between the drain and source regions. These devices are capable of delivery of ultra-low subthreshold slopes, shown in Figure 1.16b. But there are some issues, such as lower speed and mechanical nature of these devices, that need to be addressed.

1.7.4 Negative Capacitance FETs

It has been experimentally established that ferroelectric materials can provide a negative capacitance in a selected range of polarization. This property of negative capacitance has been effectively utilized by negative capacitance FET (NC-FET) to continue the process of scaling supply voltage and threshold voltage. In NC-FET, a ferroelectric layer is added with a linear dielectric layer, shown in Figure 1.17a [46]. The

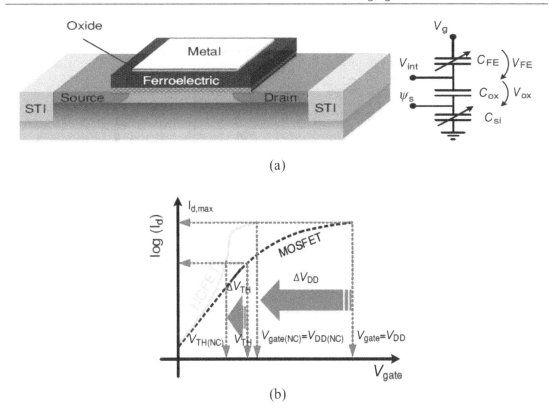

FIGURE 1.17 (a) Structure of negative capacitance FETs. (b) Reduction in V_{TH} and V_{DD} in NC-FET [46].

presence of the ferroelectric layer makes the equivalent capacitance in the gate larger than the classical capacitance of MOSFET. This enables a smaller gate voltage to induce an equivalent amount of charge, as compared to classical gate stacked MOSFET. This technique can effectively reduce the V_T and V_{DD} because the gate stack provides signal amplification and increases the surface potential, shown in Figure 1.17b [46]. These devices have been experimentally fabricated. But there are some issues, such as formation of a dead layer between the ferroelectric layer (FL) and the gate electrode. The interface properties between the FL and the semiconductor layer is also an issue.

1.7.5 Tunnel Field Effect Transistors

Tunnel field effect transistors are also called TFETs. They have the capability to outperform MOSFETs for low power applications. The structure, working, and features of TFET are discussed in the next section.

1.8 CHEMICALLY DOPED TFET

1.8.1 Theory

TFETs are gated p–i–n devices. For an N-TFET, the source region has high (10^{20} cm^{-3}) p-type doping, and the drain is moderately doped by n-type impurities. The channel is intrinsic in nature, shown in Figure 1.18a. For a P-TFET, polarities between drain and source are interchanged while maintaining the

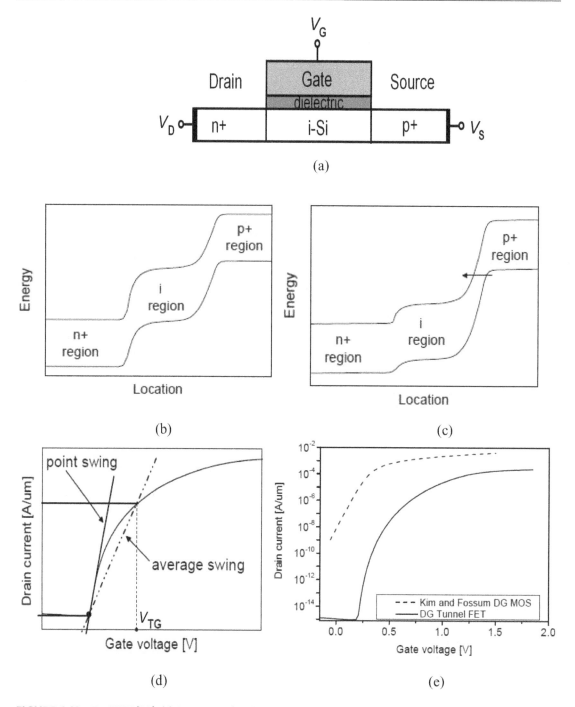

FIGURE 1.18 For TFET [39]. (a) Structure of p–i–n TFET. (b) Band diagram in ($V_{GS} = 0$ V, $V_{DS} = 1$ V) OFF-state. (c) Band diagram in ($V_{GS} = 1$ V, $V_{DS} = 1$ V) ON-state. (d) SS definition in TFET. (e) I_D–V_{GS} curves for TFET and MOSFET [38].

channel intrinsic. In the OFF-state, the electrons face a barrier to flow from source to drain through the channel region shown in Figure 1.18b. The leakage current is due to p–i–n diode leakage, which is very much less (~10^{-17} A). In the ON-state, on application of positive gate voltage the potential barrier between the channel and source region is narrowed. The electrons are capable of tunneling through the barrier instead of going over it, shown in Figure 1.18c [39]. The physical mechanism governing the carrier transport is changed as compared to MOSFET, where thermionic emission takes place. The current, due to band-to-band tunneling (BTBT), is given by equation (1.6) [47].

$$I_{BTBT} \sim T_t \sim \exp\left(-4\lambda\left(2m^*\right)^{1/2} E_g^{3/2}\right)/\left(3\hbar\left(\Delta\phi + E_g\right)\right) \tag{1.6}$$

Where T_t is the band-to-band tunneling transmission, λ is tunnel barrier width, m^* is the effective mass, E_g is the band gap of silicon, $\Delta\phi$ is the range of energy across which tunneling takes place, and $\hbar$ is the Planck's constant. The three basic requirements for tunneling of electrons from available states are the (1) availability of states where electrons can tunnel, (2) a suitably narrow energy barrier, and (3) conservation of momentum. Due to this tunneling phenomenon, it is possible for TFET to realize SS ≤ 60 mV/dec for low gate voltages at room temperatures. The SS for TFET is given by equation (1.7) [48].

$$SS = \ln(10)\left[\frac{1}{V_{eff}}\frac{dV_{eff}}{dV_{GS}} + \frac{F+b}{F^2}\frac{dF}{dV_{GS}}\right] \tag{1.7}$$

Equation 1.7 shows that SS for TFET does not have a unique value but is a function of V_{GS}. SS has the lowest value at the lowest V_{GS} and increases with V_{GS}. The different values of SS are explained in Figure 1.18d. Point SS is the lowest value of the subthreshold slope at the I_D–V_{GS} curve, and average SS is taken from where the drain current starts to increase from lowest value up to the threshold voltage. The I_D–V_{GS} curve for TFET is shown in Figure 1.18e.

1.8.2 Advantages

The salient features of TFET are listed here.

- Low SS offered by TFET is the most distinguished advantage of these devices. This can significantly reduce the power consumption of the chip and can provide a boost in low-power applications.
- Low OFF-current provided by TFET can lessen the loss of leakage current incurred in modern devices. The OFF-current in TFET is due to leakage of p–i–n diode, which is significantly lower than the OFF-current in MOSFETs.
- TFET does not follow the same scaling rules as MOSFET, where the electric field must be constant in the device [49]. TFET offers better scalability than MOSFET, as the ON-state current does not depend on TFET length.

1.8.3 Drawbacks

There are also some drawbacks or issues in these TFET devices, which are listed here.

- TFET show ambipolar behavior, which means they can also conduct in the reverse direction or negative gate voltages. For negative gate voltages, the tunnel barrier between the drain and channel gets lowered, which facilitates the electrons tunneling through, shown in Figure 1.19. This is an inherent problem with TFETs.
- Apart from all the benefits, TFET has low ON-current (~10^{-6} A) compared to MOSFET (~10^{-3} A). Many techniques, such as heterostructure TFET, strain engineered TFET, high-k dielectrics, and carbon-based TFET [49], have been proposed, but there is still room for improvement.

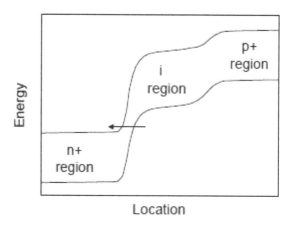

FIGURE 1.19 Ambipolar leakage in TFET [39].

- TFET also has two junctions, that is, drain-channel and source-channel junctions. A narrower tunnel barrier is required for larger ON-current in N-TFET. For a narrow tunnel barrier, the source doping should be high (~10^{20} cm^{-3}). Therefore, TFET is also affected from the problem of random dopant fluctuations and the fabrication of ultra-steep junctions at nanoscaled devices. The movement of impurity dopant atoms from drain and source regions to channel is another cause of concern.

1.9 JUNCTIONLESS TFET

As explained earlier, the presence of junctions in nanoscaled TFETs are problematic. Their fabrication and consequences are a cause of concern. The idea of junctionless TFETs was proposed by Ghosh et al. in 2014 [50]. The advent of junctionless transistors laid the foundation for JLTFET. The junctionless TFET can be a very good alternative to continue the scaling trend in TFET. The working of JLTFET is explained in the next section.

1.9.1 Working

The structure of JLTFET is shown in Figure 1.20a. In N-JLTFET, the body is composed of highly doped (10^{19} cm^{-3}) n-type material. The p-type source region is realized by an unbiased auxiliary gate of high work function, that is, 5.93 eV, shown in Figure 1.20a. Due to the high work function of the auxiliary gate, the accumulation of holes takes place under the auxiliary gate. The intrinsic channel region is generated by a gate ($L_g = 20$ nm) of work function 4.3 eV.

The gate and auxiliary gate are separated by a spacer of $L_S = 5$ nm. The drain and source contacts are on the sides. The carrier concentration in OFF-state is shown in Figure 1.20b, which shows the required barrier formation in p–i–n structure of TFET. When a positive gate voltage is applied, the carrier concentration in the channel is changed (Figure 1.20b). On the application of positive gate voltage, the barrier amid the channel and source region gets lowered, as shown in Figure 1.20c. The tunneling of electrons takes place between the valence band of the source to the conduction band of the channel region. The I_D–V_{GS} curve of JLTFET for spacer thickness $L_S = 5$ nm between the gate and auxiliary gate is shown in Figure 1.20d. The I_D–V_{GS} curve of JLTFET shows the trademark low SS of conventional TFET or TFET with chemical junctions shown in Figure 1.20e. Thus, JLTFET can provide the required tunneling phenomenon of TFET.

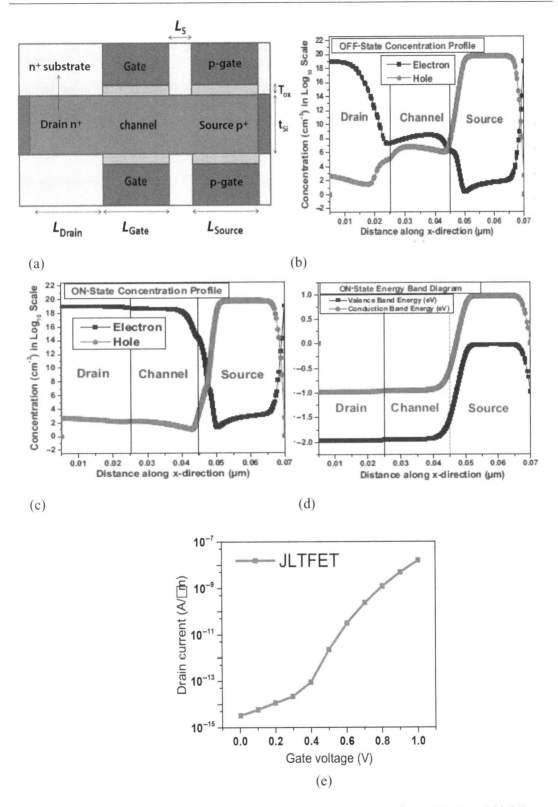

FIGURE 1.20 (a) Structure of junctionless TFET. (b) Carrier concentration in (V_{GS} = 0 V, V_{DS} = 1 V) OFF-state [50]. (c) Carrier concentration in (V_{GS} = 1 V, V_{DS} = 1 V) ON-state [50]. (d) Band diagram in (V_{GS} = 1 V, V_{DS} = 1 V) ON-state [50]. (e) I_D–V_{GS} curve for JLTFET [51].

1.10 DOPINGLESS TFET

The TFET device has also been realized on a dopingless structure [18]. The concept of a dopingless transistor (DLT) made possible with electrostatic doping has been extended to TFET structures. The use of electrodes to get the desired doping concentrations has been proposed before. The working of DLTFET is explained in the next section.

1.10.1 Working

The structure of DLTFET is shown in Figure 1.21a. The highly doped (10^{20} cm^{-3}) source region is realized by a high work function ($\phi = 5.93$ eV) electrode of platinum. The effective oxide thickness

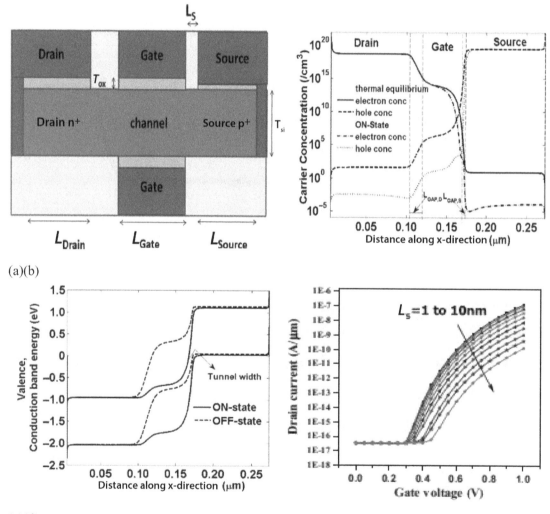

(a)(b)

(c)(d)

FIGURE 1.21 (a) Structure of dopingless TFET [51]. (b) Carrier concentration in thermal equilibrium and ON-states [18]. (c) Band diagram in ON-state ($V_{GS} = 1$ V, $V_{DS} = 1$ V) [18]. (d) I_D–V_{GS} curves for DLTFET for different spacer lengths L_S [51].

at source contact (EOT) is kept at 0.37 nm for more efficient band narrowing at the source-channel interface. The spacer SiO_2 between gate and source electrodes (L_S) must be minimum for the larger narrowing of tunnel barrier at the source-channel interface. The n-type region for the drain electrode is generated by a low work function ($\phi = 3.9$ eV) electrode of hafnium, which was also used in the DLT. To make the channel intrinsic, the work function of the gate electrode is kept at 4.5 eV, as in conventional TFETs. The thickness of oxide at the drain and gate electrodes is 3 nm. The required carrier concentration of p–i–n structure for tunneling has been realized with the use of certain electrodes, as shown in Figure 1.21b. The drain to source voltage (V_{DS}) is kept at 1 V. When a positive gate voltage of $V_{GS} = 1$ V is applied, the tunnel barrier between channel and source regions is lowered, shown in Figure 1.21c. This facilitates tunneling of electrons from the filled source region valence band to empty states in the conduction band of the channel region. The band diagrams of DLTFET in OFF- and ON-states are shown in Figure 1.20c. The I_D–V_{GS} curves for different L_S are shown in Figure 1.21d. The DLTFET I_D–V_{GS} characteristics have achieved the required TFET operation, showing low I_{OFF} and low SS.

1.11 CHALLENGES IN JLTFET AND DLTFET

These JLTFETs and DLTFETs have shown the required tunneling phenomenon and features of TFET devices. These devices can solve the problems of TFETs associated with junctions and their fabrication. But they are not free from the inherent problems of TFET. As all these devices work in a similar way, they also face similar challenges. The challenges associated with these JLTFETs and DLTFETs are of the same nature as in TFETs. Both JLTFETs and DLTFETs have low I_{ON}, like the conventional TFET. This problem of low I_{ON} hampers their ability to compete with MOSFETs. Both JLTFETs and DLTFETs suffer from the problem of ambipolarity, which needs to be worked upon [52][53]. This phenomenon of ambipolarity leads to negative conduction in these devices.

1.11.1 Low ON-State Current

Many techniques have been proposed for the improvement of ON-state behavior of JLTFETs and DLTFETs, which are charge plasma-based TFETs. The aim of our work is for the ON-state improvement of JLTFETs and DLTFETs only. The techniques for chemically doped TFETs are not mentioned here, although the strategy for ON-state improvement is common for both chemically doped and charge plasma-based TFETs, as they both are functionally the same. The use of a dual material gate with a lower work function electrode at the source side to narrow the tunnel barrier in ON-state has been proposed [54]–[56]. The use of hetrodielectric with a high-k dielectric at the source side also decreases the tunnel barrier in ON-state by enlarging the electric field due to high-k dielectric [57][58]. The conventional dielectric materials can be replaced by ferroelectric dielectric materials, which show negative capacitance. These negative capacitance dielectric have boosted the performance of JLTFETs [59]. The application of different low band-gap materials in heterojunction JLTFETs and DLTFETs, which narrow the tunnel barrier, have further enhanced their performance [60]–[62]. In these charge plasma-based JLTFETs and DLTFETs formed on an n^+ body, there is an unavoidable spacer between gate and p-gate/source electrodes. The presence of separate electrodes for channel and source regions gives rise to this spacer, which widens the tunnel barrier in ON-state [18][50]. This spacer can be eliminated by the use of a p-type body, which enriches the device performance [55]. The formation of more abrupt tunneling junctions can also be achieved by the use of a metal layer in the oxide region in the middle source and gate electrodes. This also narrows the tunnel barrier in ON-state, thereby improving the device performance. [53][63][64].

1.11.2 Ambipolarity in Charge Plasma TFETs

The issue of ambipolarity is inherent in these JLTFETs and DLTFETs. There have been many device architectures proposed for the reduction of ambipolarity. Ambipolarity is basically the conduction of electrons by tunneling from the valence band in the channel region to the conduction band in the drain region for negative V_{GS} or gate voltages. The narrowing of the tunnel barrier at the channel-drain interface for negative gate voltages is present in TFETs, which is also explained in section 1.8.3. In this section we have also focused only on techniques of ambipolar reduction for charge plasma TFETs, that is, JLTFET and DLTFET. The use of a triple material gate with low work function material on source and drain side has been proposed [20][56]. The low work function material on the source side is responsible for improvement in ON-state performance. The low work function material on the drain side widens the tunnel barrier in an ambipolar state, that is, negative gate voltages. Although this technique serves the purpose, it is very challenging from a scaling perspective. The other option of dual work function electrodes is their use on drain electrodes. The dual drain electrode also widens the tunnel barrier by changing the carrier concentration at the drain-channel interface in an ambipolar state [52][55][65][66]. The use of a low work function metal implant embedded in oxide under the spacer in the midst of the gate and drain electrodes has also been proposed, which alters the carrier concentration at the drain-channel interface [53]. This leads to broadening of the tunnel barrier in ambipolar state. The application of a thicker spacer (~15 nm) between the gate and drain electrodes also broadens the tunnel barrier by hampering the overlapping of the valence band (channel) and conduction band (drain) [64].

1.12 SIMULATION ENVIRONMENT OF JLTFET AND DLTFET

All simulations are implemented by a 2D device simulator, Silvaco Atlas [33]. The simulations for the junctionless TFET are simulated by taking the device specifications as given in [50]. To account for tunneling of carriers, the nonlocal BTBT model is activated. The effect of interface traps has also been included. Fermi–Dirac statistics are implemented in the intrinsic carrier concentration calculations needed in the SRH (Shockley–Read–Hall) expressions. The gate leakage current model has not been used, as we have presumed a high-k metal gate stack. The SRH recombination model and Lombardi mobility model are used to simulate the effect for high impurity atom. The quantum confinement model proposed by Hansch [34] is implemented in a similar manner, as in [50]. The bandgap narrowing BGN model is used to simulate the consequences of high doping. The effect of high-k spacers on silicon channel for junctionless TFET is taken similarly to that mentioned in [50], that is, by including the defects at the interface of the high-k spacer and semiconductor. The tunneling region has very fine mesh spacing to efficiently calculate the current due to tunneling.

1.13 CONCLUSION

In this chapter, we have studied the basic device physics governing the carrier transport in junctionless and dopingless devices. These junctionless or dopingless charge plasma-based devices have immense potential to replace chemically doped or junction devices at lower technology nodes. For better subthreshold performance or low-power applications, the tunnel field effect transistors based on charge plasma concept present a very good alternative to the conventional or chemically doped TFETs. However, these charge plasma-based junctionless transistors or junctionless TFET face some major challenges that need to be resolved for better device performance. The problem of GIDL in junctionless transistors can be suppressed by widening the tunnel barrier formed at the drain-channel interface or by reducing the hole concentration in the parasitic BJT. The use of an electrostatically doped drain is a very good proposed technique [67][4]. In junctionless TFETs, the problem of low ON-state current and ambipolarity have been highlighted in this chapter. These issues can be solved by the use of dual-k spacers or an oversized back gate in junctionless and dopingless devices [51][35]. Therefore, by employing these techniques, the performance of junctionless devices can be improved. These junctionless devices can be the future of microelectronic devices.

REFERENCES

[1] Behzad Razavi, *Fundamentals of Microelectronics*, Student edition. Wiley, 2006.

[2] I. Tuomi, "The Lives and Death of Moore's Law," *First Monday*, vol. 7, no. 11, 2002.

[3] "Electrical & Electronic Engineering: MOSFET, MOS structure, Threshold Voltage, MOSFET Different Rigion." [Online]. Available: http://eeeinvestigation.blogspot.com/2013/12/mosfet-mos-structure-threshold.html.

[4] M. A. Raushan, N. Alam, and M. J. Siddiqui, "Design Approach to Improve the Performance of JAMFETs," *IET Circuits, Devices & Systems*, vol. 14, no. 3, pp. 333–339, May 2020.

[5] Girish Wadhwa and B. Raj, "Design and Performance Analysis of Junctionless TFET Biosensor for High Sensitivity," *IEEE Nanotechnology*, vol. 18, pp. 567–574, 2019.

[6] "Semiconductor Device Physics | Electronics & Applied Physics Projects." [Online]. Available: http://ekim616.blogspot.com/2015/04/device-physics.html.

[7] J. Lilenfeld, "Method and Apparatus for Controlling Electric Currents," U.S Patent, 08 October 1926.

[8] Girish Wadhwa, Priyanka Kamboj, and B. Raj, "Design Optimisation of Junctionless TFET Biosensor for High Sensitivity," *Advances in Natural Sciences: Nanoscience and Nanotechnology*, vol. 10, pp. 045001, 2019.

[9] S. Gundapaneni, "Investigation of Junction-Less Transistor (JLT) for CMOS Scaling," PhD, Department of Electrical Engineering, Indian Institute of Technology, Bombay, 2012.

[10] T. K. Kim *et al.*, "First Demonstration of Junctionless Accumulation-Mode Bulk FinFETs with Robust Junction Isolation," *IEEE Electron Device Letters*, vol. 34, no. 12, pp. 1479–1481, 2013.

[11] S. Gundapaneni, S. Ganguly, and A. Kottantharayil, "Enhanced Electrostatic Integrity of Short-Channel Junctionless Transistor with High-k Spacers," *IEEE Electron Device Letters*, vol. 32, no. 10, pp. 1325–1327, 2011.

[12] Soniya, Shailendra Singh, Girish Wadhwa, and B. Raj, "An Analytical Modeling for Dual Source Vertical Tunnel Field Effect Transistor," *International Journal of Recent Technology and Engineering (IJRTE)*, vol. 8, no. 2, July 2019 (Scopus).

[13] Shailendra Singh and B. Raj, "Design and Analysis of Hetrojunction Vertical T-shaped Tunnel Field Effect Transistor," *Journal of Electronics Material*, Springer, vol. 48, no. 10, pp. 6253–6260, October 2019.

[14] C. Sahu and J. Singh, "Potential Benefits and Sensitivity Analysis of Dopingless Transistor for Low Power Applications," *IEEE Transactions on Electron Devices*, vol. 62, no. 3, pp. 729–735, 2015.

[15] Jeetendra Singh and B. Raj, "Design and Investigation of 7T2M NVSARM with Enhanced Stability and Temperature Impact on Store/Restore Energy", *IEEE Transactions on Very Large Scale Integration Systems*, vol. 27, no. 6, pp. 1322–1328, June 2019.

[16] Anil Kumar Bhardwaj, Sumeet Gupta, B. Raj, and Amandeep Singh, "Impact of Double Gate Geometry on the Performance of Carbon Nanotube Field Effect Transistor Structures for Low Power Digital Design," *Computational and Theoretical Nanoscience, ASP*, vol. 16, pp. 1813–1820, 2019.

[17] Neeraj Jain and B. Raj, "Thermal Stability Analysis and Performance Exploration of Asymmetrical Dual-k Underlap Spacer (ADKUS) SOI FinFET for Security and Privacy Applications," *Indian Journal of Pure & Applied Physics (IJPAP)*, vol. 57, pp. 352–360, May 2019 (SCI).

[18] Jeetendra Singh and B. Raj, "Tunnel Current Model of Asymmetric MIM Structure Levying Various Image Forces to Analyze the Characteristics of Filamentary Memristor," *Applied Physics A*, Springer, vol. 125, no. 3, pp. 203.1–203.11, February 2019 (SCI).

[19] Candy Goyal, Jagpal Singh Ubhi, and B. Raj, "Low Leakage Zero Ground Noise Nanoscale Full Adder using Source Biasing Technique," *Journal of Nanoelectronics and Optoelectronics*, American Scientific Publishers, vol. 14, pp. 360–370, March 2019.

[20] K. Nigam, S. Pandey, P. N. Kondekar, D. Sharma, and P. Kumar Parte, "A Barrier Controlled Charge Plasma-Based TFET with Gate Engineering for Ambipolar Suppression and RF/Linearity Performance Improvement," *IEEE Transactions on Electron Devices*, vol. 64, no. 6, pp. 2751–2757, 2017.

[21] A. Sirohi, C. Sahu, S. Member, J. Singh, and S. Member, "Analog/RF Performance Investigation of Dopingless FET for Ultra-Low Power Applications," *IEEE Access*, vol. 7, p. 1, 2019.

[22] S. Gundapaneni, M. Bajaj, R. K. Pandey, K. V. R. M. Murali, S. Ganguly, and A. Kottantharayil, "Effect of Band-to-Band Tunneling on Junctionless Transistors," *IEEE Transactions on Electron Devices*, vol. 59, no. 4, pp. 1023–1029, 2012.

[23] S. Ramaswamy and M. J. Kumar, "Raised Source/Drain Dopingless Junctionless Accumulation Mode FET: Design and Analysis," *IEEE Transactions on Electron Devices*, vol. 63, no. 11, pp. 4185–4190, 2016.

[24] S. Sahay and M. J. Kumar, "Insight into Lateral Band-to-Band-Tunneling in Nanowire Junctionless FETs," *IEEE Transactions on Electron Devices*, vol. 63, no. 10, pp. 4138–4142, 2016.

[25] S. Sahay and M. J. Kumar, "Symmetric Operation in an Extended Back Gate JLFET for Scaling to the 5-nm Regime Considering Quantum Confinement Effects," *IEEE Transactions on Electron Devices*, vol. 64, no. 1, pp. 21–27, 2017.

[26] Neeraj Jain and B. Raj, "Analysis and Performance Exploration of High-k SOI FinFETs Over the Conventional Low-k SOI FinFET toward Analog/RF Design," *Journal of Semiconductors (JoS)*, IOP Science, vol. 39, no. 12, pp. 124002–1–7, December 2018.

[27] Candy Goyal, Jagpal Singh Ubhi, and B. Raj, "A Reliable Leakage Reduction Technique for Approximate Full Adder with Reduced Ground Bounce Noise," *Journal of Mathematical Problems in Engineering*, Hindawi, vol. 2018, Article ID 3501041, 16 pp., 15 October 2018.

[28] S. Sahay and M. J. Kumar, "Physical Insights into the Nature of Gate-Induced Drain Leakage in Ultrashort Channel Nanowire FETs," *IEEE Transactions on Electron Devices*, vol. 64, no. 6, pp. 2604–2610, 2017.

[29] S. Sahay, S. Member, M. J. Kumar, and S. Member, "Nanotube Junctionless FET: Proposal, Design, and Investigation," *IEEE Transactions on Electron Devices*, vol. 64, no. 4, pp. 1851–1856, 2017.

[30] S. Sahay and M. J. Kumar, "A Novel Gate-Stack-Engineered Nanowire FET for Scaling to the Sub-10-nm Regime," *IEEE Transactions on Electron Devices*, pp. 1–5, 2016.

[31] S. Sahay and M. J. Kumar, "Spacer Design Guidelines for Nanowire FETs From Gate-Induced Drain Leakage Perspective," *IEEE Transactions on Electron Devices*, vol. 64, no. 7, pp. 3007–3015, 2017.

[32] N. Junctionless, F. Effect, S. Sahay, S. Member, M. J. Kumar, and S. Member, "Diameter Dependence of Leakage Current in Nanowire Junctionless Field Effect Transistors," *IEEE Transactions on Electron Devices*, vol. 64, no. 3, pp. 1330–1335, 2017.

[33] "Atlas User's Manual," Silvaco, Santa Clara, CA, USA, 2013.

[34] Jeetendra Singh and B. Raj, "Temperature Dependent Analytical Modeling and Simulations of Nanoscale Memristor," *Engineering Science and Technology, an International Journal*, Elsevier's, vol. 21, pp. 862–868, October 2018.

[35] Akanksha Jaiswal, R. K. Sarin, B. Raj and Shikha Sukhija, "A Novel Circular Slotted Microstrip-fed Patch Antenna with Three Triangle Shape Defected Ground Structure for Multiband Applications," *Advanced Electromagnetic (AEM)*, vol. 7, no. 3, pp. 56–63, August 2018 (Scopus).

[36] Girish Wadhwa and B. Raj, "Label Free Detection of Biomolecules using Charge-Plasma-Based Gate Underlap Dielectric Modulated Junctionless TFET," *Journal of Electronic Materials (JEMS)*, Springer, vol. 47, no. 8, pp. 4683–4693, August 2018.

[37] S. Sahay and M. J. Kumar, "Controlling the Drain Side Tunneling Width to Reduce Ambipolar Current in Tunnel FETs Using Heterodielectric BOX," *IEEE Transactions on Electron Devices*, vol. 62, no. 11, pp. 3882–3886, 2015.

[38] K. Kim and J. G. Possum, "Double-gate CMOS: Symmetrical- versus Asymmetrical-Gate Devices," *IEEE Transactions on Electron Devices*, vol. 48, no. 2, pp. 294–299, Feb. 2001.

[39] Girish Wadhwa and B. Raj, "Parametric Variation Analysis of Charge-Plasma-based Dielectric Modulated JLTFET for Biosensor Application," *IEEE Sensor Journal*, vol. 18, no. 15, 1 August 2018.

[40] Jeetendra Singh and B. Raj, "Comparative Analysis of Memristor Models for Memories Design", *JoS, IoP*, vol. 39, no. 7, pp. 074006–1–12, July 2018.

[41] Divya Yadav, Shailesh Singh Chouhan, Santosh Kumar Vishvakarma and B. Raj, "Application Specific Microcontroller Design for IoT based WSN," *Sensor Letter*, ASP, vol. 16, pp. 374–385, May 2018.

[42] Gurmohan Singh, R. K. Sarin, and B. Raj, "Fault-Tolerant Design and Analysis of Quantum-Dot Cellular Automata Based Circuits," *IEEE/IET Circuits, Devices & Systems*, vol. 12, pp. 638–664, 2018.

[43] Jeetendra Singh and B. Raj, "Modeling of Mean Barrier Height Levying Various Image Forces of Metal Insulator Metal Structure to Enhance the Performance of Conductive Filament Based Memristor Model," *IEEE Nanotechnology*, vol. 17, no. 2, pp. 268–267, March 2018.

[44] Aakash Jain, Sanjeev Sharma, and B. Raj, "Analysis of Triple Metal Surrounding Gate (TM-SG) III-V Nanowire MOSFET for Photosensing Application," *Opto-electronics Journal*, Elsevier, vol. 26, no. 2, pp. 141–148, May 2018.

[45] R. Nathanae, V. Pott, H. Kam, J. Jeon, and T. J. K. Liu, "4-Terminal Relay Technology for Complementary Logic," in *Technical Digest – International Electron Devices Meeting, IEDM*, 2009.

[46] A. Saeidi, "Exploration of Negative Capacitance Devices and Technologies," in *NANOLAB EPFL*, Lausanne, 2019.

[47] J. Appenzeller, Y.-M. Lin, J. Knoch, and P. Avouris, "Band-to-Band Tunneling in Carbon Nanotube Field-Effect Transistors," *Physical Review Letters*, vol. 93, no. 19, p. 196805, November 2004.

[48] A. Seabaugh, "Tunnel Field-effect Transistors – Status and Prospects," in *68th Device Research Conference*, vol. 3, no. 3, pp. 11–14, 2010.

[49] A. M. Ionescu and H. Riel, "Tunnel Field-Effect Transistors as Energy-Efficient Electronic Switches," *Nature* no. 479, pp. 329–337, 2011.

[50] B. Ghosh and M. W. Akram, "Junctionless Tunnel Field Effect Transistor," *IEEE Electron Device Letters*, vol. 34, no. 5, pp. 584–586, 2013.

[51] M. A. Raushan, N. Alam, M. W. Akram, and M. J. Siddiqui, "Impact of Asymmetric Dual-k Spacers on Tunnel Field Effect Transistors," *Journal of Computational Electronics*, vol. 17, no. 2, pp. 756–765, June 2018.

[52] Neeraj Jain and B. Raj, "Parasitic Capacitance and Resistance Model Development and Optimization of Raised Source/Drain SOI FinFET Structure for Analog Circuit Applications," *Journal of Nanoelectronics and Optoelectronics*, ASP, USA, vol. 13, pp. 531–539, April 2018.

[53] Shradhya Singh, S. K. Vishvakarma, and B Raj, "Analytical Modeling of Split-Gate Junction-Less Transistor for a Biosensor Application," *Sensing and Bio-sensing*, Elsevier, vol. 18, pp. 31–36, April 2018.

[54] G. Saiphani Kumar, Amandeep Singh, and B. Raj, "Design and Analysis of Gate All Around CNTFET based SRAM Cell Design," *Journal of Computational Electronics, Springer*, vol. 17, no. 1, pp. 138–145, March 2018.

[55] S. Tirkey, D. Sharma, B. R. Raad, and D. S. Yadav, "A Novel Approach to Improve the Performance of Charge Plasma Tunnel Field-Effect Transistor," *IEEE Transactions on Electron Devices*, pp. 1–8, 2017.

[56] Anju, S. Tirkey, K. Nigam, S. Pandey, D. Sharma, and P. Kondekar, "Investigation of Gate Material Engineering in Junctionless TFET to Overcome the Trade-off between Ambipolarity and RF/Linearity Metrics," *Superlattices and Microstructures*, pp. 1–9, 2017.

[57] P. Venkatesh, K. Nigam, S. Pandey, D. Sharma, and P. N. Kondekar, "Impact of Interface Trap Charges on Performance of Electrically Doped Tunnel FET With Heterogeneous Gate Dielectric," *IEEE Transactions on Device and Materials Reliability*, vol. 17, no. 1, pp. 245–252, 2017.

[58] Gurinder Pal Singh, B. S. Sohi, and B. Raj, "Material Properties Analysis of Graphene Base Transistor (GBT) for VLSI Analog Circuits," *Indian Journal of Pure & Applied Physics (IJPAP)*, vol. 55, pp. 896–902, December 2017.

[59] Neeraj Jain and B. Raj, "Impact of Underlap Spacer Region Variation on Electrostatic and Analog/RF Performance of Symmetrical High-k SOI FinFET at 20 nm Channel Length," *Journal of Semiconductors (JoS)*, IOP Science, vol. 38, no. 12, pp. 122002, December 2017.

[60] Gurmohan Singh, R. K. Sarin, and B. Raj, "Design and Analysis of Area Efficient QCA Based Reversible Logic Gates," *Journal of Microprocessors and Microsystems*, Elsevier, vol. 52, pp. 59–68, May 2017.

[61] S. Ahish, D. Sharma, Y. B. N. Kumar, and M. H. Vasantha, "Performance Enhancement of Novel InAs/Si Hetero Double-Gate Tunnel Fet using Gaussian Doping," *IEEE Transactions on Electron Devices*, vol. 63, no. 1, pp. 288–295, 2016.

[62] K. Nigam, P. Kondekar, and D. Sharma, "DC Characteristics and Analog/RF Performance of Novel Polarity Control GaAs-Ge based Tunnel Field Effect Transistor," *Superlattices and Microstructures*, vol. 92, pp. 224–231, 2016.

[63] B. R. Raad, S. Tirkey, D. Sharma, and P. Kondekar, "A New Design Approach of Dopingless Tunnel FET for Enhancement of Device Characteristics," *IEEE Transactions on Electron Devices*, vol. 64, no. 4, pp. 1830–1836, 2017.

[64] M. Aslam, S. Yadav, D. Soni, and D. Sharma, "A New Design Approach for Enhancement of DC/RF Performance with Improved Ambipolar Conduction of Dopingless TFET," *Superlattices and Microstructures*, vol. 112, pp. 86–96, 2017.

[65] B. Raj, A. K. Saxena, and S. Dasgupta, "Nanoscale FinFET Based SRAM Cell Design: Analysis of Performance Metric, Process Variation, Underlapped FinFET and Temperature Effect," *IEEE Circuits and System Magazine*, vol. 11, no. 2, pp. 38–50, 2011.

[66] S. Yadav, D. Sharma, D. Soni, and M. Aslam, "Controlling Ambipolarity with Improved RF Performance by Drain/Gate Work Function Engineering and Using High κ Dielectric Material in Electrically Doped TFET: Proposal and Optimization," *Journal of Computational Electronics*, vol. 16, pp. 1–11, 2017.

[67] M. A. Raushan, N. Alam, and M. J. Siddiqui, "Electrostatically Doped Drain Junctionless Transistor for Low-Power Applications," *Journal of Computational Electronics*, vol. 18, no. 3, pp. 864–871, 2019.

A Novel Vertical Tunnel FET and Its Application in Mixed Mode

2

Mr. K. Vanlalawmpuia and Dr. Brinda Bhowmick

Contents

DOI: 10.1201/9781003126645-2

2.1 INTRODUCTION

Since the first time the principle of the field-effect transistor (FET) was demonstrated in 1960 and the implementation of the complementary metal-oxide-semiconductor (CMOS) was realized in 1963, a tremendous persistent progress in the field of CMOS technology has been perceived [1–3]. Following the well-known Moore's law that states that the number of transistors being used in the integrated circuits (ICs) should be doubled every 18 months by scaling the dimensions of transistors [4], probably the biggest test confronting the CMOS technology is the way to produce the up-and-coming generation of superior, high efficiency, low powered devices while preserving its performance and circuit speed and reducing the cost per function, as well as its power consumption. It is noteworthy that regardless of many daunting technological difficulties, the tendency of miniaturization of the size of semiconductor devices and the subsequent exponential development and growth of the semiconductor industries have proceeded unabatedly for over many decades. The challenges accompanying the upcoming miniaturization of transistors, which at first appeared insurmountable, were eventually overcome using ingenious methods before the phenomenon could be disturbed.

However, as transistors are being continuously downscaled to the sub-50 nm regime or lower, they face a genuine threat as a result of the incapability of transistors to switch from the OFF-state to the ON-state without prompting an increase in the power dissipation, and they possess a high subthreshold swing (SS) [5], [6]. Here, the SS is described as the amount of voltage needed to lower the current by an order of magnitude and is limited to 2.3 kT/q, which is 60 mV per decade at room temperature for a metal-oxide-semiconductor field effect transistor (MOSFET) [7]. The physics relating to the main operating mechanism of the conventional CMOS transistor impose certain limitations that preclude further reduction in the supply voltage (V_{DD}) in the integrated circuit. Consequently, the existing trend of scaling down the transistors while maintaining a constant power supply voltage fails to improve the energy efficiencies in ICs that might have been acquired via reducing the power supply voltage [8]. This requires investigating and developing transistors that are dependent on new working principles, better architectural design, and even maybe innovative non-silicon-based material transistors that could surmount the shortcomings of the traditional transistors [6]. In this regard, tunnel field effect transistors (TFETs), which work based on the principle of quantum mechanical tunneling, have been shown to own a remarkable switching characteristic that goes beyond the theoretical abilities of the traditional transistors, such as the MOSFET, whose operating mechanism is due to the thermionic emission of carriers over a potential barrier [9–11]. TFETs can be implemented in ICs with low power supply voltage, which results in an exceptionally energy efficient circuit; this makes TFETs interesting for researchers and the semiconductor industries and has prompted extensive study in recent times in both theoretical and experimental fields [12], [13].

2.2 BASIC STRUCTURE AND PRINCIPLE OF OPERATION OF TFETS

Simply stated, the TFET is a gated reverse biased p–i–n device working on the principle based on the quantum mechanical band-to-band tunneling (BTBT) of carriers from the valence band (E_V) of the source region to the conduction band (E_C) of the channel region through the forbidden energy bandgap. The main distinctive feature of a TFET is the type of dopants in the source region and drain region, which is different in both the regions, where the source is doped with p+ and the drain with n+ for an

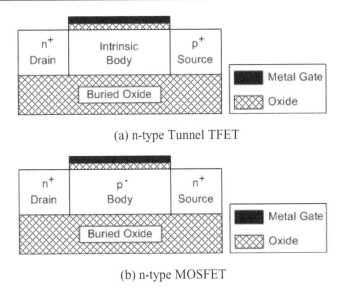

(a) n-type Tunnel TFET

(b) n-type MOSFET

FIGURE 2.1   Basic structures of (a) an n-type TFET and (b) an n-type MOSFET.

n-type TFET as opposed to n⁺ for both source and drain doping in the case of an n-type MOSFET, as illustrated in Figure 2.1. However, the channel region of the TFET is intrinsic or lowly doped. An insulator is employed to separate the gate from the channel region. A TFET is termed as n- or p-type according to the channel's dominant carrier, which is electrons for an n-type TFET or holes for p-type TFET, respectively. Here, the chapter focuses only on the n-type TFET as the principle of operation is similar for both types of TFET (p-type injection into the channel from the source is holes whereas it is electrons for n-type). However, it should be noted that regardless of the type (n-type or p-type), TFETs are ambipolar devices in comparison to MOSFETs. For instance, in an n-type TFET, if the gate is highly negatively biased, electrons being the main contribution in the current, transport may show p-type=like behavior.

For an n-type TFET, a positive gate to source voltage (V_{gs}) is applied to the gate electrode, which then induces a band bending at the source/channel junction, resulting in tunneling of electrons from E_V of the source region to E_C of the channel region. The electrons that tunneled to the channel region are then swept away to the drain terminals due to the positive biasing at the drain side. At this time, the transistor is perceived to be in the ON-state, whereas when the gate bias is zero and no tunneling occurs from the source to the channel region, it is considered as the OFF-state, as depicted in Figure 2.2. An in-depth discussion on the tunneling phenomenon is demonstrated in section 2.3. One major benefit accompanying the BTBT is the lack of the thermal (kT) dependency which makes for shaper than 60 mV/decade turn ON characteristic at T − 300 K. The resultant band to band tunneling current is mathematically defined as (detailed equations given in section 2.3) [14]:

$$I_{BTBT} \simeq A\varepsilon^2 \exp\left[-\frac{B}{\varepsilon}\right] \tag{2.1}$$

where A and B are parameters dependent on the semiconductor material. Since the p–i–n structure of the TFET is in reversed bias in the OFF-state, the diode leakage currents are the main contribution to the off current (I_{OFF}), and are substantially lower than the OFF-state current of the MOSFET [15].

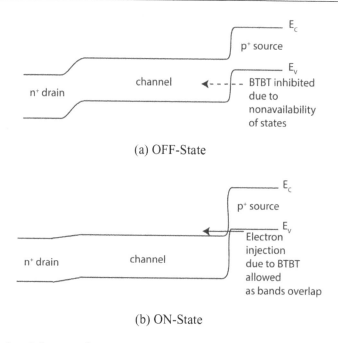

(a) OFF-State

(b) ON-State

FIGURE 2.2   Energy band diagram of an n-type TFET in the (a) OFF-state and (b) ON-state.

2.3 PHYSICS IN BAND-TO-BAND TUNNELING

The band-to-band tunneling (BTBT) is based on the phenomena of quantum mechanics in which the electrons tunneled through the forbidden energy bandgap of the semiconductor. From a physics point of view, BTBT is the snipping of electrons from the covalent bonds created among semiconductor atoms. Here, the band-to-band tunneling generation rate (G_{BTBT}) will be derived mathematically for a better understanding on how the BTBT mechanism attributes to a steeper subthreshold slope. It will start with the establishment of the general tunneling concept, along with the presumptions made for the derivations. The tunneling probability (T), current density (J) and generation rate (G) will then be determined with the help of the Wentzel–Kramers–Brillouin (WKB) approximation.

2.3.1 Presumptions Made for Derivation

The first articulation on the tunneling probability was determined by Kane utilizing time-dependent disruption hypothesis in addition to Fermi's golden rule [16], [17]. However, the section would rather utilize a rearranged along with a more instinctive way in dealing with a similar undertaking dependent on the WKB estimation [18]. The evaluation is done in accordance with tunneling of electrons through a reverse biased pn junction as shown in Figure 2.3.

So as to acquire a close form solution of G_{BTBT}, the accompanying presumptions made are essential:

1. Presume a direct bandgap semiconductor such that the momentum impact due to phonon scatterings could be neglected. In this manner, a simple analytical approach for G_{BTBT} can be achieved.

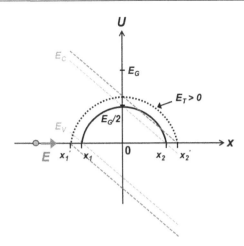

FIGURE 2.3 Representation of the energy band profile inside the depleted regions of a pn junction. The uniform electric field is represented in gray dotted lines. The imaginary distribution relations could be estimated by symmetrical two-band correlation. E denotes the energy of electron while U represents the potential barrier height. Initiation of the energy (E_T) accompanied with the transverse momentum alters the classical turning points into x_1' as well as x_2'.

2. Presume high doping rates in the semiconductor; therefore, the Fermi level is precisely aligned with the p-type semiconductor valence band ($E_V = F_P$) and the n-type semiconductor conduction band ($E_C = F_N$). In addition, we assume that states in E_V are completely filled by electrons and E_C is entirely vacant, such that the Fermi distribution is simplified ($f_v - f_c = 1$).
3. Adopt a consistent electric field through the pn junction.
4. Presume symmetrical two-band correlation for modeling the imaginary wave vector distribution relationship inside the forbidden bandgap for close form G_{BTBT} solution.

2.3.2 Evaluation of the Imaginary Dispersion Relation

According to the Wentzel–Kramers–Brillouin concept, the tunneling probability (T) is given by [18]:

$$T \simeq \exp\left[-2\int_{x_1}^{x_2} k(x)\,dx\right] \tag{2.2}$$

where x_1 and x_2 are classical turning points. To determine equation 2.2, the imaginary wave vector distribution correlation $k(x)$ should be established. In accordance with the generalized E–k relation, we have [18]:

$$k(E_x) = \sqrt{\frac{2m^*}{\hbar^2}(E_X - U)} \tag{2.3}$$

where E_X denotes the energy of the incident carriers. U determines the potential barrier heights and can be evaluated by utilizing symmetric two-band relation in [19] and assuming an even electric field throughout the pn junction:

$$(E_X - U) = c\left[x - \frac{E_G}{2q\varepsilon}\right]\left[x + \frac{E_G}{2q\varepsilon}\right] \tag{2.4}$$

From boundary conditions, that is, $U = E_G /2$ when x=0, which is accompanied by the consistent electric field, the coefficient factor c can be determined. The definitive expression for potential barrier is given as:

$$(E_X - U) = -2 \left[\frac{\left(\frac{E_G}{2}\right)^2 - (q\varepsilon x)^2}{E_G} \right] \quad (2.5)$$

To consider tunneling in 3D semiconductors, the entire momentum of the incident electron is to be considered, which includes the momentum in the tunneling path as well as perpendicular to the tunneling direction. The energy related to the transversal momentum is compensated for by adding the term E_T in equation 2.5 [14].

$$(E_X - U) = -2 \left[\frac{\left(\frac{E_G}{2}\right)^2 - (q\varepsilon x)^2}{E_G} + E_T \right] \quad (2.6)$$

Therefore, the tunneling probability is express as:

$$T(E_X, E_T) \simeq \exp \left[-2 \int_{x_1'}^{x_2'} \sqrt{\frac{2m^*}{\hbar^2} \left[2 \frac{\left(\frac{E_G}{2}\right)^2 - (q\varepsilon x)^2}{E_G} + E_T \right]} \, dx \right] \quad (2.7)$$

It should be mentioned that the integral limit changes from x_1 and x_2 to x_1' and x_2' with the addition of E_T, as the transverse momentum causes an extra damping element of the wave function and thereby alters the conventional turning positions, which results in an enhanced effective tunnel barrier width, that is, x_1' and x_2'. The new turning positions could be calculated using:

$$2 \frac{\left(\frac{E_G}{2}\right)^2 - (q\varepsilon x)^2}{E_G} + E_T = 0 \quad (2.8)$$

Therefore, $x_{1,2}' = \pm \frac{1}{q\varepsilon} \left[\left(\frac{E_G}{2}\right)^2 + \left(\frac{E_G E_T}{2}\right) \right]^{1/2} \quad (2.9)$

To calculate the integration in equation 2.7, the following coordinate transformation is carried out [19]:

$$y = \frac{q\varepsilon}{\left[\left(\frac{E_G}{2}\right)^2 + \left(\frac{E_G E_T}{2}\right) \right]^{1/2}} x \quad (2.10)$$

Using equations 2.7 and 2.10, the tunneling probability is simplified [19]:

$$T(E_X, E_T) \simeq \exp\left[-2\int_{x_1'}^{x_2'} \sqrt{\frac{2m^*}{\hbar^2}} \frac{\left(\frac{E_G}{2}\right)^2 + \left(\frac{E_G E_T}{2}\right)}{q\varepsilon E_G^{1/2}} \left(1-y^2\right)^{1/2} dy\right] \tag{2.11}$$

By solving the integration and rearranging the terms:

$$T(E_X, E_T) \simeq \exp\left[-\frac{\pi m^{*1/2} E_G^{3/2}}{2\sqrt{2}q\varepsilon\hbar}\right] \exp\left[-\frac{\sqrt{2}\pi m^{*1/2} E_G^{1/2}}{2q\varepsilon\hbar}\right] = T_o \exp\left[-\frac{E_T}{E_o}\right] \tag{2.12}$$

where, $E_o = \dfrac{\sqrt{2}\, q\varepsilon\hbar}{\pi m^{*1/2} E_G^{1/2}}$

2.3.3 Calculations of the BTBT Current and Generation Rate

By utilizing the tunneling probability expression in equation 2.12, the differential tunneling current density (dJ) could be determined as [19]:

$$dJ = q\left[\frac{1}{\hbar}\frac{\partial E_X}{\partial k_X}\right]\left[\frac{2}{(2\pi)^3} dk_X\, dk_Y\, dk_Z\right] T(E_X, E_T)(f_V - f_C) \tag{2.13}$$

where q denotes the charge, the second term is the velocity in k-space, the third term determines the amount of states/volume in k-space, the fourth term indicates the probability of tunneling, and the last term denotes the occupancy of states.

The velocities of the incident electrons were represented as group velocity of the valence band electrons wave packets [18]. Equation 2.13 could be furthermore generalized by implying that the valence band is entirely electron-filled, as well as the conduction band is vacant in the following manner: $f_V - f_C = 1$. The transverse momentum (k_Y, k_Z) can indeed be interpreted as a transverse wave vector, k_T [14].

$$k_T^2 = k_Y^2 + k_Z^2 \tag{2.14}$$

$$dk_Y\, dk_Z = 2\pi k_T\, dk_T \tag{2.15}$$

The transverse dispersion relation can then be utilized to shift the coordinates via momentum to energy space [14]:

$$E_T = \frac{\hbar^2 k_T^2}{2m^*} \tag{2.16}$$

$$2\pi k_T\, dk_T = \frac{2\pi m^*}{\hbar^2} dE_T \tag{2.17}$$

Replacing equation 2.17 in 2.13, the term for the differential tunneling current is reduced as:

$$dJ = \frac{q\,m^*}{2\pi^2\,\hbar^3} T(E_X, E_T)\, dE_X\, dE_T \tag{2.18}$$

By integrating both the normal energy and transverse energy, the tunneling current (J) could be calculated as:

$$J = \frac{q\,m^*}{2\pi^2\,\hbar^3} \int_{F_p=0}^{F_N} \int_0^E T(E_X, E_T)\, dE_X\, dE_T \tag{2.19}$$

Keep in mind that the energy integration limit along the tunnel directions ranges from the Fermi levels of n-type semiconductor material to the Fermi levels of the p-type semiconductor material, as the Fermi levels were presumed to coincide with the edges of the band. The upper limit for transverse energy integration expands to E ($E_{MAX} = E_G/2$), since $E_T > E$ corresponds to tunneling into forbidden states. Nevertheless, as the probability for tunneling decreases exponentially with E_T, it is appropriate that the upper limit should be expanded to infinity without compromising generality and for convenience.

By changing the coordinates from energy to location, the band-to-band tunneling generation rate (G_{BTBT}) can be determined from equation 2.19 and represented as:

$$G_{BTBT} = \frac{q\,m^*}{2\pi^2\,\hbar^3} \left(\frac{1}{q} \frac{\partial E_X}{\partial x}\right) \int_0^\infty T(E_X, E_T)\, dE_T \tag{2.20}$$

As a constant electrical field has been presumed, the energy is directly proportionate to the position by [20]:

$$E_X = q\,\varepsilon\,x, \quad \frac{dE_X}{dx} = q\,\varepsilon \tag{2.21}$$

Substituting equation 2.21 in 2.20, the G_{BTBT} is reduced to:

$$G_{BTBT} = \frac{q\,m^*\,\varepsilon}{2\pi^2\,\hbar^3} \int_0^\infty T_o \exp\left[-\frac{E_T}{E_o}\right] dE_T \tag{2.22}$$

By the evaluation of equation 2.22, a closed form G_{BTBT} is attained by:

$$G_{BTBT} = \frac{q\,m^*\,\varepsilon}{2\pi^2\,\hbar^3} T_o \int_0^\infty \exp\left[-\frac{E_T}{E_o}\right] dE_T = \frac{q\,m^*\,\varepsilon}{2\pi^2\,\hbar^3} T_o E_o \tag{2.23}$$

therefore, $G_{BTBT} = \dfrac{\sqrt{2}\,q^2\,m^{*1/2}}{2\pi^3\,\hbar^2\,E_G^{1/2}}\,\varepsilon^2 \exp\left[-\dfrac{\pi\,m^{*1/2}\,E_G^{3/2}}{2\sqrt{2}\,q\,\varepsilon\,\hbar}\right] = A\,\varepsilon^2 \exp\left[-\dfrac{B}{\varepsilon}\right]$ (2.24)

where $A = \dfrac{\sqrt{2}\,q^2\,m^{*1/2}}{2\pi^3\,\hbar^2\,E_G^{1/2}}$ and $B = \dfrac{\pi\,m^{*1/2}\,E_G^{3/2}}{2\sqrt{2}\,q\,\hbar}$

Even though the earlier presumptions made were appropriate for the derivation of the G_{BTBT} in a closed form, equation 2.24 offers useful knowledge on the phenomenon of tunneling. It is clear that the tunneling coefficients A and B are reliant on the semiconductor materials, and a relatively small energy bandgap and effective mass will lead to an exponential G_{BTBT}.

2.4 ELECTRICAL PARAMETERS IN TFETS

A number of electrical parameters that are of interest in TFETs are the ON-state current (I_{ON}), OFF-state leakage current (I_{OFF}), current ratio (I_{ON}/I_{OFF}), average subthreshold swing (SS_{avg}), point subthreshold swing (SS_{point}), threshold voltage (V_{th}) and total gate capacitance (C_{gg}).

2.4.1 ON-State Current

As mentioned in previous sections, when a positive gate voltage is supplied, the band bending occurs in the source/channel region, enabling the tunneling of electrons, which is then collected by the positive drain bias. In terms of the TFET transfer characteristic, the ON-state current (I_{ON}) is defined and measured at the gate to source voltage, $V_{GS} = 1$ V, at a drain to source voltage, V_{DS}, of 0.5 to 1 V. A substantial amount of tunneling current flows under this condition.

2.4.2 OFF-State Current

Here, the TFET is considered to be in the OFF-state when $V_{GS} = 0$ V with V_{DS} of 0.5 to 1 V. During this time, no band bending takes place, ensuring that only a relatively low leakage current flows of the order of around 10^{-13} A.

2.4.3 Current Ratio

Current ratio is defined as the rate of the ON-state to the OFF-state leakage currents, where the ON-state current is evaluated at $V_{GS} = 1$ V at V_{DS} of 0.5 to 1 V and OFF current at $V_{GS} = 0$ V at V_{DS} of 0.5 to 1 V. Typically, the I_{ON}/I_{OFF} ratio of a TFET is ~10^8.

2.4.4 Subthreshold Swings

The subthreshold swing (SS) of a transistor is determined as the amount of gate voltage needed to lower the drain current by an order of magnitude. It is a vital parameter in determining the switching characteristics of a transistor and crucial in circuit designing. Mathematically, it is expressed as:

$$SS = \frac{\partial V_{gs}}{\partial \left(\log I_D \right)} \text{ mV / decade} \tag{2.25}$$

In the subthreshold region, it is noteworthy that traditional MOSFETs differ from TFETs. A MOSFET's subthreshold slope is constrained by the device's diffusion–current dynamics in weak inversion; hence, the lowest theoretical SS for an ideal MOSFET is around 60 mV per decade at room temperature (300^0 K). The current conduction mechanism in TFETs, however, is dependent on the tunnel barrier length instead of depending upon inversion layers and does not have the same physical constraint as MOSFETs. The SS in TFETs is not dependent on the thermal factor (kT/q), as the tunneling current is only slightly temperature dependent. Mathematically, the SS in TFETs is expressed as:

$$SS_{TFET} = \frac{V_{gs}^2}{5.75 \left(V_{gs}^2 + Const \right)} \text{ mV / decade} \tag{2.26}$$

where the *Const* is defined by the TFET dimension and material parameter. It is seen from equation (2.26) that the SS_{TFET} depends greatly on the gate to source bias as well as the TFET geometrical parameters. There are two types of subthreshold swing in TFET: point subthreshold swing (SS_{point}) and average subthreshold swing (SS_{avg}). SS_{point} is specified as the least value of the subthreshold swing anywhere on the transfer characteristic graph, generally measured at the steepest point in the curve as soon as the tunneling current starts flowing, while the SS_{avg} is measured from the point where the TFET begins to turn on up to the threshold current (I_T) (typically at 10^{-7} A). It is usually measured using constant current method.

2.4.5 Threshold Voltage

Previously, the TFET's threshold voltage (V_{TH}) had been defined on the basis of a constant current method, where the gate voltage for which the drain current is 10^{-7} A is widely employed. Nevertheless, this technique utilizes a random value and possesses virtually no practical significance. However, a recent concept of the V_{TH} of the TFETs has emerged. It is specified as the gate voltage for which the energy band narrowing begins to get saturated with the application of the gate bias.

2.4.6 Total Gate Capacitance

In TFETs, the contributions to the total gate capacitance (C_{gg}) are mainly due to the gate to drain capacitance (C_{gd}), also known as the Miller capacitance, while the gate to source capacitance (C_{gs}) attributes remarkably less to C_{gg} as a result of the existence of the source/channel tunnel barriers. Consequently, the C_{gs} increases when a positive gate voltage is applied, which reduces the channel/drain potential barriers. For a silicon-based TFET, it is seen that when the gate to source as well as the drain to source voltages are equal, the C_{gg} is solely dominated by the C_{gd}. The pinch-off point in TFET is at higher values of the drain to source voltage at higher gate to source voltage. The primary explanation is that for higher gate to source voltage, the band bending at the source/channel side is much higher, implying that a huge portion of the drain to source voltage appears on the source end. Therefore, the drain bias for a certain gate to source voltage tends to affect the barrier on the source side of the tunnel barrier until a decent voltage (say 1 V) is attained; beyond this point, the gate to drain capacitance tends to reduce gradually.

2.5 A FEW TECHNIQUES FOR IMPROVING THE PERFORMANCE OF TFETS

Although TFETs are deemed to replace the traditional MOSFET technologies, as discussed earlier, due to the physical operating principle, that is, BTBT of electrons, which is accountable for the current transport in TFETs, essentially give a low ON-state current (I_{ON}). Hence, the ON-state current acquired by the conventional TFET is remarkably low and therefore generally it does not satisfy the criteria by several orders of magnitude (I_{ON} should be at least ~100 µA/µm, making the I_{ON}/I_{OFF} ratio ~ 10^6 for $V_{DD} = 0.5$ V or lower). This is the major drawback of TFETs, and until this issue is resolved, TFETs are not suitable to be implemented for modern standard CMOS technologies. In this regard, much work on TFETs was devoted toward enhancing the transistor ON-state current.

The drive current or the tunneling currents in TFETs, which is directly proportional to the tunneling probability, is given by the approximated equation of Wentzel–Kramers–Brillouin (WKB), which is expressed as [21], [22]:

$$I_D \; \alpha \; \exp\left[-\frac{4\sqrt{2m^*}\,E_g^{*3/2}}{3\,|e|\,\hbar\left(E_g^* + \Delta\Phi\right)}\sqrt{\frac{\varepsilon_{Si}}{\varepsilon_{ox}}}\,t_{ox}\,t_{Si}\right]\Delta\Phi \tag{2.27}$$

where e designates the electronic charge, the effective carrier mass is denoted by m^*, E_g^* represents the bandgap of the semiconductor material at the tunneling region, the dielectric constant of the semiconductor and the gate oxide are ε_{Si} and ε_{ox}, respectively, and t_{Si} and t_{ox} denote the depth of the device (which is for silicon material in this context) and the thickness of the gate oxide, correspondingly. The energy extends for which tunneling takes place is denoted by $\Delta\Phi$, and $\hbar$ denotes the reduced Planck's constant. It is worth noting that the WKB approximation for the drain current (I_D) in equation 2.27 in all cases may not be correct. For instance, in the case of a direct bandgap semiconductor such as indium arsenide and heterojunction such as silicon/germanium or when quantum effects and Phonon-assisted tunneling turn out to be dominant [23]. From equation 2.27 it can be concluded that the tunneling current could be improved using the methods shown here:

 i. Reducing the gate insulator thickness (t_{ox}).
 ii. Using an oxide material that has a high dielectric constant (ε_{ox}).
 iii. Using different materials that have an effective carrier mass (m^*) and energy bandgap (E_g^*) lower than silicon.

These methods are also combined to give the optimum performance of the device, not only for enhancing I_{ON} but also for reduction in the leakage current, and thus for improvement in the subthreshold swing, which is a crucial electrical characteristic in any transistor devices. Apart from the techniques mentioned, many researchers have been developing and exploring new strategies and techniques in terms of both modifying the structural architectures (TFET device structure design engineering) as well as materials engineering. In a more extensive manner, several methods to enhance the device performance of the TFET can be categorized accordingly.

2.5.1 Gate Engineering

As the gate controls the band bending and the carrier injection density in TFETs, gate engineering is found to have a tremendous impact on the transistor ON-state current. This implies using multiple gates, including using a double gate instead of the traditional single gated TFET. The double gate enhances the coupling of the gate voltage with the channel potential because the tunneling takes place from two areas, underneath each of the gates at the source/channel junction [24]. For an ultra-thin body double gate TFET (DG-TFET), the increment in the I_{ON} is more than twice as that of a single gate TFET because of the impact of both the gates [25]. In addition to this, the gate all around TFET (GAA TFET) is another method to control the channel and boost I_{ON} [26]. Another technique is the dual-material gate (DMG) TFET, in which different work function material gates are used and the existence of low work function gate material close to the source enhances the electric field, while high work function gate material near the channel/drain junction reduces the ambipolar current [27].

2.5.2 Materials Engineering

As silicon-based manufacturing technologies have advanced and improved over recent decades, TFETs dependent on silicon materials are profoundly attractive, as this enables TFETs to exploit the previously developed technologies of silicon processing and makes the implementation of TFETs in the latest CMOS process flow much simpler. However, as seen from the fundamental mechanics of tunneling in equation 2.27,

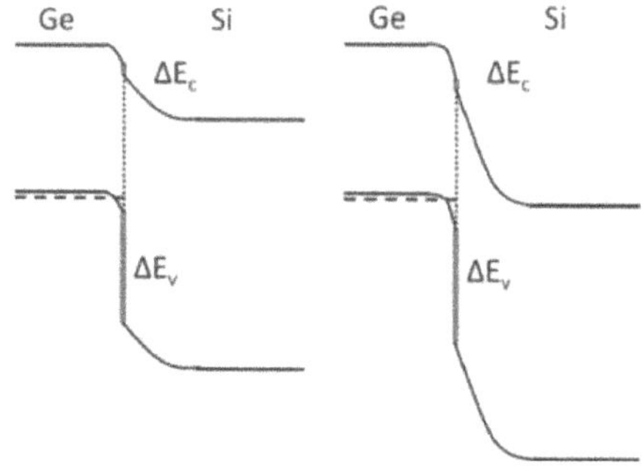

FIGURE 2.4 Energy band alignment in heterojunction Ge/Si n-type TFET.

semiconductor materials with a smaller bandgap are preferred for a greater tunnel probability and higher drive current. Owing to its wide bandgap and the indirect tunneling mechanism, silicon attributes to only low tunneling current in TFETs. As the silicon-based TFETs fail to meet the required ON-current for advanced technologies, another method of improving tunneling in the TFET is therefore to use a smaller bandgap material. This include using germanium, III–V semiconductors, carbon nanomaterials, etc. [28–31].

Germanium (Ge) is chemically identical to silicon (Si) material and is compatible to the current CMOS processes to a significant degree. In recent decades, the higher versatility and carrier mobility of germanium in contrast to silicon has drawn the interest of researchers. Germanium possesses a bandgap of 0.66 eV, which makes it a suitable candidate for improving the TFET performances. I_{ON} can be incremented by ~2,700 times by implementing germanium in the channel of the TFET instead of the conventional Si-based method [32]. A silicon-germanium (SiGe) alloy in which the mole fraction is carefully estimated is also suitable, as the effective bandgap in SiGe can be lesser than that of Si and Ge. An alternate way is to implement a heterojunction structure TFET in the tunneling junction in which germanium is employed at the source while silicon is used at the channel region. As the electron affinity differs from each other (4.0 eV for germanium and 4.05 eV for silicon), they form a staggered (type-II) heterojunction. The energy band showing the junction profile for ON-state and OFF-state of the germanium/silicon heterojunction n-type TFET is shown in Figure 2.4.

In Figure 2.4, the effective energy barrier for tunneling is calculated to be 0.61 eV that is, $E_g(Ge) - \Delta E_C$, which is somewhat lower than in pure germanium. Furthermore, as mentioned before, the process of incorporating of germanium in silicon technology is no issue, since in modern CMOS technologies, germanium is in fact utilized to cause compressive strains to improve hole mobility in p-type transistors [33]. The mathematical expression for the tunneling barrier of type-II heterojunctions among the materials x and y is given by:

$$E_{g,\,eff} = E_{g,\,x} - \Delta E_C = E_{g,\,y} - \Delta E_V \tag{2.28}$$

In particular, strained silicon and strained germanium of 2% and −2.7%, respectively grown on 42% of a silicon-germanium buffer can attain an effective bandgap of 122 milli-electron volt. Apart from these type-II heterojunctions, some III–V materials such as indium arsenide (InAs), having a bandgap of 0.36 eV, and an indium gallium arsenide (InGaAs) compound semiconductor alloy with a bandgap of 0.58 eV for $In_{0.7}Ga_{0.3}As$ and a bandgap of 0.74 eV for $In_{0.53}Ga_{0.47}As$ could be used to further enhance the performances of the TFET, depending on its application. These III–V material-based TFETs have shown great performance in terms of both I_{ON} and linear output, which have been proven both theoretically and practically [34], [35]. Equation 2.28 can also be used to find other staggered hetero-junctions for TFETs applications.

2.5.3 Doping Engineering

Another technique for enhancing the TFET in terms of I_{ON} is focusing on the doping techniques (especially the source region doping). Increasing the source doping concentration in TFETs reduces the

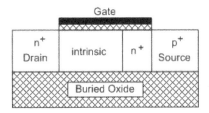

FIGURE 2.5 Pocket doping in TFET.

tunneling-barrier width due to the increase in the built-in electric field between the source and channel, owing to the bandgap narrowing. This increases the ON-state current substantially. It is noteworthy to mention that varying the source doping concentration does not in any way necessarily affect the OFF-state leakage currents in TFETs. Without changing the semiconductor material, the TFET ON-state current can also be enhanced by using a technique called pocket doping, as depicted in Figure 2.5.

The pocket region gets depleted and leaves an ionized positive charge. Here, the pocket region does not alter the potential of the intrinsic region for a gate voltage as well as drain voltage. Large electric fields around the tunneling region tend to increase the generation rate, thereby enhancing the tunneling current. Concurrently, this extra field often ensures that the current is raised directly after the inter-band tunneling is switched on. In consequence of this, the point subthreshold swing is additionally lessened. Optimization of the pocket doping concentration, width, and thickness and activating of the dopants have been explored and established for the pocket coped TFET, where an average subthreshold slope of 46 mV/decade with an I_{ON} = 1.4 µA/µm is attained [36].

A different method of doping technique, yet an effective option, is to implement a delta doped layer in the source region. The device structure is illustrated along with the energy band profiles in Figure 2.6.

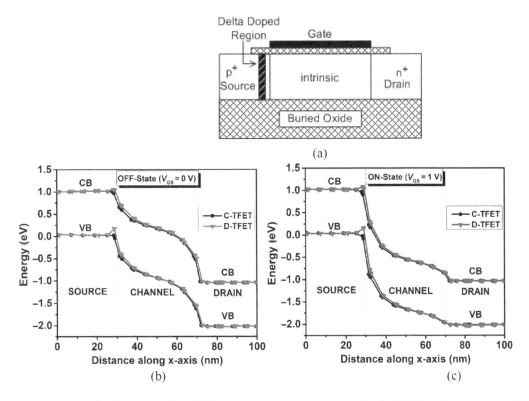

FIGURE 2.6 (a) Delta doped TFET (D-TFET) structure. Energy band profile of D-TFET and conventional TFET (C-TFET) in (b) OFF-state and (c) ON-state [37].

The incorporation of a heavily doped delta layer in the source area lays the groundwork for an increase in the tunneling capability and thereby offers higher drain current. In the same manner as the pocket doping, the length and the distance of inserting the delta doped layer inside the source should be optimized. The heavily doped delta layer (about 10^2 orders higher in doping concentration) is incorporated by using metal-organic chemical vapor deposition (MOCVD). As compared to the conventional silicon-based TFET, the tunneling length of the delta doped TFET(D-TFET) is reduced, which aids in more effective tunneling of electrons, seen in Figure 2.6c. The D-TFET can acquire a high I_{ON}/I_{OFF} current ratio of 10^{11} and a reasonable subthreshold swing of ~52 mV/decade [37].

2.6 LATERAL AND VERTICAL TFETS

Over the past years, researchers have developed novel TFET architectures to improve performance [38–41]. As the TFET principal operating mechanism is based on band-to-band tunneling (BTBT), depending on the design of the TFET structure, there are two distinctive types of tunneling modes that can be employed. The first one is the lateral tunneling, where, due to the application of a positive bias to the gate, the band starts to bend, and the electrons are injected from the source region into the channel region, primarily in a direction parallel to the semiconductor/gate oxide dielectric interface. Lateral tunneling is also sometimes stated to point tunneling [42] which is the fundamental BTBT mechanism and operating mode for a typical conventional TFET. Even though there is no exact definition of a vertical TFET, the convention adopted here is that a device is termed a vertical tunnel FET if the electron BTBT direction is predominantly associated with the direction of the electric field of the gate. Here the source region should be overlapped by the gate. As the direction of gate electric field and the electron tunneling are in the same direction or in alignment,

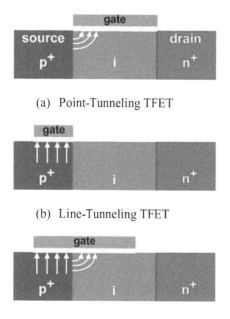

(a) Point-Tunneling TFET

(b) Line-Tunneling TFET

(c) Point + Line-Tunneling TFET

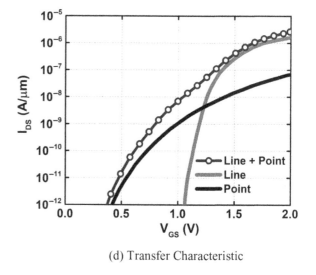

(d) Transfer Characteristic

FIGURE 2.7   TFET structures employing (a) point-tunneling, (b) line-tunneling, and (c) point plus line-tunneling TFET along with (d) the transfer characteristic curve of point, line, and line+point tunneling TFETs (arrows in the TFET structures indicate the electron tunneling paths) [45].

vertical tunneling is also referred to as line tunneling. In a vertical-based TFET, the structural dimensions and arrangements play a crucial role in determining the tunneling path length. The junction depletion width can also be lowered effectively. The gate to source overlap region can also determine the cross-section area of the tunneling junction. Due to this, line tunneling TFETs can attain a much higher ON current of about 2–3 orders of magnitude and a steeper subthreshold slope than point tunneling TFETs. Another advantage is that Vertical TFETs are also more immune to leakage currents as compared to Lateral TFETs [43], [44]. However, TFET based on lateral tunneling gives an advantage when it comes to ease of fabrication of the device. Compared to the regular CMOS process flow, a distinct as well as a converse ion implantation for the source region and drain region is the only additional step necessary. However, when considering the effect of physical parameters on electrical characteristics, the designing window decreases substantially. Another way of designing the TFET is to implement a structure that employs both line and point tunneling. This is done by partially overlapping of the gate and the source region. This can offer a sufficiently high ON current depending on the configuration, but the optimization of the overlap region should be precise. The basic TFET structures employing line tunneling, point tunneling, and both point + line tunneling along with the transfer characteristics are depicted in Figure 2.7 [45]. It should be noted that the central point of the current curve on the transfer characteristic can be shifted to the left by using a lower metal gate work function. The reduction of gate work function enhances the tunneling of carriers in the source/channel region and reduces the tunneling in the drain/channel region.

2.7 VARIOUS VERTICAL TFET ARCHITECTURES

In this section, various TFET structures utilizing vertical tunneling or both vertical and lateral tunneling are presented. The device structures are designed and optimized to give the optimum TFET performance by using different methods, including the techniques discussed in the previous sections.

2.7.1 Germanium Source with Overlap TFET

A germanium source TFET employing both line and point tunneling is shown in Figure 2.8a. A gate sidewall spacer of silicon nitride (Si_3N_4) that is 8 nm wide is used. An oxide thickness of 3 nm is used. The thickness of the silicon body is 70 nm, while the germanium source thickness, which the gate region overlaps, is of 21 nm. At the ON-state, an inversion surfaced layered electrons appears on the silicon body and a strong vertical electric field in the germanium source allows electrons to be tunneled through the germanium to the inverted layer of the germanium region to supply electrons and pass throughout the silicon channel region. The gate overlaps the source region, and this region is not sufficiently highly doped to allow the region to be inverted for acceptable gate voltages. In addition, appropriate functionality of the transistor ensures that the silicon body channel inversion threshold voltage is less than that of the onset of BTBT, such that drain voltage could be coupled to the germanium source surface potential, splitting quasi-Fermi levels to activate BTBT. At this point, the device possesses a small tunneling bandgap due to the overlap region (it induces vertical tunneling), which attributes to an off-state leakage current of 0.1 pA/μm for gate bias of 0.5 volt and an ON current of 0.4 μA/μm. Hence an I_{ON}/I_{OFF} current ratio of the order 10^6 at a low voltage operation of 0.5 V is achieved for the germanium source with overlap TFET [46].

7.2 L-Shaped Gate TFET

As the name suggest, the L-shaped gate TFET (LG-TFET) [47] is named thus because the gate region resembles the letter L. An n$^+$ pocket doping that is also the same shape as the gate is introduced

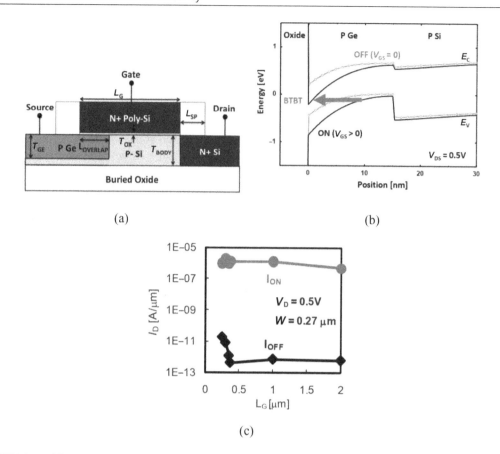

FIGURE 2.8 (a) Germanium source with overlap TFET, (b) energy band diagram, and (c) the ON- and OFF-state current of the germanium source with overlap TFET [46].

between the source and channel region; this allows the BTBT to take place in a direction both parallel and perpendicular to the channel. The utilization of both types of BTBT (i.e., line and point tunneling) due to the device design increases the gate electric field from the top of the tunneling junction when the transistor is in the ON-state. As the band-to-band generation rate is strongly dependent on the electric field, the enhanced electric field enhances the tunneling current significantly. The n^+ pocket is of 5 nm thickness, while the thickness of the silicon dioxide (SiO_2) is 2 nm. The height of the source region and drain region are 30 nm and 10 nm, correspondingly. The p^+ source is doped with doping concentrations of $1 \times 10^{20}/cm^3$; a doping concentration of $1 \times 10^{19}/cm^3$ is used for the drain and the pocket doped regions. The substrate is lightly doped with doping concentrations of $1 \times 10^{17}/cm^3$. The height and width of the L-shaped gate is 40 nm and 6 nm, respectively, with a work function of $\phi M = 4.17$ electron volt.

As seen from the transfer characteristic of the LG-TFET in Figure 2.9b, the inclusion of the n^+ gate pocket has a slight improvement in I_{ON} from 1.54×10^{-6} to 2.52×10^{-6} A/μm at a drain to source bias of 0.5 V. Here, a high average subthreshold swing of 47.21 mV per decade and a low point subthreshold swing of 38.5 mV per decade is achieved. Since the overlap length plays a crucial role in the LG-TFET, therefore the length of the overlapped region is optimized in Figure 2.9c, which shows the optimized length is at 5 nm. When the overlap is less than 5 nm, I_{ON} is decremented due to the decrease in the electric field near the tunneling junction, which is justified from the energy band diagram in Figure 2.9d.

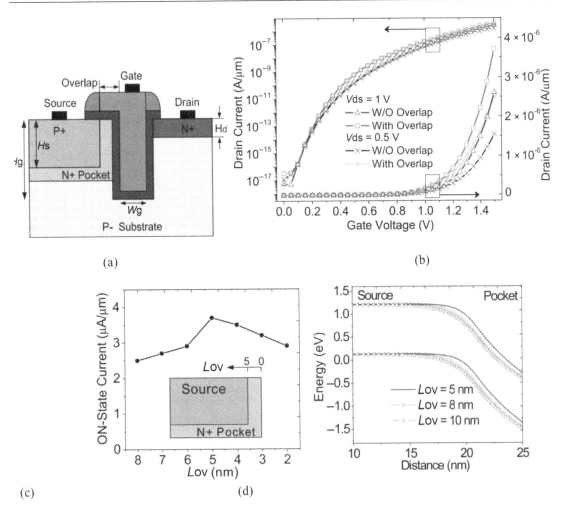

FIGURE 2.9 (a) Structure of the L-shaped gate TFET (LG-TFET). (b) Transfer characteristic with and without lateral gate to pocket overlapping. (c) ON-state current at different overlap length. (d) Energy band corresponding to the overlap length of the LG-TFET [47].

2.7.3 Heterojunction TFET with T-Shaped Gate

The heterojunction TFET with a T-shaped gate (HTG-TFET) employs the recess path at the substrate to change point tunneling into line tunneling perpendicular to the p- silicon region that enhances the tunneling region and I_{ON}. The structure uses two source – source1 and source2 – to maximize the tunneling region, the drain region is situated at the bottom of the transistor to minimize the gate to drain capacitances. As the metal gate shape looks like the letter T, the TFET is named after it. The T-shaped gate region is overlapped with n$^+$ pocket doping in both directions, vertically and laterally. As mutual line and point tunneling on both ends of the gate occur concurrently, the ON-state current is greatly increased. Additionally, the gate pocket overlapping region increments the electric field and the tunnel region. As higher electric field attributes to higher band-to-band generation rate, this overlap region is responsible for achieving a higher ON-state current. Here, a silicon germanium (SiGe) alloy is implemented at the pocket region, as opposed to silicon. The advantage of using SiGe is that, depending on the percentage of silicon or germanium required, the mole fraction x can be adjusted. The BTBT parameters of the SiGe are set as A = 0.295 × 10^{15}/cm^3 sec and B = 18.9 × 10^6 V/cm, which is given in [49]. The thickness of the pocket doped region is 5 nm, with an optimized overlap length of 7 nm. The pocket region is doped with a concentration of 1 × 10^{19}/cm^3. Hafnium oxide (HfO$_2$) of 2 nm thickness is utilized as the gate insulator. A metal work function of 4.33 eV is utilized. The length and height of the T-shaped gate are 10 nm and 60 nm, respectively. The height of both the source regions are 40 nm, with a drain region height of 20 nm.

From the transfer characteristic curve in Figure 2.10b, it is witnessed that a higher germanium composition of the SiGe compound semiconductor in the pocket region is beneficial for the device in terms

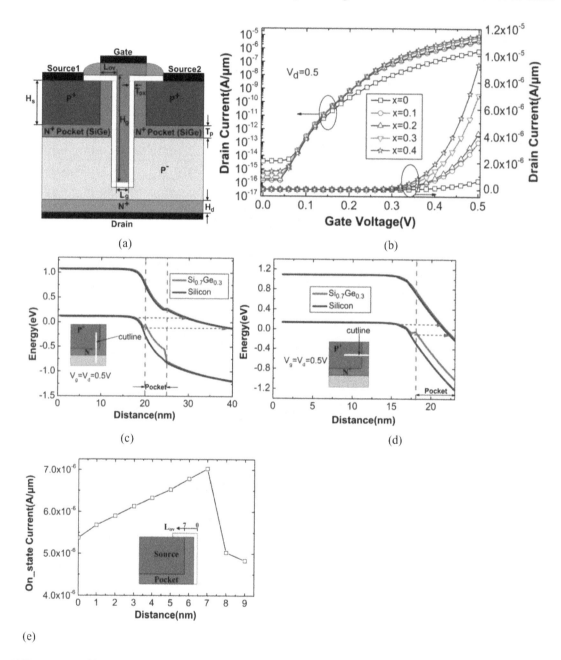

(a) (b)

(c) (d)

(e)

FIGURE 2.10   (a) Heterojunction TFET with a T-shaped gate structure (HTG-TFET). (b) Transfer characteristic. (c) Energy band diagram for point tunneling. (d) Energy band diagram for line tunneling. (e) ON-state current for variations of pocket doped region for the HTG-TFET [48].

of I_{ON}. However, the same effect is seen in terms of the OFF current, as the OFF current degrades for higher mole fraction value. The reason is that the bandgap of the $Si_{1-x}Ge_x$ semiconductor reduces with an increase in the germanium composition (i.e., $x = 0$ to 0.4, which is 40% of the germanium composition). Here, a mole fraction of $x = 0.3$ is considered as the optimum germanium composition for the HTG-TFET. An average subthreshold swing of 44.64 mV/decade along with a point subthreshold swing of 36.59 mV/decade is attained for the HTG-TFET [48].

For better understanding of the HTG-TFET, the energy band profile, with the inset showing the location of the cut-line (point or line), is illustrated in Figures 2.10c and d. As stated earlier, the tunneling width for the pocket doped region of $Si_{0.7}Ge_{0.3}$ is lesser in comparison with using a silicon pocket region owing to the smaller bandgap of 30% germanium composition in the SiGe alloy. The SiGe at $x = 0.3$ gives better overall device performance. It is also visualized that the transistor ON current varies for a variation in the pocket overlap length (L_{ov}), as seen in Figure 2.10e. Here, as the overlap length is increased from 0 to 7 nm (the lateral gate and the source/pocket interface are aligned), the drain current is increased and attains an $I_{ON} = 7.02$ $\mu A/\mu m$. That being said, the ON current tends to degrade when the L_{ov} is lower than 7 nm, as a result of decrement in the gate electric field. The decrement in I_{ON} at L_{ov} greater than 7 nm stems from the fact that the electric field close to the tunnel junction does not rise further while the energy band lowering in the source region attributes to a lesser, steeper band bending.

2.7.4 Germanium Source with Delta Doped Layer Vertical TFET

The germanium source with delta doped layer vertical TFET structure (Ge Source DD Vert TFET) [50] is portrayed in Figure 2.11a. Unlike the TFETs discussed earlier, here the electron BTBT from the source to the channel region takes place only vertically (line tunneling). This means that the path of the electron BTBT is directly aligned with the path of the metal gate electric field. This is possible since the entire source region (including the channel region) is being overlapped by the metal gate. Hence, stronger electric fields from the gate are encountered by the electrons, thereby allowing more tunneling, which enhances the device performance along with its ON current [43]. The energy band profile in Figure 2.11b depicts the electron tunneling in the ON-state, and the cut-line where the measurement is taken is denoted by A–B in Figure 2.11a. The gate length is measured at 30 nm, with a metal work function of 4.18 eV. The thickness of the germanium source, silicon channel, and gate insulator (HfO_2) are 12, 3, and 2 nm, respectively. The

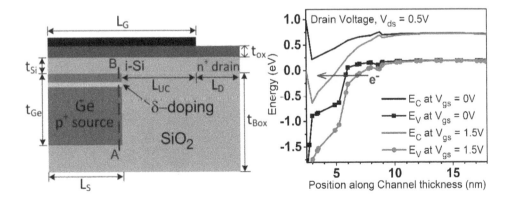

(a) (b)

FIGURE 2.11 (a) Germanium source with delta doped layer vertical TFET structure (Ge Source DD vert TFET) (b) Energy band profile in the ON- and OFF-state.

length of the source, drain, and the silicon channel under-cut layer are 15, 10, and 15 nm, respectively. The germanium source is doped with doping concentrations of $1 \times 10^{20}/cm^3$ along with a delta (δ) doped region of concentrations $5 \times 10^{16}/cm^3$, which is implemented within the source at 2 nm below the channel and the source junction. This layer is of 1 nm thickness, with an exponential decay profile.

The Synopsis technology computer-aided design simulator (TCAD) is utilized in performing these simulations [51]. The physics models activated are as follows: Fermi–Dirac statistics, bandgap narrowing model with the Old-Slotboom model, Masetti-mobility model, Shockley-Read-Hall recombination, and a dynamic nonlocal BTBT model where the values of the coefficients are properly calibrated with experimental data [46]. Here the parameters of silicon are calibrated as A =$3.29 \times 10^{15}/cm^3$ sec and B = 23.8 $\times$ 10^6 V/cm and germanium parameters are A = $1.67 \times 10^{15}/cm^3$ sec and B = 6.55×10^6 V/cm, respectively.

It is observed that the channel under-cut length plays a vital role in optimizing the TFET performances in terms of its subthreshold slope, which is a crucial device parameter. However, the ON-state current is not affected by varying the channel under-cut length but has a major effect on I_{OFF}. This leads to a well-improved subthreshold slope and a lower I_{OFF} at a channel under-cut length of 15 nm. As a positive drain voltage is applied in the OFF-state, the channel/drain potential is reduced. Accordingly, as the channel under-cut length increases, the influence of the drain at the souce/channel region is reduced. This decreases the electron BTB generation rate in the OFF-state. Hence, an improvement in I_{OFF} from 2.40 $\times$ 10^{-10} A/μm to 2.77×10^{-14} A/μm is attained for a channel under-cut length from 0 to 15 nm, respectively. This leads to a huge increment in the I_{ON}/I_{OFF} current ratio of up to four orders of magnitude.

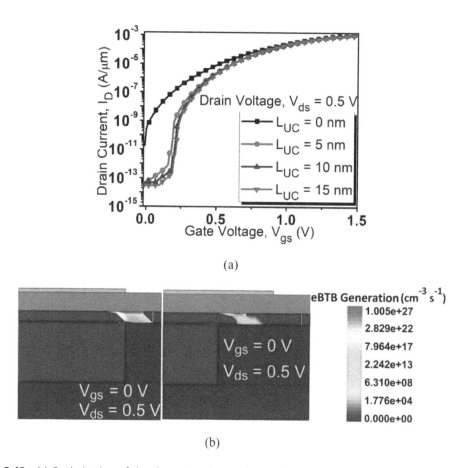

(a)

(b)

FIGURE 2.12 (a) Optimization of the channel under-cut length (L_{UC}). (b) BTB electron generation rate in the OFF-state for L_{UC} = 0 and L_{UC} = 15 nm.

As mentioned earlier, the employment of a delta doped layer within the germanium source region is further investigated. The analysis is conducted by investigating two different forms of doping in the source region, one being the uniformly doped germanium source while the other is the aforementioned germanium source with a delta doped layer. It is seen from the transfer characteristic given in Figure 2.13a that the delta doped layer contributes to an additional decrement in the OFF-state leakage current by an order of magnitude. This can be explained with the help of Figure 2.13c, which shows that the electron band-to-band generation rate is much lower, by two orders of magnitude, in the OFF-state when a delta doped layer is inserted. This means that the electrons are confined within the delta doped layer, which then improved the gate electrostatic control [52]. The optimization of the doping concentration in the delta layer is also carried out where $5 \times 10^{16}/cm^3$ is taken to be the optimized value in Figure 2.13b. The subthreshold swing (SS) of the Ge Source DD Vert TFET is determined using equation (2.29):

$$SS = \frac{\left(V_{th} - V_{OFF}\right)}{\left[\log\left(I_{th}\right) - \log\left(I_{OFF}\right)\right]} \tag{2.29}$$

where V_{th} and V_{OFF} denote the threshold voltage and the gate to source voltage at OFF current (I_{OFF}), respectively. The threshold current (I_{th}) is taken at the drain current value of 10^{-7} A. When the doping concentration is kept very low, the built-in electrical fields within the source and channel tunnel junction would be

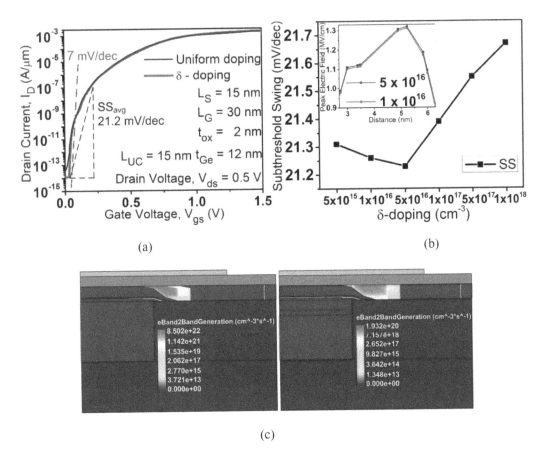

(a)

(b)

(c)

FIGURE 2.13 (a) Transfer characteristics of the optimized Ge Source DD Vert TFET with and without delta doping. (b) Optimization of the delta doping concentration. (c) Electron band-to-band generation rate in the OFF-state with and without delta doping.

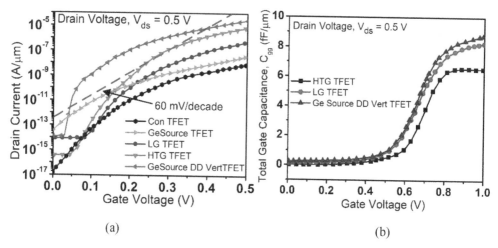

FIGURE 2.14 (a) Comparison of the transfer characteristic of the conventional TFET, germanium source TFET with vertical TFETs (LG-TFET, HTG-TFET and Ge Source DD Vert TFET) (b) Comparison of the total gate capacitance for the vertical TFETs (LG-TFET, HTG-TFET and Ge Source DD Vert TFET).

reduced, resulting in a high average subthreshold swing. On the other hand, when the doping is kept high, there is no significant electrostatic enhancement, since the doping profile reaches the normal continuous uniform doping. Hence, a substantially low OFF-state current of 1.175×10^{-14} A/µm and a high ON current of 1.98 mA/µm is achieved in the optimized vertical structure with the addition of a delta doped region. The current ratio is found out to be 1.68×10^{11}. A very low average subthreshold swing of 21.24 mV/decade and a relatively low point subthreshold swing of 7 mV/decade is attained for the Ge Source DD Vert TFET [50].

The benchmarking of the vertical TFETs (point and line tunneling) disccused in the previous section along with a lateral tunneling based conventional TFET and germanium source TFET is visualized in Figure 2.14a. From the transfer characteristic curve, it shows that the Ge Source DD Vert TFET possess the steepest subthreshold slope in addition to the highest ON-state current as compared to the diverse line and point tunneling TFETs. The total capacitance-voltage characteristics curve of the LG-TFET, HTG-TFET, and the Ge Source DD Vertical TFET is also portrayed in Figure 2.14b. The total gate capacitance comprises the Miller capacitances, including the gate to source capacitance and the gate to drain capacitance, in which the gate to drain capacitance is the main contributor, as discussed in the previous sections. This indicate that the depletion capacitance at the drain for the Ge Source DD Vert TFET is the highest among the structures. However, as the magnitudes are all in the same order (femto farad) and only a small difference is observed, this will not induce severe Miller capacitance problems in circuit designing. This shows that with proper implementation and a well-designed TFET structure, TFETs employing line tunneling can attain a much better overall device characteristic. Itis worth noting that the threshold voltage is measured at a drain current value of 10^{-7} A. The ON- and OFF-state current is taken at 0.5 V and 0 V of the gate voltage. The total gate capacitance is measured at gate voltage of 0.8 V. A summarized comparison of the parameters of the TFETs is given in Table 2.1.

TABLE 2.1   Benchmarking of the Vertical TFETs.

PERFORMANCE PARAMETERS	LG-TFET	HTG-TFET	GE SOURCE DD VERT TFET
I_{OFF} (A/µm)	8.7×10^{-15}	3.6×10^{-16}	1.2×10^{-14}
I_{ON} (A/µm)	4.7×10^{-7}	6.8×10^{-6}	2.76×10^{-5}
I_{ON}/I_{OFF}	5.4×10^{7}	1.8×10^{10}	2.3×10^{9}
V_T (V)	0.40	0.28	0.20
SS_{avg} (mV/dec)	47.21	44.64	21.24
C_{gg} (fF/µm)	7.39	6.32	7.73

2.8 MIXED MODE GE SOURCE DD VERTICAL TFET

Due to various attributes such as its low leakage currents, the sub-60 mV per decade subthreshold slope and the controlability of ambipolarities, TFETs are suited for various ultra-low power applications. A digital inverter, the basic building blocks of any modern circuit, is the best method to determine the performances of a TFET in circuit application. In TFET inverters, the power capacitance is 2.6 times higher than the total gate capacitances, unlike the MOSFET inverters, in which the output capacitances and the total gate capacitances are equal. The evaluation of the effective capacitances as well as drive currents for delay of TFETs have been performed [53]. The digital inverter realization showing the implementation of a complementary TFET (C-TFET) inverter circuit and comparison of the transient characteristic for the conventional TFET inverter and the germanium source delta doped vertical TFET inverter are illustrated in Figures 2.15a and b, correspondingly. The mixed-mode complementary TFET for both transistors is analyzed where the load capacitance (C_L) is held constant at 0.65 femtofarad with an input voltage of

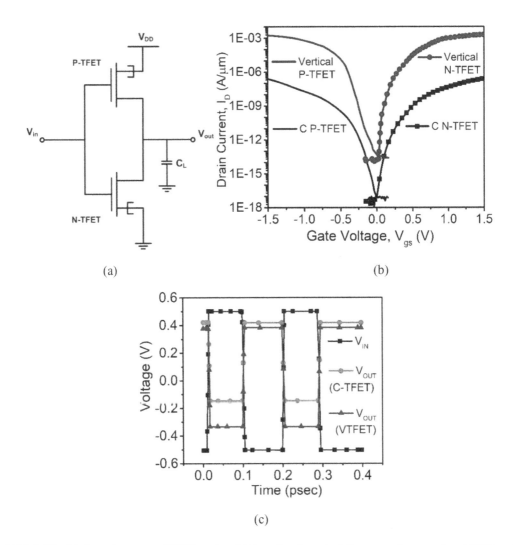

FIGURE 2.15 (a) Complementary TFET inverter. (b) Transfer characteristic curve of conventional TFET and Ge Source DD Vert TFET for both p-type and n-type. (c) Transient characteristic of the conventional TFET and Ge Source DD Vert TFET.

TABLE 2.2 Delay Parameters for Transient Characteristics of Conventional Tunnel and Ge Source DD Vert TFET.

DELAY PARAMETERS (PS)	CONVENTIONAL TFET	GE SOURCE DD VERT TFET
τ_{PLH}	0.129	0.121
τ_{PHL}	0.915	0.025
τ_p	0.587	0.073

ramp −0.5 to 0.5 volt applied. Figure 2.15c shows that the overshoot as well as the undershoot in the transient characteristic for both the TFETs in the comparison are irrelevant. That being said, the low level output of the Ge Source DD Vert TFET is increased in contrast with the conventional TFET owing to higher values of the total gate capacitances and Miller capacitances of vertical TFETs. The average delay parameter (τ_p) from Figure 2.15c is calculated by:

$$\tau_p = \frac{\tau_{PLH} + \tau_{PHL}}{2} \tag{2.30}$$

where τ_{PLH} signifies the rising edge delay and τ_{PHL} indicates the falling edge delay. The calculations in Table 2.2 reveal that the Ge Source DD Vert TFET has the superiority over the conventional TFET from the perspective of the average delay parameter, which is crucial in circuit performance.

2.9 CONCLUSION

In this chapter, we established a fundamental understanding of the TFET structure and its operating principle along with some important electrical characteristics of TFETs. We looked into the physics behind the band to band tunneling phenomenon. We listed several strategies researchers have explored to solve the low ON-state current problem as well as to improve various characteristics in TFET. Various TFET device structure implementations that utilize point tunneling and line tunneling as well as both point and line tunneling together were also compared and discussed. The TFET structure that adopted vertical tunneling (line tunneling) possesses a better subthreshold swing as well as better overall device performance. Among the vertical TFETs discussed, the Ge Source DD Vert TFET has attained a better ON-state current of 2.76×10^{-5} A/μm, a total gate capacitance of 7.73 fF/μm, and a point and average subthreshold swing of 7 mV/dec and 21.24 mV/dec, respectively. Finally, the mixed mode Ge Source DD Vertical TFET was implemented, and the transient characteristic showed an average delay parameter of 0.073 ps, which is 155.76 % lower compared to the conventional TFET, which is crucial in TFET circuit performance. The Ge Source DD Vert Tunnel is indeed considered to be a feasible candidate for future ultra-low power and high-performance technology.

REFERENCES

[1] D. Kahng and M. M. Atalla, "Silicon-Silicon Dioxide Field Induced Surface Device," *Solid State Device Research Conference*, June 1960.
[2] F. Wanlass and C. Sah, "Nanowatt Logic Using Field-Effect Metal-Oxide Semiconductor Triodes," *IEEE International Solid-State Circuits Conference Digest of Technical Papers*, pp. 32–33, Feb. 1963.
[3] W. F. Brinkman, D. E. Haggan, and W. W. Troutman, "A History of the Invention of the Transistor and Where It Will Lead Us," *IEEE Journal of Solid-State Circuits*, vol. 32, pp. 1858–1865, Dec. 1997.

[4] G. E. Moore, "Cramming More Components Onto Integrated Circuits," *Electronics*, vol. 87, pp. 114–117, Apr. 1965.

[5] C. A. Mack, "Fifty Years of Moore's Law," *IEEE Transactions on Semiconductor Manufacturing*, vol. 24, pp. 202–207, May 2011.

[6] International Technology Roadmap for Semiconductors (ITRS), "Emerging Research Devices," www.itrs.net, 2013.

[7] S. M. Sze and K. K. Ng, *Physics of Semiconductor Devices*, Hoboken, NJ: Wiley, 315 pp., 2007.

[8] Shailendra Singh and B. Raj, "A 2-D Analytical Surface Potential and Drain current Modeling of Double-Gate Vertical T-shaped Tunnel FET," *Journal of Computational Electronics*, Springer, vol. 19, pp. 1154–1163, Apr. 2020.

[9] W. Y. Choi, B.-G. Park, J. D. Lee, and T.-J. K. Liu, "Tunneling Field-Effect Transistors (TFETs) with Subthreshold Swing (SS) Less than 60 mV/dec," *IEEE Electron Device Letters*, vol. 28, no. 8, pp. 743–745, Aug. 2007.

[10] Soniya, Shailendra Singh, Girish Wadhwa, and B. Raj, "Design and Analysis of Dual Source Vertical Tunnel Field Effect Transistor for High Performance," *Transactions on Electrical and Electronics Materials*, Springer, vol. 21, pp. 74–82, Oct. 2019.

[11] A. C. Seabaugh and Q. Zhang, "Low-Voltage Tunnel Transistors for beyond CMOS Logic," *Proceedings of the IEEE*, pp. 2095–2110, Dec. 2010.

[12] Girish Wadhwa and B. Raj, "Design and Performance Analysis of Junctionless TFET Biosensor for High Sensitivity," *IEEE Nanotechnology*, vol. 18, pp. 567–574, 2019.

[13] B. Bhushan, K. Nayak, and V. R. Rao, "DC Compact Model for SOI Tunnel Field-Effect Transistors," *IEEE Transactions on Electron Devices*, vol. 59, no. 10, pp. 2635–2642, Oct. 2012.

[14] Girish Wadhwa, Priyanka Kamboj, and B. Raj, "Design Optimisation of Junctionless TFET Biosensor for High Sensitivity," *Advances in Natural Sciences: Nanoscience and Nanotechnology*, vol. 10, pp. 045001, 2019.

[15] S. H. Kim, Z. A. Jacobson, and T.-J. K. Liu, "Impact of Body Doping and Thickness on the Performance of Germanium-Source TFETs," *IEEE Transaction on Electron Devices*, vol. 57, no. 7, pp. 1710–1713, Jul. 2010.

[16] Soniya, Shailendra Singh, Girish Wadhwa, and B. Raj, "An Analytical Modeling for Dual Source Vertical Tunnel Field Effect Transistor," *International Journal of Recent Technology and Engineering (IJRTE)*, vol. 8, no. 2, Jul. 2019.

[17] Shailendra Singh and B. Raj, "Design and Analysis of Hetrojunction Vertical T-shaped Tunnel Field Effect Transistor," *Journal of Electronics Material*, Springer, vol. 48, no. 10, pp. 6253–6260, Oct. 2019.

[18] Candy Goyal, Jagpal Singh Ubhi, and B. Raj, "A Low Leakage CNTFET based Inexact Full Adder for Low Power Image Processing Applications," *International Journal of Circuit Theory and Applications*, Wiley, vol. 47, no. 9, pp. 1446–1458, Sept. 2019.

[19] C. Kittel, *Introduction to Solid State Physics*, Hoboken, NJ: Wiley, 317 pp., 1971.

[20] R. F. Pierret, *Semiconductor Device Fundamentals*. Reading, MA: Addison-Wesley, pp. 91, 1996.

[21] J. Knoch and J. Appenzeller, "A Novel Concept for Field-Effect Transistors: The Tunneling Carbon Nanotube FET," in *IEEE Proceedings of 63rd DRC*, pp. 153–156, June 2005.

[22] Anil Kumar Bhardwaj, Sumeet Gupta, B. Raj, and Amandeep Singh, "Impact of Double Gate Geometry on the Performance of Carbon Nanotube Field Effect Transistor Structures for Low Power Digital Design," *Computational and Theoretical Nanoscience*, ASP, vol. 16, pp. 1813–1820, 2019.

[23] Neeraj Jain and B. Raj, "Thermal Stability Analysis and Performance Exploration of Asymmetrical Dual-k underlap Spacer (ADKUS) SOI FinFET for Security and Privacy Applications," *Indian Journal of Pure & Applied Physics (IJPAP)*, vol. 57, pp. 352–360, May 2019.

[24] Jeetendra Singh and B. Raj, "Tunnel Current Model of Asymmetric MIM Structure Levying Various Image Forces to Analyze the Characteristics of Filamentary Memristor," *Applied Physics A*, Springer, vol. 125, no. 3, pp. 203.1–203.11, Feb. 2019 (SCI).

[25] Candy Goyal, Jagpal Singh Ubhi, and B. Raj, "Low Leakage Zero Ground Noise Nanoscale Full Adder using Source Biasing Technique," *Journal of Nanoelectronics and Optoelectronics*, American Scientific Publishers, vol. 14, pp. 360–370, Mar. 2019.

[26] G. V. Luong, S. Strangio, A. Tiedemann, S. Lenk, S. Trellenkamp, K. K. Bourdelle, Q. T. Zhao, and S. Mantl, "Experimental Demonstration of Strained Si Nanowire GAA n-TFETs and Inverter Operation with Complementary TFET Logic at Low Supply Voltages," *Solid-State Electronics*, vol. 115, Part B, pp. 152–159, 2016.

[27] S. Saurabh and M. J. Kumar, "Novel Attributes of a Dual Material Gate Nanoscale Tunnel Field-Effect Transistor," *IEEE Transactions on Electron Devices*, vol. 58, no. 2, pp. 404–410, Feb. 2011.

[28] G. Dewey, B. Chu-Kung, J. Boardman, J. M. Fastenau, J. Kavalieros, R. Kotlyar, W. K. Liu, D. Lubyshev, M. Metz, N. Mukherjee, P. Oakey, R. Pillarisetty, M. Radosavljevic, H. Then, and R. Chau, "Fabrication, Characterization, and Physics of III-V Heterojunction Tunneling Field Effect Transistors (H-TFET) for Steep Sub-threshold Swing," in *IEEE International Electron Devices Meeting (IEDM)*, pp. 33.6.1–33.6.4, IEEE, 2011.

[29] H. Zhao, Y. Chen, Y. Wang, F. Zhou, F. Xue, and J. Lee, "InGaAs Tunneling Field-Effect-Transistors with Atomic-Layer Deposited Gate Oxides," *IEEE Transactions on Electron Devices*, vol. 58, pp. 2990–2995, Sept. 2011.

[30] M. H. Lee, S. T. Chang, T.-H. Wu, and W.-N. Tseng, "Driving Current Enhancement of Strained Ge (110) p-type Tunnel FETs and Anisotropic Effect," *IEEE Electron Device Letters*, vol. 32, pp. 1355–1357, Oct. 2011.

[31] Neeraj Jain and Balwinder Raj, "Analysis and Performance Exploration of High-k SOI FinFETs Over the Conventional Low-k SOI FinFET toward Analog/RF Design," *Journal of Semiconductors (JoS)*, IOP Science, vol. 39, no. 12, pp. 124002–1–7, Dec. 2018.

[32] Candy Goyal, Jagpal Singh Ubhi, and Balwinder Raj, "A Reliable Leakage Reduction Technique for Approximate Full Adder with Reduced Ground Bounce Noise," *Journal of Mathematical Problems in Engineering*, Hindawi, vol. 2018, Article ID 3501041, 16 pp., 15 Oct. 2018.

[33] Anuradha, Jeetendra Singh, Balwinder Raj, and Mamta Khosla, "Design and Performance Analysis of Nanoscale Memristor-based Nonvolatile SRAM," *Journal of Sensor Letter*, American Scientific Publishers, vol. 16, pp. 798–805, Oct. 2018.

[34] Jeetendra Singh, Sanjeev Sharma, Balwinder Raj, and Mamta Khosla, "Analysis of Barrier Layer Thickness on Performance of In1-xGaxAs based Gate Stack Cylindrical Gate Nanowire MOSFET," *JNO*, ASP, vol. 13, pp. 1473–1477, Oct. 2018.

[35] S. Mookerjea, D. Mohata, R. Krishnan, J. Singh, A. Vallett, A. Ali, T. Mayer, V. Narayanan, D. Schlom, A. Liu, and S. Datta, "Experimental Demonstration of 100 nm Channel Length In0.53Ga0.47As-based Vertical Inter-band Tunnel Field Effect Transistors (TFETs) for Ultralow-Power Logic and SRAM Applications," in *IEDM Technical Digest*, pp. 949–952, 2009.

[36] A. Tura, Z. Zhang, P. Liu, Y. Xie, and J.C.S. Woo, "Vertical Silicon p–n–p–n Tunnel nMOSFET with MBE-grown Tunneling Junction," *IEEE Transactions on Electron Device*, vol. 58, no. 7, pp. 1907–1913, 2011.

[37] S. Panda, S. Dash, S. K. Behera, and G. P. Mishra, "Delta-doped Tunnel FET (D-TFET) to Improve Current Ratio (ION/IOFF) and ON-Current Performance," *Journal of Computational Electronics*, vol. 15, pp. 857–864, 2016, DOI:10.1007/s10825-016-0860-z.

[38] N. Bagga, A. Kumar, and S. Dasgupta, "Demonstration of a Novel Two Source Region Tunnel FET," *IEEE Transactions on Electron Devices*, vol. 64, no. 12, pp. 5256–5262, Dec. 2017.

[39] C. Liu, Q. Ren, Z. Chen, L. Zhao, C. Liu, Q. Liu, W. Yu, X. Liu, and Q. T. Zhao, "A T-Shaped SOI Tunneling Field-Effect Transistor with Novel Operation Modes," *IEEE Journal of the Electron Devices Society*, vol. 7, pp. 1114–1118, Oct. 2019.

[40] S. Yang, H. Lv, B. Lu, S. Yan, and Y. Zhang, "A Novel Planar Architecture for Heterojunction TFETs with Improved Performance and Its Digital Application as an Inverter," *IEEE Access*, vol. 8, pp. 23559–23567, Jan. 2020.

[41] Balwinder Raj, A. K. Saxena, and S. Dasgupta, "Nanoscale FinFET Based SRAM Cell Design: Analysis of Performance metric, Process Variation, Underlapped FinFET and Temperature Effect," *IEEE Circuits and System Magazine*, vol. 11, no. 2, pp. 38–50, 2011.

[42] W. G. Vandenberghe, A. S. Verhulst, G. Groeseneken, B. Sorée, and W. Magnus, "Analytical Model for Point and Line Tunneling in a Tunnel Field-Effect Transistor," in *Proceedings of International Conference on Simulation of Semiconductor Processes and Devices*, pp. 137–140, Sep. 2008.

[43] S. Blaeser, S. Glass, C. S. Braucks, K. Narimani, N. V. D. Driesch, S. Wirths, A. T. Tiedemann, S. Trellenkamp, D. Buca, S. Mantl, and Q. T. Zhao, "Line Tunneling Dominating Charge Transport in SiGe/Si Heterostructure TFETs," *IEEE Transactions on Electron Devices*, vol. 63, no. 11, pp. 4173–4178, Nov. 2016.

[44] A. Archarya, A. B. Solanki, S. Glass, Q. T. Zhao, and B. Anand, "Impact of Gate – Source Overlap on the Device/Circuit Analog Performance of Line TFETs," *IEEE Transactions on Electron Devices*, vol. 66, no. 9, pp. 4081–4086, Sept. 2019.

[45] R. Rooyackers, "Trends and Challenges in Tunnel-FETs for Low Power Electronics," in *2019 34th Symposium on Microelectronics Technology and Devices (SBMicro)*, Aug. 2019.

[46] S. H. Kim, H. Kam, C. Hu, and T.-J. K. Liu, "Germanium-source Tunnel Field Effect Transistors with Record High I_{ON}/I_{OFF}," in *Proceedings of Symposia on VLSI Technology*, Jun. 2009, pp. 178–179.

[47] Z. Yang, "Tunnel field-effect Transistor with an L-shaped Gate," *IEEE Electron Device Letters*, vol. 37, no. 7, pp. 839–842, Jul. 2016.

[48] Girish Wadhwa and B. Raj, "Label Free Detection of Biomolecules using Charge-Plasma-Based Gate Underlap Dielectric Modulated Junctionless TFET," *Journal of Electronic Materials (JEMS)*, Springer, vol. 47, no. 8, pp 4683–4693, Aug. 2018.

[49] Girish Wadhwa and B. Raj, "Parametric Variation Analysis of Charge-Plasma-based Dielectric Modulated JLTFET for Biosensor Application," *IEEE Sensor Journal*, vol. 18, no. 15, 1 Aug. 2018.

[50] Jeetendra Singh and B. Raj, "Comparative Analysis of Memristor Models for Memories Design", *JoS, IoP*, vol. 39, no. 7, pp. 074006–1–12, Jul. 2018.

[51] *Sentaurus User's Manual*, Synopsys, Inc., Mountain View, CA, USA, 2013.

[52] G. Scappucci, G. Capellini, W. C. T. Lee, and M. Y. Simmons, "Ultradense Phosphorus in Germanium Delta-Doped Layers," *Applied Physics Letters*, vol. 94, no. 16, Apr. 2009, Art. no. 162106.

[53] S. Mookerjee, R. Krishnan, S. Datta, and V. Narayanan, "Effective Capacitance and Drive Current for Tunnel FET (TFET) CV/I Estimation," *IEEE Transactions on Electron Devices*, vol. 56, no. 9, pp. 2092–2098, 2009.

Vertical T-Shaped Heterojunction Tunnel Field-Effect Transistor for Low Power Security Systems

3

Shailendra Singh, Balwant Raj, and Balwinder Raj

Contents

DOI: 10.1201/9781003126645-3

3.1 INTRODUCTION

In the last four decades, semiconductor scaling, novelty with the structural design, and fabrication enhancement technology are the main governing points that significantly developed complementary metal oxide semiconductor (CMOS) technology. This will result in a faster integrated circuit (IC) that comes with superior performance. However, continuous scaling of the device leads to short channel effects (SCEs) such as gate-induced drain leakage (GIDL), drain-induced barrier lowering (DIBL), the hot electron effect, and a subthreshold slope restricted to 60 mV/decade [1–3]. So in order to maintain better performance, we have to open the doors for new technology in which it will be a challenge to suppress the SCEs. In this new era of advanced electronics gadgets, a large amount of memory space is required with ultra-low power consumption in order to achieve high performance. Initially, the first transistor was invented by the Bell laboratory in the 1940s [4–5]. The journey of device dimensions moved from centimeters to the nanoscale of today's devices. The first IC was introduced by Jack Kilby in 1958 at Texas Instruments [6]. In the last four decades, continuous miniaturization of the transistor has taken place in silicon-based CMOS technology, which has led to the growth in the semiconductor industry. This miniaturization in silicon-integrated circuits continues to advance on an exponential scale defined by the law of Moore, as shown in Figure 3.1, which is driven by the cost and its functionality [7, 51, 52]. Continuous scaling of the MOS transistor's physical feature size is the key factor for building high-density integrated electronic systems and increasing efficiency (by increasing device agility, decreasing intrinsic delay by 30% per technology generation, and lowering the cost). However, the transistor's miniaturization faces an imminent power crisis and insurmountable barriers to static power dissipation due to fundamental limitations of metal-oxide semiconductor field-effect transistors (MOSFETs) (i.e., high leakage current in OFF-state (I_{OFF}), gate leakage, short channel effects (SCEs), and high subthreshold swing [8–10]). Therefore, in the technological advancement of the nanoscale regime leading to increased switching performance, improved energy efficiency, and reduced supply voltage, new nanoscale devices are expected to arrive [53–56].

3.1.1 Scaling: A Historical Perspective

The scaling factor allows manufacturing of nanoscale devices at the same cost. However, traditional scaling based on reduction of physical dimensions of metal–oxide–semiconductor transistors, with simultaneous reduction of supply voltages and dissipated power, is reaching its limits [9–10]. Therefore, it is required that the researchers explore possible solutions to enable the next technology nodes, making transistors smaller and faster. This will require further innovations in device architectures and new materials and inventive techniques of patterning features of nanometer size. Scaling is a method to reduce the device dimension in the regulated manner, so that device occupied area will be the minimum possible without any changes in the output characteristics [57–62].

Issues such as high power consumption and less packing density can efficiently be solved with the aid of scaling, thus leading to the creation of an effective device that can be further operated with a low biasing voltage.

One of the scaling theory's most relevant principles is that vertical and horizontal dimensions should be measured by a common scaling factor in order to prevent the leakage current and to deliver efficient channel reliability. This helps to decrease the supply voltages effectively. To retain the similar features of the devices during scaling process, the doping concentration should be increased. In 1925, Lilienfeld discovered the insulated gate field effect transistor (IGFET) for the first time in the world of innovation technology. It was an effective substitute for vacuum tube technology and paved way for the production of low-power semiconductor transistor devices of small scale [11]. Kahng and Atilla were the first to provide a MOSFET made up of silicon, which was practically demonstrated in 1960 [12]. In 1958, an IC evaluation was done by Jack Rabaey and Robert Noyce at Texas Instruments. The first IC was used for SET-RESET (SR) flip-flop memory application [13]. In 1959, Richard Feynman found in his research that "there is plenty of space at the edge" that can be used efficiently in further size scaling in small scale semiconductor devices [14]. Finally, in 1965, Gordon Moore came up with the new law regarding the packing density called Moore's law. He is also the cofounder of the Intel company. According to him, the device density on a microchip will double every year. In addition, the overall production cost will be halved. Moore also predict that in the future much technology will be enhanced following the scale in linearity. In 1962, the first logic family with TTL units was formed [14]. Intel introduced the first microprocessor, which consisted of more than 1,800 PMOS transistors, in 1972. After Intel's microprocessor, many different series of microprocessors also entered the market using NMOS transistor technology, but they soon disappeared due to their heavy power consumption because of the exponential rise in chip density. However, to resolve the issue of high power consumption, CMOS technology was introduced, which was well suited for low power consumption devices. Therefore, the device technology scaling jumped from small scale integration (SSI) to very large scale integration (VLSI). With CMOS technology, one of the most beneficial aspects was to decrease the scaling of the device to nanoscale dimensions. The International Development Roadmap for Semiconductors (ITRS) has established a scaling concept with respect to complicated power consumption and manufacturing cost at the production stage. The roadmap by the ITRS is shown in Figure 3.1.

In addition, a wafer having the capacity of 1 TB, having more than 800 billion transistors, made by Intel was called static random access memory (SRAM) [15]. Different researchers have come together in terms of reducing the dimension. In 1972, Dennard and his colleagues proposed the scaling measurements [16]. New projects such as the International Roadmap for Devices and Systems (IRDS) have gained publicity following the dissolution of ITRS, as shown in Figure 3.2.

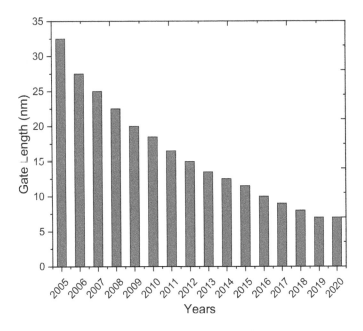

FIGURE 3.1   Scaling for the gate length variations with respect to the ITRS years.

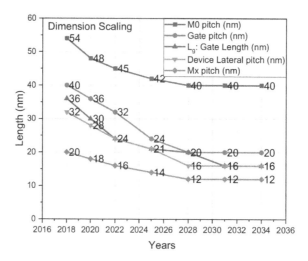

FIGURE 3.2   IRDS 2018 projected scaling of key ground rules.

To overcome these limits, more disruptive technologies have to be developed based on unconventional mechanisms and novel materials. In addition, creative circuit and system designs may take advantage of some unique features of novel devices unavailable in Si MOSFETs.

3.1.2 Challenging Issue of Scaling

The numerous explanations for carrying out miniaturization of the devices are as follows:

- Cost reduced per transistor because nano devices occupy a small region.
- Added transistors that perform complex functions can be integrated on the same chip.
- Another very low capacitance value, which reduces the power consumption required to turn on the device.

Despite these benefits, scaling poses significant challenges during product manufacture.

- Initially, the power consumption at high operating frequency is rising, which causes the chip to generate a significant amount of heat. Due to this, it is so challenging for the manufacturer to kept the IC cool for low power application. Scaling down the supply voltage (V_{dd}) is the solution for rising power consumption. However, this will reduce the operating speed of the device. Consequently, the key goal of the system is to reduce the power consumption in a regulated way [17].
- MOSFET switching delay can be minimized by synchronizing the subthreshold voltage (V_t) of the device with the supply voltage (V_{dd}). However, this leads to an increase in leakage current. Therefore, scaling in the oxide thickness layer (t_{ox}) is one of the ways of reducing subthreshold voltage. Ultimately, we have to maintain the relation between the operating speed and the power consumption [18].
- Second, the scaling drawbacks arise due to characteristics such as subthreshold voltage, subthreshold slope, and drain induced barrier lowering (DIBL). It also is observed from a literature survey that by reducing oxide thickness, the gate-controlled-depletion-depth (d) will be influenced [17–19]. Below 90 nm channel, the device will not regulate the t_{ox} length less than 1.2 nm, which is approximate 6 atomic layers of SiO_2.

3.1.2 Next to more than Moore's Law

With reference to the International Technology Roadmap for Semiconductors (ITRS) data, the collective graph of digital specification and nondigital specification is shown in Figure 3.3, called 'More-than-Moore'. Downscaling offers benefits; thus, this phenomenon, known as 'More Moore', must be sustained. Therefore, the potential of nondigital technologies is significantly improved in both processing technologies and design growth.

It will also cover the nondigital analogue circuits application into a single package. The fully integrated systems can therefore communicate skillfully with people as well as with the world. This drifting is beneficial in so-called More-than-Moore device functional sustainability [20].

3.1.3 Channel Engineering Methods of the Device

3.1.3.1 Shallow at Source and Drain Junction

In order to reduce the barrier coupling of source side and drain side, the junction depth adjacent to the gate side is reduced. However, to sustain the sheet resistance, the doping concentration of the source and drain side will be increased. On the other hand, doping density cannot be increased after a limit because the solid solubility of doping particles is $\sim 10^{20}$ cm^{-3}. After execution of an annealing process to gain very low resistivity, it is not possible to have the source–drain junction by triggering the dopants [21]. The leakage current will be increases due to the abrupt junction of source and drain due to band-to-band tunneling (BTBT), which will degrade the overall performance of device.

3.1.3.2 Halo Doping

Several researchers are working upon the single halo (SH) and double halo (DH) technique for device channel modification. Although this technique improves the device performance, it is more suitable for lateral asymmetric channel (LAC) devices [22]. By embedding this technique, a pinch-off region is formed

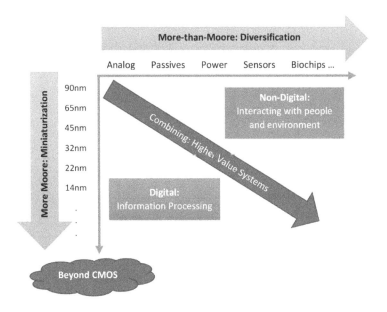

FIGURE 3.3 Graph of digital device minimization called 'More Moore' with respect to 'More-than-Moore'.

near the drain region of the device with greater subthreshold voltage. This will improve the device drain current.

3.1.3.3 Strain Method

Due to scaling, the source and drain will overlap each other due to reductions in the device dimensions. However, it will increase the electric field drastically. In order to overcome this problem, doping needs to be increased. But with the increase in the doping concentration in channel, the charge particle scattering will lead to reducing the mobility inside the channel. On the other hand, the threshold voltage of the device will continuously vary because of random fluctuation of dopant. In order to reduce the mobility degradation due to charge carrier scattering, one technique, called strain technology, improves the mobility of carriers. This will also help in modifying the lattice constant of the material. Second, it ensures that carrier traps are formed by band structure alteration, which ultimately increases the mobility [9]. The lower band gap materials are used to increase the mobility like silicon-germanium. By increasing the germanium content, mobility will increase.

3.1.3.4 Multi-Material Gate

A multi-material gate is one of the most effective techniques in reducing the hot electron effect (HEE) used for the channel engineering method. The two different materials of different work function will cascade after each other; these are called tunneling gate length (L_1) and auxiliary gate length (L_2). The M_2 (material) of the drain side has lower work function, and the M_1 toward the source side has higher work function, as shown in Figure 3.4. This technique is effective for channel length less than 25 nm. In 1999, the double-material-gate design (DMG) was proposed, and in 2008, the triple-material-gate design (TMG) was proposed [23]. Dissimilar work functions at the source side and the drain side will lead to a high electric field. This will result in improving the transport efficiency of the charges of device.

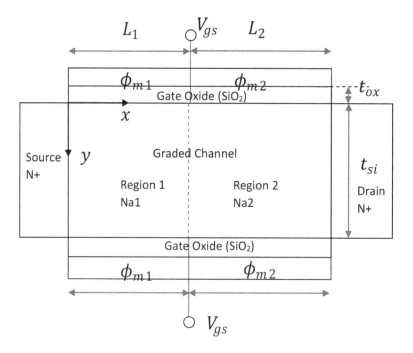

FIGURE 3.4 Dual material gate structure, where $\phi_{m1} > \phi_{m2}$.

3.1.4 Gate Engineering Methods

3.1.4.1 *High-k Dielectric-Material Gate Dielectric*

The high k-dielectric constant material is used to significantly reduce the leakage current and gate tunneling with low gate oxide thickness. In most of the semiconductor fabrication industry, a substitute to the SiO_2 will be replaced by a material with a high k-dielectric constant like HfO_2, with the value of $k = 21$. The combination of low dielectric k with high dielectric k stacking is a famous method that easily scales the device further without actually influencing the device performance. The equivalent oxide thickness (EOT) can be calculated by using equation 3.1. This will calculate the actual thickness of oxide for low dielectric k (t_{ox1}) with high-k dielectric (t_{ox2}) [24].

$$EOT = t_{ox1} + \frac{\varepsilon_{r1}}{\varepsilon_{r2}} t_{ox2} \hspace{4cm} (3.1)$$

Where ε_{r1} is absolute permittivity for SiO_2 and ε_{r2} is absolute permittivity for another high k-dielectric constant.

3.1.4.2 *Metal Gate*

High k-dielectric oxide stacking is an efficient method of reducing the leakage current; however, loss of mobility still takes place to a large degree with the use of a high-k dielectric. So the metal gate is established for providing protection against mobility degradation. Work function variation occurs due to the inconsistency of different dielectric gate alignments in the band [25].

3.1.4.3 *Multiple Gates*

Multiple gates are an effective technique used for THE reduction of SCE for a low scale dimension device. It provides good control over the channel, which prevents the device form SCE and improves the device's ability [26]. Among all structures, cylindrical structure that surround the channel will provide a great pathway from charge carriers. Due to introduction of the tri-gate MOSFETs, fabrication is now feasible up to 22 nm. It is surrounded on three sides to provide high controllability over the channel at very low dimensions. This will also improve the drain current due to suppression of the leakage current.

Various researches have also evaluated new devices such as gate-all-around (GAA), SOI, and DG-MOSFETs as shown in Figure 3.5, to increase the degree of gate-over-channel control. Of all

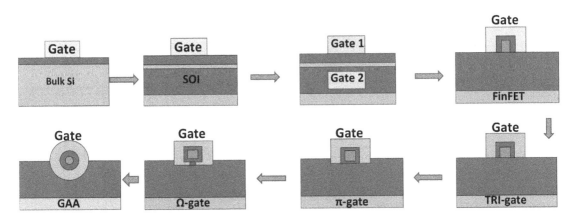

FIGURE 3.5 Various evaluations of multiple gate structures.

multi-gate systems, the system structure of gate-all-around (GAA) and nanowires (NWs) comes with the minimum channel length, which offers a better scaling device.

3.1.5 TFET (A Low Power Device)

A TFET works on the breakdown mechanism for reverse biased p–i–n junction semiconductor devices. This mechanism effectively works under a high electric field that induces quantum mechanical inter-band tunneling phenomena (IBT) at the tunnel junction, resulting in large current at a small voltage level [27–29]. TFETs have been proposed as a suitable candidate for the sharp transition between OFF-and ON-states, have obtained the lowest OFF-state current, and potentially will scale the supply voltage in a technology compatible with the existing CMOS technology platforms [30–31].

3.1.5.1 TFET Structure and Operation

TFET is a gated p–i-structure that is based upon the working principle of quantum mechanical band-to-band tunneling (BTBT). This is made up of a solid-state semiconductor that is operated in a reverse bias gate control mechanism. The operating principle is clearly understood by device schematic and energy band diagram, as shown in Figure 3.6. In the reverse bias, the electron will flow according to the Zener tunneling mechanism [32] from the source valance band to the channel conduction band. In the OFF-state condition, the channel region is fully depleted due to high p+ source doping and n+ drain doping. The depletion regions get reduced by applying the gate voltage biasing. This will turn ON the transistor mode due to the increase in the electric field. The minority charge carries will transmit from the source to the channel and

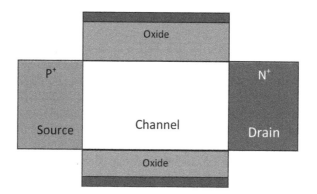

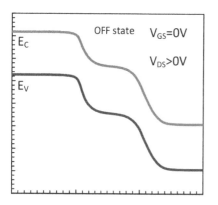
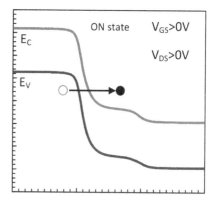

FIGURE 3.6   Double gate TFET device schematics and energy band diagram.

are regulated by the gate biasing voltage. However, in the case of the MOSFET, thermionic emission will oppose the injected charges, due to which high voltage is required to suppress the barrier. This process of tunneling charge carriers from source to channel medium of the device is called the BTBT process [33].

The main advantage of the tunneling phenomena is that the energy bandgap cuts the 'tail' of the Boltzmann electrons in the p-type source region [28–29]. This reduces thermal dependency, which allows a steeper subthreshold slope (SS) below the 60 mV/decade. The tunneling current thus obtained can be expressed as in equation 3.2 [34].

$$I_{BTBT} \cong A\varepsilon^2 exp\left[\frac{B}{\varepsilon}\right] \qquad (3.2)$$

Where A and B parameters are material dependent. The p–i–n structured TFET device is reverse biased for the OFF-state region, and hence the leakage current is smaller than in MOSFET. Therefore, TFET has ability to avoid SCEs, can achieve high I_{ON}/I_{OFF} ratio, and is independent of geometrical scaling.

3.1.5.2 The WKB Approximation Method

The Wentzel–Kramer's–Brillouin (WKB) approximation enables the probability tunneling current to be determined using arbitrary shaped potential barriers. A formal rigorous derivation is detailed in [35]. A TFET will employ the BTBT modes (lateral and vertical tunneling). The multiple advantages of perpendicular vertical tunneling to the gate-dielectric interface semiconductor makes it attractive compared to a traditional TFET design. Obtaining high ON current in TFETs is a challenging task, as ION is critically dependent on the transmission probability, T_{WKB}, of the barrier to tunneling between bands. As indicated in Figure 3.7, the potential barrier can be estimated by a triangular shape, so that T_{WKB} can be measured using the WKB approximation [35].

$$T_{WKB} \approx exp\left(\frac{4\lambda\sqrt{2m^*}\sqrt{E_G^3}}{3q\hbar\left(E_G + \Delta\phi\right)}\right) \qquad (3.3)$$

Where the bandgap is E_G and the effective mass is m^*. The spatial extent of the transition region at the source–channel interface is defined here by screening tunneling length (λ) depending on the specific geometry of the device, shown in Figure 3.7. In a TFET, the gate voltage (V_{GS}) at constant drain voltage (V_{DS})

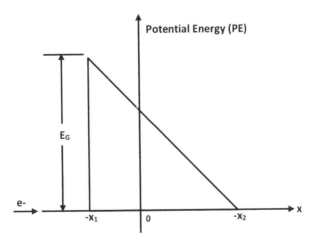

FIGURE 3.7 BTBT can be calculated by approximating the energy barrier width by a triangular potential energy barrier, where the electrons must tunnel through the widest distance at the base of the triangle.

increases; as a result, λ reduces, and the energetic difference ($\Delta\Phi$) between the valence band in the channel and conduction band at the source increases. Thus, in a first approximation, the drain current is a super-exponential function of V_{GS}. As a result, the TFET's point slope, in comparison to the MOSFET, is no longer a constant and is directly reliant on V_{GS}. The minimum value subthreshold swing is obtained at the minimum gate voltage. Whenever the source tunneling barrier transmission probability approaches unity, even for slighter V_{GS} variation, the high rate of tunneling current and a steeper slope are obtained.

3.1.5.3 Subthreshold Swing in TFET

The characteristic of the device below threshold voltage (V_{TH}) is known as subthreshold slope of the device. The subthreshold slope (SS) is defined as the minimum potential of gate-source voltage required to lesser the subthreshold current by a factor of 10. It is defined as the minimum amount of the gate biasing voltage required for lowering the subthreshold current by a factor of 10.

The subthreshold swing expression is defined as follows.

$$SS = \left(\frac{\partial log I_D}{\partial log V_G}\right)^{-1} \tag{3.4}$$

The unit of the subthreshold swing is defined as mV/decade.

In the case of MOSFET, the slope is independent of gate biasing voltage (V_{GS}) below the threshold voltage.

$$SS = \frac{\partial log V_G}{\partial \varphi_s} \frac{\partial \varphi_s}{\partial log I_D} = \left(1 + \frac{C_{DEP}}{C_{OX}}\right)\left(\frac{kT}{q} ln10\right) \tag{3.5}$$

Where

$$\left(\frac{kT}{q} ln10\right) \cong 60mV / Decade$$

Where at $T = 300$ k and C_{DEP} capacitance occur at the depletion region and C_{OX} is oxide capacitance. φ_s is the surface potential. At the equilibrium condition, the Boltzmann mobile charge carriers' distribution at the source region explain that a MOSFET can switch very steeply from OFF- to ON-state and vice versa.

The TFET has an inter-band electron tunneling process in which the charge carrier injection from the source side, in contrast to the MOSFET, which has thermionic carrier emission, so that the TFET can achieve a sub-60 mV/decade subthreshold slope. Wang, Hilsenbeck et al. reported that there is the possibility of low power applications in the case of TFETs [31]. Zhang et al., through a theoretical analysis, showed that the subthreshold swing of TFETs can be diminished to under 60 mV/decade [36].

Bhuwalka et al. achieved the subthreshold slope with a value of 44 mV/decade in the vertical structure of TFET [37] (Figure 3.8). Based on carbon nanotube (CNT) technology, Appenzeller et al. implemented the tunneling device to achieve SS of less than 60 mV/decade [38]. Knoll et al. demonstrated experimentally a novel and convenient method for making dopant segregation (DS) tunneling junctions in the strained-Si nanowire TFETs [39]. Both n-type and p-type TFETs displayed strong ON currents. For n-type TFETs, a minimum SS of 30 mV/decade was achieved at 300 K.

In a recently published paper, a subthreshold slope has been reported less than 60 mV/decade [40]. However the TFET reported a very low SS value, the MOSFET still had a better ON/OFF current ratio. This happens due to a low ON current outcome from TFET characteristics. The low ON current is due to the disparity between the TFET and the MOSFET carrier injection mechanisms. The band-to-band tunneling (BTBT) current is well known to become less sensitive to the electric field. This happens due to the

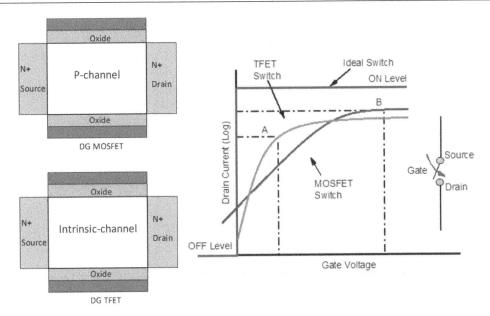

FIGURE 3.8   (a) Schematic diagram of double gate n-type tunnel TFET and MOSFET. (b) Subthreshold slope comparison for conventional MOSFET (blue) and a TFET (green) for ON- and OFF-state.

difference in potential source region and channel region increases. Again, it is difficult to attain significant transconductance beyond the subthreshold region. The TFET subthreshold swing (*SS*) can be obtained for the TFET by taking the first derivative of tunneling current with respect to gate source voltage (V_{GS}) [37].

$$S = log \left[\frac{1}{V_R} \frac{dV_R}{V_{GS}} + \frac{\varepsilon + B}{\varepsilon^2} \frac{d\varepsilon}{dV_{GS}} \right]^{-1} \tag{3.6}$$

It can be predicted from the equation that the subthreshold slope is almost independent of the temperature (KT/q).

3.1.5.4 Techniques for TFETs Performance Enhancements

From the previous discussions, it can be analyzed that the TFET subthreshold slope and ON-state current are regulated by the junction electric field, which is control by gate biasing voltage. For short channel length device, the characteristics length is an effective parameter. The scaling of the characteristics length is created by use of structural engineering such as a nanowire, dual gate, equivalent oxide thickness (EOT), and smaller body thickness, which can be used to enhance the performance of the TFET. Another method is the doping engineering, which fruitfully enhance the performance of a silicon TFET by implanting pocket/halo doping of different polarity near to source region. The pocket region will be depleted, leaving the ionized positive charges, which results in electric field to the intrinsic region at a given biasing voltage. Due to this tunneling, the generation rate will be improved with a steeper subthreshold slope.

3.1.5.5 Leakage Current in TFET

There are five major kinds of leakage current that effect the TFET device even in the OFF-state region. The leakage current arises from thermionic emission and high-k-gate stacks over the source and a drain in built potential. Among all options, the gate-induced drain leakage current (GIDL) is found to an effective drain leakage current in OFF-state medium. A hot charge carrier on GIDL arises due to induced interface traps and oxide trapped charges. Previously, the thermionic emission was a parallel process to tunneling in order to transmit the change carriers of holes and electrons form valence band to conduction band. Shockley-Read-Hall recombination holds the thermal effect of the device because at low biasing voltage thermal effect becomes prominent. In addition to the material of narrower bandgap material like InAs, a new drain leakage current will become dominant via band-to-defect tunneling. In this process, the electron will become thermally excited and form a valance band to interface traps. This situation need to prevented the biasing voltage will be exceeded from E_G/q, reverse mode of the biasing will arise due to the minority charge carrier at the drain side, which results in the ambipolar conduction. This ambipolar conduction can be reduce by building asymmetry into the drain side. This can be done by the gate underlap and gate overlap condition. Another way is to use the hetero-junction material of a lower bandgap at the source side, in comparison to the channel and drain.

3.1.6 Simulation Tools

Numerical simulations are becoming more important in modern scientific research work. Using technology computer aided design (TCAD) simulation tools provides a prototype of designed structured or characteristics that can be approximately predicted in a relatively short time. Therefore, it saves our time, cost, and rigorous experimental work. It also proves useful information for analytical and experimental work. TCAD tools are performed as 2D or 3D simulations, which fundamentally depends on finite element methods (FEM), where the device fabrication and its electrical behavior can be assessed based on the physical processing mechanisms or electrical mechanisms in the device via electric field, doping profile, potential biasing, charge carrier density, etc. [40–41].

The Sentaurus TCAD tool is a device simulation tool that is based upon the industry's standards with several fabricator collaborators, and the prototypes are tuned to the experimental details. This technology has been found to provide precise prediction of the future system with its characteristics to fulfill the demands of the International Technology Road Map for Semiconductors (ITRS) [19].

3.3.1.6.1 TFET Models

In a TCAD simulation tool, TFET devices are basically divided into two types of band-to-band tunneling (BTBT) models:

- Local BTBT model
- Nonlocal BTBT model

The Kane model, Klassen model and Hurkx model are the local BTBT models, while the WKB model is a nonlocal BTBT model. In the local model, a simple equation is used to drive the electric field, which is a significant parameter to find the tunneling width and tunneling current. The local model parameters easily fit the simulation of a TFET to find its effective parameters. However, these models are inappropriate for the TFET simulation because BTBT tunneling is fundamentally a physical process that can be incorporated by nonlocal BTBT process. The nonlocal BTBT process physically depends on the width of the tunneling barrier and the energy states. This model estimates the tunneling path, volume effect, and direction to the flow of charges. Although commercially many of the device simulation tools under TCAD, such as Sentaurus, Silvaco, Medici, and HyENEXSS, inherit the properties of the TFET models for execution purpose. Among them, we used Sentaurus simulation software for our results analysis because most of the manufacturers follow this tool for their fabrication purpose. In realty, BTBTs have multidimensional effects. The tunneling current is determined by taking into account the profile of the energy band along each tunneling slice with reverse bias voltage applied to the junction of the device. The schematic diagram of an energy band profile in which the range of valence band electron permitted for tunneling current is shown in Figure 3.9. Table 3.1 reports the default value of the main parameter in the model.

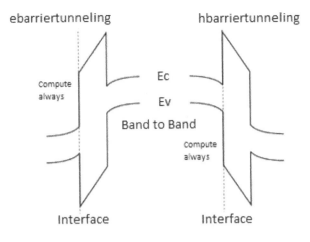

TABLE 3.1 Nonlocal BTBT Parameters for Sentaurus TCAD Simulation Tool

PARAMETER USED	DEFAULT VALUE
MC	0.160
MV	0.240
ME-TUNNEL	0.160
MH-TUNNEL	0.250

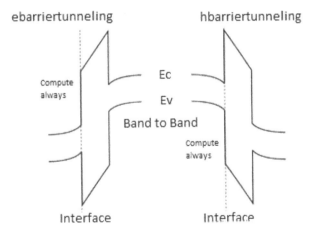

FIGURE 3.9 Schematic diagram of nonlocal band-to-band tunneling (BTBT) in reverse bias junction in the Sentaurus TCAD tool [40].

3.1.7 Major Device Parameters

The device performance and its functionality were determined by various parameters. The following sections discuss only a few descriptive parameters.

3.1.7.1 Subthreshold Slope

The subthreshold slope is basically the variation in the gate biasing voltage dropping the tunneling current by the factor 10. The drive characteristics below the subthreshold voltage (V_t) are known as the subthreshold regime. Ideally, the MOS device should have zero subthreshold slope, that is, below the subthreshold voltage, the device current must be zero, which in fact can never be attained. However, in reality, there is always some current flowing below the threshold voltage. The SS is an important parameter for evaluating the subthreshold properties of MOSFET devices on the short-channel nanoscale regime. It provides for how many volts the gate bias should be increased to increase the current subthreshold by 10 times and is provided by [42].

$$SS = \left(\frac{\partial log I_D}{\partial log V_G} \right)^{-1}$$

(3.7)

3.1.7.2 Drain Induced Barrier Lowering

Drain induced barrier lowering (DIBL) mainly predicts the V_{ds} effect on subthreshold characteristics. Due to a large amount of drain bias, the source–channel barrier will be reduced. The DBIL basically occurs in the short channel devices in which the device is controlled by both the gate biasing and the drain biasing. It can be measured by lateral shifting of transfer curves in the subthreshold region, where change in threshold voltage (ΔV_t) is divided by V_{dd} variation of two curves [43].

$$DIBL = \frac{\Delta V_t}{\Delta V_{dd}}$$

(3.8)

3.1.7.3 Cut-Off Frequency

When evaluating the devices for analog use, cut-off frequency (f_T) is the significant parameter. The measured value of the cut-off frequency occurs when the current gain is unity. It is given by equation 3.9, where g_m represents the transconductance and the total capacitance of the device is C_T [44].

$$f_T = \frac{g_m}{2\pi C_T}$$

(3.9)

3.1.7.4 Transconductance

Transconductance (g_m) is the reciprocal of resistance and is defined by equation 3.10. It is the electrical characteristic that relates the current through a device's output to the voltage at the device's input [45].

$$g_m = \frac{\Delta I_d}{\Delta V_{gs}}$$

(3.10)

3.1.7.5 Early Voltage

Early voltage (V_{EA}): This parameter effectively converts the DC power into the AC frequency with the performance gain, and the term is given by equation 3.11. It is the main figure of merit for the analogy application [46].

$$V_{EA} = \frac{I_d}{g_d}$$

(3.11)

3.1.7.6 Responsivity and Quantum Efficiency

In addition to response time and detectivity, the different parameters for evaluating photosensor efficiency are responsivity (R_e) and quantum efficiency (Q_e). The Q_e gain is defined as the number of charge carriers that pass for each photon absorbed per second between the electrodes and the responsivity of a photosensor. R_e is defined as the photocurrent generated on the effective area of the sensors per unit of light power incident. The expression for R_e and Q_e is shown by equations 3.12 and 3.13, where h represents the Planck constant, q is the electron charge, and incident radiation wavelength is represented by λ [47].

$$Responsivity\,(R_e) = \frac{Available\ Photocurrent, I_{ph}}{Incident\ optical\ power, P_o} \tag{3.12}$$

$$Quantum\,Efficiency\ (Q_e) = R_e\,\frac{h_c}{q\lambda} \tag{3.13}$$

3.1.8 Proposed Device Strucuture and Its Specifications

A 3D vertical p–i–n structure of silicon material with source, channel, and drain with different doping concentration is given in Figure 3.10a. To shorten the effective tunneling length for transmission of charges, a layer of 10 nm SiGe material is imposed between source and channel. This will reduce the energy bandgap form 1.1 eV to 0.7 eV and enhance the device ON current. The next Figure 3.10b will show the cross-section view with channel length of 50 nm and source drain length of 30 nm each. The work function used for the device (ϕ_m) is 4.5 eV (n-type) with the HfO$_2$ as the gate metal work function, with the width of 2 nm. Table 3.2 shows the description of the credentials used in the device structure.

Figures 3.11a and b show the equivalent representation of vertical T-shaped TFET hMobility and eMobility concentration of holes and electron (cm^2 V^{-1}s^{-1}) at source-channel region and channel-drain region, respectively. Both of the charge concentrations, eMobility and hMobility, are high at the center of the interface region. This is due to the high electric field at the center of the device as compared to the side dimensions.

3.1.9 Vertical T-Shaped Device as an Inverter Device

The mixed-mode technique is used to club the n-type and p-type vertical T-shaped device to make an inverter device using TCAD Sentaurus simulation software, as shown in Figure 3.12. The only difference between the p-type and the n-type device is the work function used and the area factor. Both the values will be greater for the p-type device, that is, the area factor will be twice the value of the n-type and the work function will be 4.5 eV for n-type and 5.3 eV for the p-type. The change concentration for the n-vertical T-shaped TFET will be a p–i–n structure while for p-vertical T-shaped TFET is a n–i–p structure. The oxide thickness of HfO$_2$ will be taken as 2 nm, with the EOT value of the 0.37 nm with respect to the SiO$_2$. The Si$_{1-x}$Ge$_x$ mole fraction is taken as 0.8 throughout the simulation.

Figure 3.13 shows the device current characteristics of the configuration for both n-type and p-type at the single display graph. Both devices achieve approximately similar subthreshold voltage, with the value of 0.357 V and 3.49 V at V_{ds} = 1 V. The current ratio of ON current and OFF current will be approximate of 10^{-11} with the suppressed ambipolar current, with the symbolic representation of each device with its graph, respectively.

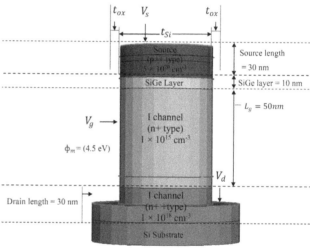

FIGURE 3.10 (a) 3D equivalent diagram of the proposed device vertical T-shaped TFET. (b) 2D equivalent schematic diagram of the vertical T-shaped TFET with SiGe layer with mentioned specifications.

TABLE 3.2 Device Specifications

CREDENTIALS USED	STANDARDS
Source (p++)	5×10^{20} cm^{-3}
Channel (n+)	1×10^{15} cm^{-3}
Drain (n++)	1×10^{18} cm^{-3}
Gate oxide thickness	2 nm
Work function used (n-type)	4.5 eV
Work function used (p-type)	5.6 eV
Channel length	50 nm
Source/drain length	30 nm

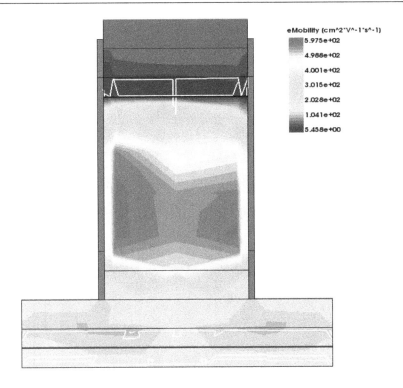

eMobility (cm^2*V^-1*s^-1)

5.975e+02
4.988e+02
4.001e+02
3.015e+02
2.028e+02
1.041e+02
5.458e+00

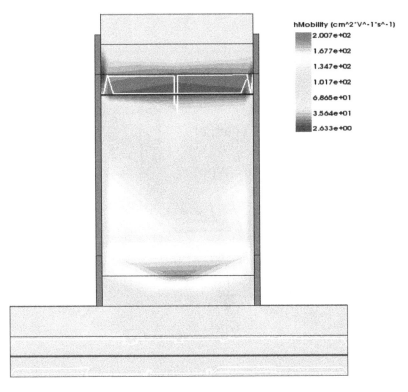

hMobility (cm^2*V^-1*s^-1)

2.007e+02
1.677e+02
1.347e+02
1.017e+02
6.865e+01
3.564e+01
2.633e+00

FIGURE 3.11   A 2D representation of the (a) eMobility and (b) hMobility concentration rate of charge carrier of holes and electrons (cm^2 V^{-1}s^{-1}) with the SiGe of 10 cm with the value of $x = 0.8$.

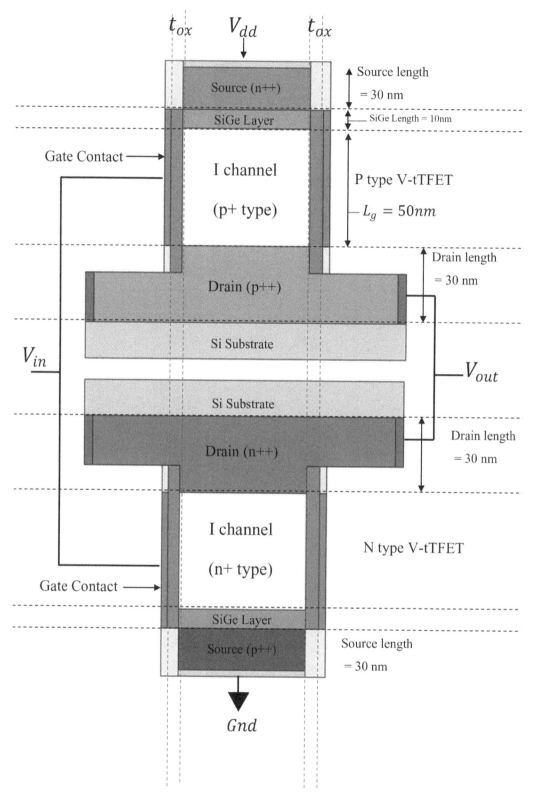

FIGURE 3.12   Mixed-mode vertical T-shaped heterojunction inverter circuit of p-type and n-type.

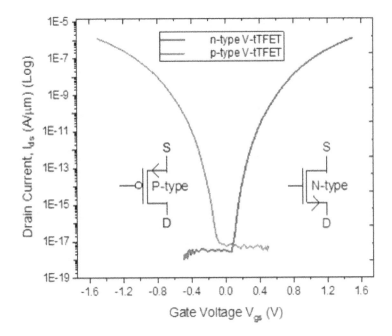

FIGURE 3.13   Drain current characteristics of the n-type and p-type vertical T-shaped TFET.

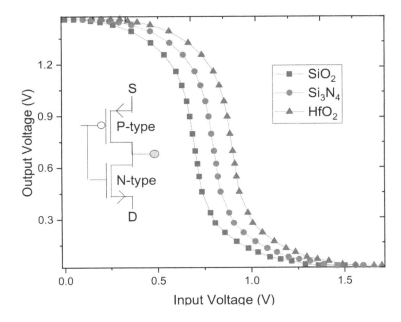

FIGURE 3.14   Transfer characteristics of vertical T-shaped inverter TFET with respect to different oxide material.

Figure 3.14 shows the simulated voltage transfer characteristics with different dielectric constant value. The material used as a dielectric constant is SiO_2 ($k = 3.9$), Si_3N_4 ($k = 7.0$), and HfO_2 ($k = 21$), respectively. From the graph, it can be seen that the HfO_2 material will show the maximum output voltage transfer characteristics from transition 0 to transition 1, with comparison to the Si_3N_4 and HfO_2.

3.1.10 Vertical T-Shaped TFET as a Low Power Application

Moving forward to the device application as a security challenge for low power application devices with respect to cost production and IC fabrication are continuous to outsources. However, this will increase the overall cost for the manufacturer, along with the security concerns. To overcome this situation currently, most of the new emerging techniques have been studied in terms of hardware security in designing the physical unclone function (PUF) and random number generation (RNG) [48]. So in order to alleviate this threat, we review the following PUF method. This method has been used frequently for the authentication process [49]. Therefore, many modeling methods have been design to estimate the response of the PUF function.

3.1.10.1 Device Characteristics

In the Figure 3.12, we saw that complementary circuit of a vertical T-shaped TFET has been developed using mixed mode technique. Now to corelate it with the hardware security concern, it has been divided into four cells: (i) security requirements at the top level; (ii) hardware primitives for the next level; (iii) drain current characteristics for the effective execution of those primitives; and (iv) the biasing voltage and other specifications. Formation of the given device will lead successfully to dealing with the security threats. In the case of nanoelectronics devices, the subthreshold slope and subthreshold voltage will leads to efficient low power devices. This implementation could lead to the current mode logic (CML) to defend the differential power analysis (DPA). So this chapter mainly focuses on the CML circuits with a vertical T-shaped TFET-based inverter. In many of the nanoscale devices, such as TFET, ambipolarity is one of the unwanted conductions of the negative biasing. This can be reduced with tunable polarity biasing also being suppressed in the proposed device.

3.1.10.2 Current Mode Logic of Vertical T-Shaped TFET

The current mode logic comes under the family of digital logic [50]. This gate basically consists of a tail current source and steering current core with differential attached load. With the different input trasistor network, the switching of the constant current and the voltage variation at the output load device occur. With the overall operation, the power consumption will be stable of the CML circuit. This will reduce the effect of DPA outbreak. Figure 3.15a shows the scematic diagram of the CML circuit, which include the pull-up and pull-down network. Figure 3.15b represents the details of the vertical T-shaped TFET device inverter buffer circuit.

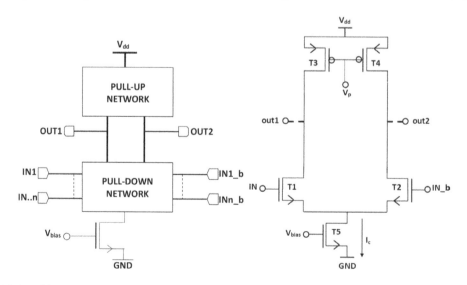

FIGURE 3.15 (a) Current mode logic universal circuit. (b) Schematic diagram of CML-based vertical T-shaped TFET inverter circuit.

3.1.11 Conclusion

This chapter deals with the basic concept and evaluation of the TFET as an exchange for the conventional MOSFET for low power application devices. A new vertical p–i–n junction device is proposed and analyzed using TCAD simulation. The vertical tunnel FET features a source channel distribution that will enhance the scalability of the simulated device. Further, both vertical p–i–n junction and n–i–p junction devices have been clubbed using mixed-mode technique to form an inverter circuit. A transfer current characteristic for the gate oxide materials of SiO$_2$ ($k = 3.9$), Si$_3$N$_4$ ($k = 7.0$) and HfO$_2$ ($k = 21$) has been compared, out of which HfO$_2$ comes out with the best output for the transition from 0 to 1. Further, the proposed devices have been analyzed for the hardware security system as lower power circuit device using current mode logic.

REFERENCES

[1] Tanaka, Junko, Toru Toyabe, Siego Ihara, Shin'ichiro Kimura, Hiromasa Noda, and Kiyoo Itoh. "Simulation of sub-0.1-mu m MOSFETs with completely suppressed short-channel effect." IEEE Electron Device Letters 14, no. 8 (1993): 396–399.

[2] Bricout, P.-H. and Emmanuel Dubois. "Short-channel effect immunity and current capability of sub-0.1-micron MOSFET's using a recessed channel." IEEE Transactions on Electron Devices 43, no. 8 (1996): 1251–1255.

[3] Young, Konrad K. "Short-channel effect in fully depleted SOI MOSFETs." IEEE Transactions on Electron Devices 36, no. 2 (1989): 399–402.

[4] Shockley, William. "The path to the conception of the junction transistor." IEEE Transactions on Electron Devices 31, no. 11 (1984): 1523–1546.

[5] Wadhwa, G. and B. Raj. "Design and performance analysis of junctionless TFET biosensor for high sensitivity." IEEE Nanotechnology 1, no. 8 (2019): 567–574.

[6] Instruments, Texas. "The chip that Jack built." www.ti.com/corp/docs/kilbyctr/jackbuilt.shtml [Online], 2011.

[7] Moore, Gordon E. "Cramming more components onto integrated circuits." IEEE Solid-State Circuits Society 38, no. 8 (1965): 114–117.

[8] Frank, David J., Robert H. Dennard, Edward Nowak, Paul M. Solomon, Yuan Taur, and Hon-Sum Philip Wong. "Device scaling limits of Si MOSFETs and their application dependencies." Proceedings of the IEEE 89, no. 3 (2001): 259–288.

[9] Wadhera, T., D. Kakkar, G. Wadhwa, and B. Raj. "Recent advances and progress in development of the field effect transistor biosensor: A review." Journal of Electronic Materials 48, no. 12 (2019): 7635–7646.

[10] Zhirnov, Victor V., Ralph K. Cavin, James A. Hutchby, and George I. Bourianoff. "Limits to binary logic switch scaling: a Gedanken model." Proceedings of the IEEE 91, no. 11 (2003): 1934–1939.

[11] Edgar, Lilienfeld Julius. "Method and apparatus for controlling electric currents." U.S. Patent 1,745,175, issued January 28, 1930.

[12] Sharma, S. K., B. Raj, and M. Khosla. "Enhanced photosensitivity of highly spectrum selective cylindrical gate In1-xGaxAs nanowire MOSFET photodetector." Modern Physics Letter-B 33, no. 12 (2019): 1950144.

[13] Rabaey, Jan M., Anantha P. Chandrakasan, and Borivoje Nikolic. "The devices." In Digital Integrated Circuits: A Design Perspective, 2nd edn. Pearson Education, New York (2003).

[14] Hoefflinger, B. "IRDS – International roadmap for devices and systems, rebooting computing, S3S." In B. Murmann and B. Hoefflinger (eds.), Nano-Chips 2030. The Frontiers Collection. Springer, Cham (2020). https://doi.org/10.1007/978-3-030-18338-7_2.

[15] Jain, N. and B. Raj. "Thermal stability analysis and performance exploration of Asymmetrical Dual-k Underlap Spacer (ADKUS) SOI FinFET for security and privacy applications." Indian Journal of Pure & Applied Physics (IJPAP) 57 (2019): 352–360.

[16] Lo, S.-H., Douglas A. Buchanan, and Yuan Taur. "Modeling and characterization of quantization, polysilicon depletion, and direct tunneling effects in MOSFETs with ultrathin oxides." IBM Journal of Research and Development 43, no. 3 (1999): 327–337.

[17] Singh, Shailendra, and Balwinder Raj. "Study of parametric variations on hetero-junction vertical t-shape TFET for suppressing ambipolar conduction." Indian Journal of Pure & Applied Physics 58 (June 2020): 478–485.

[18] Timp, G., J. Bude, K. K. Bourdelle, J. Garno, A. Ghetti, H. Gossmann, M. Green et al. "The ballistic nano-transistor." In International Electron Devices Meeting 1999. Technical Digest (Cat. No. 99CH36318), pp. 55–58. IEEE, 1999.

[19] "More-than-Moore." White Paper. www.itrs.net/papers.html [Online], accessed August 25, 2015.

[20] Naguib, Hussein M., Iain D. Calder, Vu Q. Ho, and Abdalla A. Naem. "Process of fabricating MOS devices having shallow source and drain junctions." U.S. Patent 4,683,645, issued August 4, 1987.

[21] Singh, Shailendra, and Balwinder Raj. "Analytical modelling and simulation of Si-Ge hetero-junction dual material gate vertical T-Shaped tunnel FET." Silicon 13 (2021): 1139–1150.

[22] Goel, Ekta, Sanjay Kumar, Balraj Singh, Kunal Singh, and Satyabrata Jit. "Two-dimensional model for subthreshold current and subthreshold swing of graded-channel dual-material double-gate (GCDMDG) MOSFETs." Superlattices and Microstructures 106 (2017): 147–155.

[23] Gusev, E. P., Douglas A. Buchanan, Eduard Cartier, Arvind Kumar, Don DiMaria, Supratik Guha, A. Callegari et al. "Ultrathin high-K gate stacks for advanced CMOS devices." In International Electron Devices Meeting. Technical Digest (Cat. No. 01CH37224), pp. 20–21. IEEE, 2001.

[24] Singh, Shailendra, and Balwinder Raj. "Two-dimensional analytical modeling of the surface potential and drain current of a double-gate vertical t-shaped tunnel field-effect transistor." Journal of Computational Electronics (2020): 1–10.

[25] Goyal, C., J. S. Ubhi, and B. Raj. "Low leakage zero ground noise nanoscale full adder using source biasing technique." Journal of Nanoelectronics and Optoelectronics, American Scientific Publishers 14 (2019): 360–370.

[26] Colinge, Jean-Pierre. "Multiple-gate SOIMOSFETS." Solid-State Electronics 48, no. 6 (2004): 897–905.

[27] Jain, A., S. Sharma, and B. Raj. "Analysis of Triple Metal Surrounding Gate (TM-SG) III-V nanowire MOSFET for photosensing application." Opto-electronics Journal, Elsevier 26, no. 2 (2018): 141–148.

[28] Singh, Shailendra, and Balwinder Raj. "Analysis of ONOFIC technique using SiGe heterojunction double gate vertical TFET for low power applications." Silicon (2020): 1–10.

[29] Choi, Woo Young, Byung-Gook Park, Jong Duk Lee, and Tsu-Jae King Liu. "Tunneling field-effect transistors (TFETs) with subthreshold swing (SS) less than 60 mV/dec." IEEE Electron Device Letters 28, no. 8 (2007): 743–745.

[30] Wang, P.-F., K. Hilsenbeck, Th Nirschl, M. Oswald, Ch Stepper, M. Weis, D. Schmitt-Landsiedel, and W. Hansch. "Complementary tunneling transistor for low power application." Solid-State Electronics 48, no. 12 (2004): 2281–2286.

[31] Singh, Shailendra, and Balwinder Raj. "Design and analysis of I ON and ambipolar current for vertical TFET." In Manufacturing Engineering, pp. 541–559. Springer, Singapore (2020).

[32] Gurinder Pal Singh, B. S. Sohi, and Balwinder Raj. "Material properties analysis of Graphene Base Transistor (GBT) for VLSI analog circuits." Indian Journal of Pure & Applied Physics (IJPAP) 55 (2017): 896–902.

[33] Moll, J. L. Physics of Semiconductors, pp. 249–253. McGraw Hill, New York (1964).

[34] Sze, Simon M. and Kwok K. Ng. Physics of Semiconductor Devices. John Wiley & Sons, Hoboken, NJ (2006).

[35] Singh, Shailendra, and Balwinder Raj. "Modeling and simulation analysis of SiGe heterojunction Double Gate Vertical t-shaped tunnel FET." Superlattices and Microstructures (2020): 106496.

[36] Singh, Shailendra, and Balwinder Raj. "Vertical tunnel-FET analysis for excessive low power digital applications." In 2018 First International Conference on Secure Cyber Computing and Communication (ICSCCC), pp. 192–197. IEEE, 2018.

[37] Bansal, Priya and B. Raj. "Memristor: A versatile nonlinear model for dopant drift and boundary issues." JCTN, American Scientific Publishers 14, no. 5 (2017): 2319–2325.

[38] Singh, Shailendra and Balwinder Raj. "Design and analysis of a heterojunction vertical t-shaped tunnel field effect transistor." Journal of Electronic Materials 48, no. 10 (2019): 6253–6260.

[39] Jain, Neeraj and Balwinder Raj. "Device and circuit co-design perspective comprehensive approach on FinFET technology – A review." Journal of Electron Devices 23, no. 1 (2016): 1890–1901.

[40] Sentaurus User's Manual, Synopsys, Inc., Mountain View, CA, USA, 2017.09.

[41] Singh, Shailendra, and Balwinder Raj. "Modeling and simulation analysis of SiGe heterojunction Double Gate Vertical t-shaped tunnel FET." Superlattices and Microstructures (2020): 106496.

[42] Bansal, P. and B. Raj. "Memristor modeling and analysis for linear dopant drift kinetics." Journal of Nanoengineering and Nanomanufacturing, American Scientific Publishers 6 (2016): 1–7.

[43] Sharma, Savitesh M., S. Dasgupta, and M. V. Kartikeyan. "FinFETs for RF applications: A literature review." In 2018 Conference on Emerging Devices and Smart Systems (ICEDSS), pp. 280–287. IEEE, 2018.

[44] Badgujjar, Soniya, Girish Wadhwa, Shailendra Singh, and Balwinder Raj. "Design and analysis of dual source vertical tunnel field effect transistor for high performance." Transactions on Electrical and Electronic Materials 21, no. 1 (2020): 74–82.

[45] Jain, Aakash, Sanjeev Sharma, and Balwinder Raj. "Design and analysis of high sensitivity photosensor using cylindrical surrounding gate MOSFET for low power sensor applications." Engineering Science and Technology, An International Journal, Elsevier's 19, no. 4 (2016): 1864–1870.

[46] Xie, Xing, So-Ying Kwok, Zhenzhen Lu, Yankuan Liu, Yulin Cao, Linbao Luo, Juan Antonio Zapien et al. "Visible – NIR photodetectors based on CdTe nanoribbons." Nanoscale 4, no. 9 (2012): 2914–2919.

[47] Wu, Chi-Woo, Alexei N. Nazarov, Isabelle Ferain, Nima Dehdashti Akhavan, Ran Yan, Pedram Razavi, Ran Yu, Rodrigo T. Doria, and Jean-Pierre Colinge. "Low subthreshold slope in junctionless multigate transistors." Applied Physics Letters 96, no. 10 (2010): 102106.

[48] Singh, Shailendra, and Balwinder Raj. "Analytical modeling and simulation analysis of T-shaped III-V heterojunction vertical T-FET." Superlattices and Microstructures (2020): 106717.

[49] Singh, K. and B. Raj. "Comparison of temperature dependent performance and analysis of SWCNT bundle and copper as VLSI interconnects." Research Cell: An International Journal of Engineering Sciences 17, no. 1 (2016): 583–595.

[50] Verma, Sachin K., Shailendra Singh, Girish Wadhwa, and Balwinder Raj. "Detection of biomolecules using charge-plasma based gate underlap dielectric modulated dopingless TFET." Transactions on Electrical and Electronic Materials 21, no. 5 (2020): 528–535.

[51] Kumar, Sunil and B. Raj. "Analysis of ION and ambipolar current for dual-material gate-drain overlapped DG-TFET." Journal of Nanoelectronics and Optoelectronics, American Scientific Publishers, USA 11 (2016): 323–333.

[52] Kumar, Sunil, Sumit Kumar, Karamveer, Keshav Kumar Gupta, and B. Raj. "Analysis of double gate dual material TFET device for low power SRAM cell design." Quantum Matter, American Scientific Publishers 5 (2016): 762–766.

[53] Aggarwal, Raghav, B. Raj, Lakhwinder Singh Teji, and Ritesh Kumar. "Analysis of leakage current and power dissipation in 5T SRAM cell." Quantum Matter, American Scientific Publishers 5 (2016): 792–797.

[54] Khandelwal, S., B. Raj, and R. D. Gupta. "Design and optimization of 6T FinFET SRAM in 45 nm technology." Quantum Matter, American Scientific Publishers 5 (2016): 552–556.

[55] Kumar, Sunil and B. Raj. "Compact channel potential analytical modeling of DG-TFET based on evanescent – mode approach." Journal of Computational Electronics, Springer 14, no. 2 (2015): 820–827.

[56] Singh, K. and B. Raj. "Temperature dependent modeling and performance evaluation of multi-walled CNT and single-walled CNT as global interconnects." Journal of Electronic Materials, Springer 44, no. 12 (2015): 4825–4835.

[57] Singh, K. and B. Raj. "Performance and analysis of temperature dependent multi-walled carbon nanotubes as global interconnects at different technology nodes." Journal of Computational Electronics, Springer 14, no. 2 (2015): 469–476.

[58] Raj, B., A. K. Saxena, and S. Dasgupta. "Nanoscale FinFET based SRAM cell design: Analysis of performance metric, process variation, underlapped FinFET and temperature effect." IEEE Circuits and System Magazine 11, no. 2 (2011): 38–50.

[59] Raj, B., S. K. Vishvakarma, A. K. Saxena, and S. Dasgupta. "Analytical modeling of nanoscale double gate FinFET device." International Journal of Intelligent Electronics System 1 (2007): 633–637.

[60] Vishvakarma, S. K., V. Agarwal, Balwinder Raj, A. K. Saxena, and S. Dasgupta. "Two dimensional analytical potential modeling of nanoscale symmetric double gate (SDG) MOSFET with Ultra Thin Body (UTB)." Journal of Computational and Theoretical Nanoscience (JCTN), American Scientific Publishers (ASP), USA 4, no. 5 (2007): 1144–1148.

[61] Pattanaik, Manisha, B. Raj, Shashikant Sharma, and Anjan Kumar. "Diode based trimode multi-threshold CMOS technique for ground bounce noise reduction in static CMOS adders." Advanced Materials Research, Trans Tech Publications, Switzerland 548 (2012): 885–889.

[62] Bhushan, S., S. Khandelwal, and B. Raj. "Analyzing different mode FinFET based memory cell at different power supply for leakage reduction." Seventh International Conference on Bio-inspired Computing: Theories and Application (BIC-TA 2012), Advances in Intelligent Systems and Computing 202 (2013): 89–100.

Silicene and Germanene Nanoribbons for Interconnect Applications

4

Varun Sharma and Pankaj Srivastava

Contents

4.1 INTRODUCTION

In the past few decades, ubiquitous implementation of integrated circuits (IC) implies their importance and utility in every day-to-day electronic product. These varied and diverse applications aroused an imperative urge for multiple functionalities on the same IC, which resulted in the incessant miniaturization of the devices on the IC. Traditionally, field effect transistors (FET) embedded on the IC are most commonly comprised of bulk or 3D semiconductor channel materials, mainly silicon, germanium, and the other III-V semiconductors. Through these bulk materials, to date, researchers and technologists have been able to successfully scale down the devices to sub-nanoscale dimensions over the last five decades, which was in good agreement with Gordon Moore's famous prediction (popularly coined as Moore's law) stating that the density of transistors on a chip will double every two years [1–4]. Following this prediction, researchers, technologists, and industries set up their goals and strategies and devised a roadmap termed the International Technology Roadmap for Semiconductors (ITRS) [5]. These strategies and predictions have recently started to falter as the active channel lengths started hitting the nanometers regime. This has been the case for the following primary reasons: (a) At these dimensions, the transport of the carrier over the channel is not principally governed by the laws of statistical mechanics (typically known as the Boltzmann equation of transport [6]). (b) At these dimensions, the electrostatic control of the controlling terminal, usually termed as the gate, becomes inappropriate. (c) At these dimensions, the short channel effects in the devices become prominent (short channel effects are commonly referred as current leakage and the resulting heat dissipation caused by an increase in static power) [7]. (d) At these dimensions, the significance of contacts and interconnects between devices becomes more important, as comparatively significant power is dissipated at the interface rather than the channel itself. As a result, researchers are looking for various possible solutions at every level of abstraction (viz. circuit, device, and materials). To address these short channel effects, devices such as multiple gate transistors [8, 9], FinFETs [10], TFETs [11–13], gate-all-around FETs [14], and ultrathin body transistors (UTB) [15, 16] are actively being looked into. Further, various circuit level techniques such as power gating [17] and use of multi-threshold transistors also have been popularly used. These approaches are often coined as top-down approaches to mitigate scaling issues. Ever since the famous talk by Richard Feynman – "There is plenty of room at the bottom" [18] – the 'bottom-up viewpoint' has always been considered as one of the unique approaches to understand various fundamental properties. Recently, with the advances in computational and synthesis techniques, the bottom-up approach is serving as a foundation stone for new-era nanoscale devices [19–21]. However, both the approaches, namely top-to-bottom and bottom-up, have their own advantages and limitations. For example, on the one hand, it is hard to capture the exact physics in a nanometer regime in the top-to-bottom approach. On the other hand, it is difficult to culminate the complete quantum (bottom-up) viewpoint for a bigger circuit and system. Thus, the optimal solutions at each system, circuit, device, and material level are consistently being proposed to achieve the required overall solution. In recent decades, fundamental bottlenecks posed due to the intrinsic materials properties such as electron mobility (due to scattering in bulk materials) and mechanical and thermal instability raised a severe motivation for the research community to look for promising new materials [22]. Nanomaterials for varied applications are the consequence of such an incessant unquenched thrust of the scientific community. Nanomaterials constitute the genre of materials where at least one of the effective dimensions is in the range of nanometers (nm). Unlike bulk materials, nanomaterials offer a lot of scope not just to overcome existing issues but also to widen the applicability horizon. Among these nanomaterials, low dimensional nanomaterials such as C-based nanostructures have procured special attention. These structures include 0D fullerenes (bulky balls of carbon), 1D carbon nanotubes (CNT), graphene nanoribbons (GNRs), and the 2D graphene which is monolayer of graphite [23]. The exotic properties of graphene are backed by the active research by in this area, which eventually rejuvenated the overall research interest in the field of layered materials possessing such peculiar electronic and optical behavior [24]. Moreover, their excellent mechanical strength, optical transparency, and unique direct bandgap make them suitable for a wide variety of applications [25–33]. From the device perspective, it is also evident that for channels comprising a 2D layered semiconductor, the carriers will remain confined in the atomically thin layered channels, ensuring a better electrostatic control. Such an excellent gate control ensures reduced leakage current and reduced short channel effects. Other than the typical semiconducting applications, these materials also possess enhanced mobility of carriers, tunable electronic properties, and intrinsic spin orbit coupling (SOC), which widens their pool of applications [34]. Thus, immediately after the phenomenal success of graphene synthesis, researchers have come up with a number of exciting materials such as transition metal dichalcogenides (TMDs) [35, 36], hexagonal-boron nitride (h-BN), etc. [37, 38]. Apart from these compounded materials, other group-IV elemental analogues of graphene, namely silicene germanene and phosphorene [39–42], have also gained an enormous research interest. Silicene and germanene exhibited several astonishing properties enabling them to not only serve as active material for state-of-art devices but also to append entirely new dimensions to the field [43, 44]. These materials have high carrier mobility, and due to strong SOC, the substantial bottleneck of zero bandgap is greatly reduced compared to graphene [45–49]. Similar to graphene, in order to further modulate the bandgap, silicene and germanene can further be stripped off into 1D nanoribbons with definite edge states. Unlike two-dimensional versions, nanoribbons (NR) are confined in two directions and are allowed to grow only along a third direction. Due to such confinement, these materials exhibit different and unique electronic properties compared to their 2D form. Based on the edge states, the nomenclature follows as armchair silicene/germanene nanoribbon (ASiNR/AGeNR) and zigzag silicene/germanene nanoribbon (ZSiNR/ZGeNR), as shown in Figure 4.1.

Similar to graphene nanoribbons, these structures also have unique ribbon width-dependent electronic and magnetic properties [50]. Further, it must be mentioned that based on the direction of cutting, the 2D sheet armchair or the zigzag edge states can be produced. There is a 30° cutting direction difference that creates ASiNR/AGeNR and ZSiNR/ZGeNR, respectively. In nanoribbons, the edge atoms are attached with only the two nearest neighboring atoms. Therefore, an additional unsaturated chemical bond (the dangling bond) is associated to each edge of the Si/Ge atom at both the edges. The presence of these dangling bonds at the ribbon edges provides a convenient way to alter the electronic properties and the stability of NRs through termination or the chemical edge modification [45, 51–53]. Further, like GNRs in the recent past, research pertaining to substitutional doping in these SiNR/GeNRs has led to exhibition of some exceptional properties [54–57]. Apart from typical semiconductor applications, these materials (specifically the ones having zero bandgaps) can also have other prospective applications such as nanoscale interconnects [58–60]. Spintronics is another such field in which precise utilization of the spin degree of freedom of an electron, rather than its electronic charge, serves as the fundamental quantity for the flow of information [61] and is being looked at as a promising low power alternative. Silicene and germanene possess strong SOC and can serve as promising candidates for these applications, too [62, 63]. Furthermore, applications such as contact engineering and designing

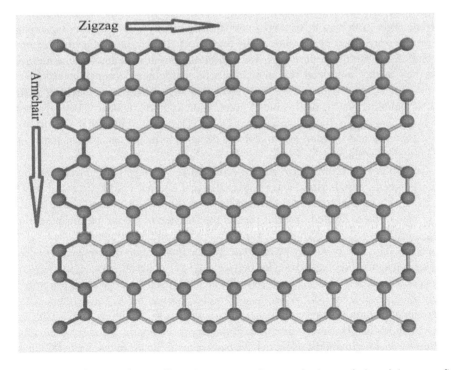

FIGURE 4.1   Directions of cutting the 2D silicene/germanene sheet to obtain armchair and zigzag configurations.

metal interconnects have also garnered significant scientific attention, as they can bring down the power consumption to a great extent [58, 64]. Hence, it is imperative to study and analyze the important properties of these atomically thin layered materials so as to address such diverse applications.

4.2 THEORETICAL AND EXPERIMENTAL INVESTIGATIONS ON SILICENE AND GERMANENE

Ever since the idea of flat structure of C-atoms packed in a hexagonal arrangement came into existence, the researchers have been continuously eyeing new materials. Both theoreticians and experimentalist have continuously explored new evidences for the existence of different 2D materials. Studies suggest that not only new materials but their novel synthesis methods also play a crucial role in differentiating the intrinsic properties of these materials. Here are few of the seminal works primarily focused on theoretical and experimental evidences of silicene and germanene.

Takeda et al. [65] have, for the first time, theoretically predicted the existence of silicon- and germanium-based corrugated graphene-like structures based on the first-principle calculations. Their studies show that unlike the graphene, the Si- and Ge-based structures exist in slightly buckled form. Further, they have also calculated the electronic band structures that indicated the same Dirac cones near the Fermi level as was earlier predicted for graphene. A more rigorous first-principle-based study conducted by Cahangirov et al. [66] suggested that indeed silicon- and germanium-based structures can possess honeycomb structure like graphene. However, they strongly emphasized that low buckling structures are energetically more favorable compared to the planar and high buckling forms. They have also shown that Dirac cones still exist in these structures that ensure the promise of high electron mobility with these structures. Studies for a one-dimensional confined structure, namely nanoribbons for both silicene and germanene, were also carried out. Lalmi et al., [67] have presented the experimental evidence for a ($\sqrt{3} \times \sqrt{3}$) silicene sheet grown on a Ag (111) substrate. Atomically resolved scanning tunneling microscopic (STM) images presented in the work imply that direct condensation of a silicon atomic flux onto the single-crystal Ag substrate in ultra-high vacuum conditions is pragmatically achievable. Through STM images, they further presented two distinct silicon sub-lattices occupying positions at different heights (0.02 nm), indicating possible sp^2–sp^3 hybridizations. In another report, Davila et al. [68] proposed a novel synthesis technique for the realization of germanium akin to graphene. They reported the growth of a ($\sqrt{3} \times \sqrt{3}$) germanene superstructure on a ($\sqrt{7} \times \sqrt{7}$) Au (111) configuration. In the report, they proposed several other phases, too, and the aforementioned was deduced to be the most stable and almost flat compared to other proposed structures.

Several other reports (both theoretical and experimental) covering silicene layer growth on Ag (100), Ag (110), ZrB_2, graphene, h-BN, MoS_2, Ca, Pb, Mg. etc., and alternatives like Pt (111), Au (111), etc. for germanene have also been presented in the recent past [69, 70–81]. The purpose of exploring different substrate for growth of these layered materials lies in the fact that based on the substrate interaction and deposition rate, different silicene/germanene superstructures exhibit different physical, electronic, and magnetic properties. This can primarily be attributed to the modulation of physical parameters such the lattice, buckling constant, and charge transfer between the 2D material surface and substrate material [88]. Other theoretical studies also suggest several nonmetallic substrates such as h-boron nitride, SiC, MX_2 (M = transition metal atom (Mo, W, etc.) and X = chalcogen atom (S, Se, Te, etc.)). These materials are predicted to induce minimal impact on the Dirac cones of grown 2D layered materials, ensuring their utility for device application. However, since most of the reports on nonmagnetic substrates are theoretical in nature and very few experimental evidences have been reported so far for the same [82], it is imperative to look for the substrate material (be it metallic or nonmetallic) that induces minimal impact on these 2D nanomaterials or, alternatively, modulates their properties for the desired utilization.

4.3 TAILORING THE ELECTRONIC/MAGNETIC PROPERTIES OF SILICENE AND GERMANENE

From the studies, it is evident that like graphene, silicene and germanene are also predicted to have a semimetallic nature [65]. Thus, for any plausible switching device application, it is imperative to ensure the semiconducting behavior of these materials. Along the lines of graphene, researchers have focused on several ideas pertaining to the realization of semiconducting properties in these materials. This includes application of an external electric field, saturation of dangling bond through a different functionalization group, and introducing external dopants. Apart from these measures, one-dimensional confinement of these materials (creating nanoribbons) also induces several interesting and useful properties to them. Like graphene nanoribbons (GNRs), silicene and germanene also possess interesting properties when cut from a sheet in a particular direction, that is, zigzag and armchair. It is revealed through studies that these nanostructures have highly susceptible edges that can undergo reconstruction or functionalization that can modulate the properties of these nanostructures. Literature also suggests that doping these structures can play a vital role in modulating the properties of these nanostructures. Below are some crucial landmarks from the literature that suggest how key properties of silicene and germanene nanoribbons can be modulated for varied vital usage of these materials.

Ding and Ni [89] have investigated the electronic and magnetic properties of silicon nanoribbons (SiNR). Based on the first-principles calculation, they showed that armchair nanoribbons can exhibit a metallic or a semiconducting behavior based on the width of the nanoribbon. However, based on the energy state calculated, they demarked that zigzag nanoribbons exhibit a magnetic ground state (antiferromagnetic). Under the application of a transverse electric field, these zigzag structures tend to exhibit the half-metallicity that is widely utilized in spintronic device applications. Other studies by Pang et al. [53] have carried out studies on zigzag and armchair germanene nanoribbons. They revealed that, similar to silicene nanoribbons, armchair GeNR exhibit a nonmagnetic semiconducting behavior, whereas zigzag structures exhibit an antiferromagnetic semiconducting ground state. In this work, they also studied the effects of substitutional doping, which showed that both N-doping and B-doping is preferred at the edge site rather than center. Ding et al. [90] suggested that on the account of the sp^2–sp^3 edge bipolar magnetic semiconducting behavior can be induced in zigzag silicene nanoribbons. They also suggested the measures to tune them to act as half-metals using suitable dopants, hence predicting their utility for spintronic devices. Zhang et al. [91] have studied the geometric, electronic, and magnetic properties of armchair silicene nanoribbons. They predicted that zigzag silicene nanoribbons possess a magnetic ground state when terminated asymmetrically with hydrogen. They considered various configurations and suggested that, although asymmetric edge terminations induce a magnetic character in the nanoribbons, they are not the most stable ones. They also studied the effect of a transverse electric field on these structures and found them suitable for spintronic device applications under appropriate electric field effects. Recently, Tao et al. [44] grew silicene on a Ag (111) film on mica; following this, a unique encapsulated delamination transfer and native Ag contact were developed for the first time, thereby preserving the crucial properties of buckled silicene during the device fabrication and characterization process. This work for the first time demonstrates and proves a silicene device can be in solid agreement with predictions of a Dirac-like ambipolar charge transport.

On the same lines in recent years, several works have come to light that suggest different methods to alter the electronic and magnetic properties of these novel materials. Interaction with halogen, transition metals, and coinage metals (Ag, Au) are prime among those [54, 92–107]. These studies suggest a profound impact of foreign atoms on various properties of these materials. The vitality of such studies also lies in the fact that these elements, such as hydrogen, oxygen, halogens, transition metals, and coinage metals (Ag, Au), can be very commonly induced as the impurities in the cleanroom environment, henceforth making all such cases a practical scenario.

With this chapter, we present the prospects of SiNR and GeNR as metal interconnects under various functionalization and doping scenarios. We have systematically analyzed the performance of various configurations by studying their structural, electronic, and transport properties. Here we will present the key performance parameter based on density functional theory (DFT) calculations and nonequilibrium Green's function (NEGF) formulism-based two-probe analysis that includes quantum resistance (R_Q), kinetic inductance (L_K), and quantum capacitance (C_Q) for the considered structures.

4.4 EDGE FUNCTIONALIZED GENRS FOR INTERCONNECT APPLICATIONS [109]

In this section, we present the analysis of impact of edge oxidation and hydrogenation on electronic and transport properties of zigzag and armchair germanene nanoribbons (ZGeNR and AGeNR); we have modeled all possible terminations of hydrogen and oxygen for AGeNR and ZGeNR. Figures 4.2 and Figure 4.3 present all possible termination cases for AGeNR and ZGeNR, along with the appropriate names, as presented in the caption for each case.

Here blue, white and red spheres represent the germanium, hydrogen and oxygen atom respectively.

First, to comment on structural stability of each configuration, we have calculated the binding energy (E_b) as per equation 4.1.

$$E_b = E_{Total} - E_{Bare} - xE_H - yE_O \qquad (4.1)$$

Here, E_{Total}, E_{Bare}, E_H, E_O, indicate the total energy of functionalized AGeNR/ZGeNR, bare AGeNR/ZGeNR, and the H- and O- atoms, respectively. Also x, y represents the total count of these atoms respectively. Further, based on the DFT calculations, we also calculated the respective bandgaps and Fermi velocity (v_f) in each of the cases. These are tabulated in Table 4.1

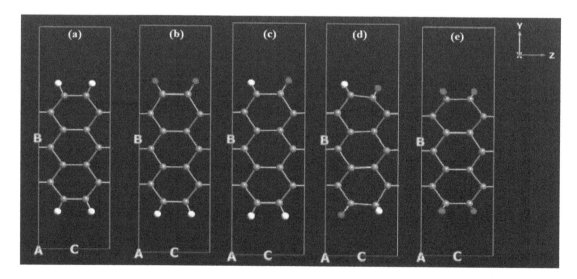

FIGURE 4.2 The relaxed configurations for functionalized AGeNR (ribbon width N_A = 7): (a) H-AGeNR-H, (b) O-H-AGeNR-H-H, (c) O-H-AGeNR-H, (d) O-H-AGeNR-O-H, (e) O-AGeNR-O.

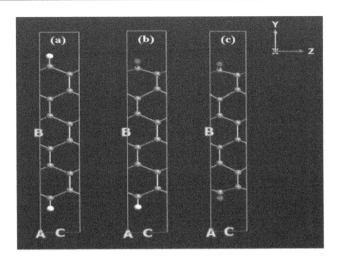

FIGURE 4.3 The relaxed configurations for functionalized ZGeNR (ribbon width N_z = 6): (a) H-ZGeNR-H, (b) O-ZGeNR-H, (c) O-ZGeNR-O.

TABLE 4.1 Calculated Values of Binding Energy, Bandgap, and Fermi Velocity for Armchair and Zigzag Germanene Nanoribbons with Considered Edge Terminations

EDGE STATES	CONFIGURATIONS (AS PER TERMINATIONS)	BINDING ENERGY (EV)	BANDGAP (EV)	FERMI VELOCITY (M/SEC)
Armchair	H-AGeNR-H	−12.03	0.36 (direct)	6.61×10^3
	O-AGeNR-H	−18.97	0.18 (indirect)	1.89×10^4
	O-H-AGeNR-H-H	−15.39	Metallic	2.47×10^5
	O-H-AGeNR-O-H	−18.67	Metallic	2.95×10^5
	O-AGeNR-O	−26.07	0.089 (indirect)	9.14×10^2
Zigzag	H-ZGeNR-H	−6.47	Semimetallic	9.19×10^4
	O-ZGeNR-H	−10.08	Metallic	4.19×10^5
	O-ZGeNR-O	−13.65	Metallic	5.21×10^5

The tabulated values for E_b show that O- termination strengthens the stability of the GeNR structures (both AGeNR and ZGeNR cases) in comparison to the pristine GeNRs. The E_b values also suggest that the higher the concentration of the O-atom at the edge of the GeNRs, the more stable is the overall system. As a result, both O-AGeNR-O and O-ZGeNR-O are reported to have the highest values of the E_b (viz. −26.07 eV, −13.65 eV, respectively). The electronic properties were then gauged for AGeNR and ZGeNR based on the calculated bandgap via E–k diagrams. The tabulated values suggest that the bandgap typically reduces as the concentration of elemental O-atom at the edges increases. This typical behavior, on one hand, can be attributed to the symmetry of edge terminating elements. But also, on the other hand, since the charge induced at the edges of AGeNR from O-atom is much larger compared to the H-atom, the band edges appear very near to the Fermi level, resulting in much narrower bandgap (for O-AGeNR-O case). Hence, we can infer that O-functionalized AGeNRs can be semiconducting or metallic depending upon the site of O-termination. However, for ZGeNR, the case is quite straightforward, where all the terminations are metallic in nature with almost zero bandgap. This implies that higher concentration of O-atoms at the edges of ZGeNR improves the metallicity of the structure to greater extent, which underscores its utility for interconnect applications.

To gauge their performances toward interconnect applications, a two-probe analysis has been performed to calculate transmission channel (N_{ch}), kinetic inductance (L_K), and quantum capacitance (C_Q) for considered structures. The quantum resistance can then be modeled by the equation shown below [58]:

$$R_Q = \frac{1}{G_Q} = \frac{h}{2q^2} \frac{1}{N_{ch}} \qquad (4.2)$$

Here, R_Q is defined as the quantum resistance for very small dimensions, N_{ch} defines the number transmission channels. Equation 4.2 highlights the fact that quantum resistance varies inversely with the value of the number of transmitting channels, N_{ch}. To comment on the value of N_{ch}, transmission eigenvalues are then calculate. Transmission eigen values represents the probability (or the possibility) with which the carrier (electron wave) can pass through the central region; higher probability indicates less scattering and a more smooth flow of carriers. Transmission mode/N_{ch} around the Fermi level is then calculated, and the results are tabulated in Table 4.2. Tabulated values makes it evident that only O-H-AGeNR-H-H and O-H-AGeNR-O-H exhibit non-zero values of N_{ch} ($N_{ch} = 1$ and 2, respectively). Similar analysis can be performed for the zigzag counterparts, where the tabulated values suggest that R_Q will be more or less same as the value of N_{ch} and is the same for all the three cases. However, a detailed analysis for the transmission eigen spectra shows that due to presence of c2-symmetry (right at the edges) in the case of O- ZGeNR-O, the channel tends to offer smooth flow of carrier (or least scattering) in this case. It must be mentioned here that this symmetry is a strong function of bias voltage, and it collapses with the increase in bias voltage leading to an undesired NDR effect here. Further, the dynamic performance is governed by inductance and capacitance, which is typically modeled through equations for L_K, C_Q with respect to 1D nanoscale devices [58] thus:

$$L_K = \frac{h}{4q^2 v_f} \frac{1}{N_{ch}} \qquad (4.3)$$

$$C_Q = \frac{4q^2}{h v_f} N_{ch} \qquad (4.4)$$

The calculated results are tabulated in Table 4.2. As it is implied from equations 4.3 and 4.4 that both, L_k and C_Q depend inversely on the v_f, which means the greater the v_f, lesser the inductive coupling (L_k) and parasitic delay (C_Q) will be. Hence, it is essential to calculate the v_f for each of functionalized configuration. The v_f values for each case are calculated, and the values were presented in Table 4.1. Values

TABLE 4.2   The Calculated Transmission Channel (N_{ch}), Kinetic Inductance (L_k), and Quantum Capacitance (C_Q) for Armchair and Zigzag Germanene Nanoribbons with Considered Edge Terminations

EDGE STATES	CONFIGURATIONS (AS PER TERMINATIONS)	TRANSMISSION CHANNELS (N_{ch})	KINETIC INDUCTANCE (L_k) (nH/μm)	QUANTUM CAPACITANCE (C_Q) (pF/cm)
	H-AGeNR-H	NIL	NIL	NIL
	O-AGeNR-H	NIL	NIL	NIL
Armchair	O-H-AGeNR-H-H	1	26.20	6.26
	O-H-AGeNR-O-H	2	109.67	10.48
	O-AGeNR-O	NIL	NIL	NIL
	H-ZGeNR-H O-ZGeNR-H	3	23.47	50.45
Zigzag	O-ZGeNR-O	3	5.15	11.06
		3	4.14	8.89

in Table 4.1 indicate that with increase in concentration of the oxygen atom, –all AGeNR configurations – except for O-AGeNR-O –the magnitude of v_f increases. Further, in the case of ZGeNR, the significance of the value of v_f is even more, as the value of N_{ch} is constant for each configuration of ZGeNR. The results indicate that with the increase in concentration of O- atoms at the edges of ZGeNRs, the value of v_f increases. This leads to significant improvement in the value of L_k and C_Q for O-ZGeNR-H and O-ZGeNR-O cases compared to the pristine ones (H-ZGeNR-H). However, as mentioned previously, for the higher bias voltage, as N_{ch} tends to reduce, L_k and C_Q may also ruin the performance of interconnect. These results are further matched with the already reported results for silicene [108].

Hence, through calculations of R_Q, L_k, and C_Q, it is revealed that, based on the significantly higher values of v_f and N_{ch}, the O-ZGeNR-O configuration offers minimum quantum resistance and delay and can be a potential candidate for future nanoscale interconnect applications. Similar analysis can also be done for the silicene nanoribbons, and it can be shown that O- functionalization enhances the metallicity of the SiNRs, making them useful for interconnect applications.

4.5 SUBSTITUTIONALLY DOPED GENRS, SINRS FOR INTERCONNECTS APPLICATIONS WITH COMMENTS ON RELIABILITY AND SECURITY [110]

Here we present the analysis of the impact of substrate induced impurities (as dopants) on GeNR. For proper comparative and exhaustive study, both AGeNR and ZGeNR have been investigated, with substitutional Au doping at different sites. We have presented all possible stable configuration, both doped and pristine, that is, undoped AGeNR/ZGeNR cases (for quick comparison), as shown in Figure 4.4.

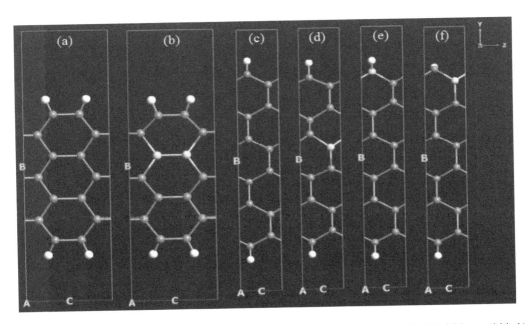

FIGURE 4.4   The relaxed configurations for AGeNR (ribbon width N_A = 7) and ZGeNR (ribbon width N_Z = 6): (a) pristine AGeNR, (b) center-doped (C-AGeNR), (c) pristine ZGeNR, (d) center-doped (C-ZGeNR), (e) edge-doped (E-ZGeNR), and (f) near-edge doped (NE-ZGeNR). Here blue, white and yellow spheres represent the germanium, hydrogen, and gold atoms, respectively.

The naming conventions and associated abbreviations used for different Au-doping configurations are as follows: (i) center-doped AGeNR (C-AGeNR), (ii) center-doped ZGeNR (C-ZGeNR), (iii) edge-doped ZGeNR (E-ZGeNR), and (iv) near-edge doped ZGeNR (NE-ZGeNR).

To precisely ensure the thermodynamical stabilities for these configurations formation energy per atom calculations have been performed as per the relation given below:

$$E_{FE} = \frac{E_{Total} - pE_{Ge} - qE_{Au} - rE_{H}}{p + q + r} \tag{4.5}$$

Here, E_{Total}, E_{Ge}, E_{Au}, and E_{H} represent the total energy of the doped AGeNR or ZGeNR configurations, Ge- atom, Au- atoms, and H- atom (in their isolated form). Also, p,q,r indicates the total number of Ge-, Au-, H- atoms, respectively. The calculated formation energy (E_{FE}) for each of the doped AGeNR and ZGeNR cases are tabulated in Table 4.3.

The calculated value of formation energy clearly shows that stability of the Au-doped AGeNR/ ZGeNR is a strong function of doping site (in particular in the case of ZGeNR). From tabulated results of E_{FE} in Table 4.3, it is evident that with $E_{FE(NE-ZGeNR)} = -3.46$ eV, NE-ZGeNR outperforms others in terms of stability, with others following the order E-ZGeNR < C-ZGeNR < NE-ZGeNR. Further, based on the values of the bandgap listed in the table, it can be said that all the doped configurations exhibit strong metallic behavior, making them suitable for interconnect applications.

To examine the performance of these doped structures as interconnects, we again use the same performance parameter metric that is R_Q, L_K, C_Q. Based on the two-probe analysis and the equations listed earlier, we again calculated the parameters for these doped configurations. These are listed in Table 4.4.

TABLE 4.3 Calculated Values of Formation Energy per Atom, Bandgap, and Fermi Velocity for Armchair and Zigzag Germanene Nanoribbons with Considered Doping Sites

EDGE STATES	CONFIGURATIONS (AS PER DOPING)	FORMATION ENERGY (eV)	BANDGAP (eV)	FERMI VELOCITY (m/sec)
Armchair	Pristine AGeNR	−3.10	0.36 (direct)	6.61×10^3
	C-AGeNR	−3.29	Metallic	4.05×10^5
Zigzag	Pristine ZGeNR	−3.32	Semimetallic	9.19×10^4
	C-ZGeNR	−3.45	Metallic	4.88×10^5
	E-ZGeNR	−3.42	Metallic	7.34×10^5
	NE-ZGeNR	−3.46	Metallic	7.87×10^5

TABLE 4.4 Calculated Values of Transmission Channels (N_{ch}), Kinetic Inductance (L_k), and Quantum Capacitance (C_Q) for Various Doped and Pristine ZGeNR and AGeNR Configurations

CONFIGURATIONS	TRANSMISSION CHANNELS (N_{CH})	KINETIC INDUCTANCE (L_k) IN nH/μm	QUANTUM CAPACITANCE (C_Q) IN pF/cm
Pristine AGeNR	NIL	NIL	NIL
C-AGeNR	2	7.99	7.63
Pristine ZGeNR	3	23.47	50.45
C-ZGeNR	4	3.32	12.66
E-ZGeNR	2	4.41	4.21
NE-ZGeNR	3	2.74	5.89

The values of N_{ch} for each configuration are listed out in Table 4.4. The absence of N_{ch} for pristine AGeNRs versus $N_{ch} = 2$ for C-AGeNR is a clear indication of the origin of N_{ch} due to Au doping. However, this is not true for doped ZGeNR, as the value of N_{ch} here is dependent on the site of doping. Tabulated values in Table 4.4 clearly show that maximum (minimum) value of N_{ch} is observed in case of C-ZGeNR (E-ZGeNR). For better insights into this, we have also calculated the values of N_{ch} for different biasing voltage ranges, as mentioned in Table 4.5. The trends from the values listed in table indicate that N_{ch} is indeed a function of bias voltages as well. This can be simply attributed to the fact that with increased bias voltages, electrodes tend to decouple with the central region. Hence, it can be deduced that seamless coupling is a strong function of the bias voltage. This observation holds true in every case (except E-ZGeNR); the coupling between electrodes and the central region tends to get modulated with the applied bias voltage.

In the light of the previously mentioned important parameters, such as R_Q, L_K, and C_Q, for every possible configuration, we investigated the possible dependence of these parameters on bias voltages as well. Our results highlight that E-ZGeNR (with $L_K = 4.41$ nH/μm and $C_Q = 4.21$ pF/cm) is one of the most reliable and secure nanoscale interconnects among various considered Au-doped GeNR configurations. Further, similar calculations and projections have also been done for the Ag-doped SiNRs, and the key findings are tabulated in Tables 4.6, 4.7, and 4.8.

TABLE 4.5 Calculated Values of Transmission Channels N_{ch} for All Doped and Undoped ZGeNR/AGeNR Configurations at Different Bias Voltage

CONFIGURATIONS	PRISTINE AGeNR	C-AGeNR	PRISTINE ZGeNR	C-ZGeNR	E-Z GeNR	NE-ZGeNR
Bias Voltage = 0.0 V	NIL	2	3	4	2	3
0.5 V	NIL	1	1	3	2	3
1.0 V	1	1	2	2	2	4

TABLE 4.6 Calculated Values of Formation Energy per Atom, Bandgap, and Fermi Velocity for Armchair and Zigzag Silicene Nanoribbons with Considered Doping Sites

EDGE STATES	CONFIGURATIONS (AS PER DOPING)	FORMATION ENERGY (eV)	BANDGAP (eV)	FERMI VELOCITY (m/sec)
Armchair	Pristine ASiNR	−3.90	0.50 (direct)	4.05×10^3
	C-ASiNR	−4.30	Metallic	4.59×10^5
Zigzag	Pristine ZSiNR	−4.09	Semimetallic	6.92×10^4
	C-ZSiNR	−4.65	Metallic	7.42×10^5
	E-ZSiNR	−4.61	Metallic	7.84×10^5
	NE-ZSiNR	−4.64	Metallic	9.59×10^5

TABLE 4.7 Calculated Values of Transmission Channels (N_{ch}), Kinetic Inductance (L_k), and Quantum Capacitance (C_Q) for Various Doped and Undoped ZSiNR/ASiNR Configurations

CONFIGURATIONS	TRANSMISSION CHANNELS (N_{ch})	KINETIC INDUCTANCE (L_k) IN nH/μm	QUANTUM CAPACITANCE (C_Q) IN pF/cm
Pristine ASiNR	NIL	NIL	NIL
C-ASiNR	2	7.05	6.73
Pristine ZSiNR	2	46.75	44.66
C-ZSiNR	5	1.74	10.41
E-ZSiNR	2	4.13	3.94
NE-ZSiNR	3	2.24	4.83

TABLE 4.8 Calculated Values of Transmission Channels N_{ch} for Various Doped and Undoped ZSiNR/ASiNR Configurations at Different Bias Voltage

CONFIGURATIONS	PRISTINE ASiNR	C-ASiNR	PRISTINE ZSiNR	C-ZSiNR	E-ZSiNR	NE-ZSiNR
Bias Voltage = 0.0 V	NIL	2	2	5	2	3
0.5 V	NIL	1	1	3	2	3
1.0 V	1	1	2	2	2	3

4.6 CONCLUSION AND FUTURE OUTLOOK

In a nutshell, it can be stated that both functionalization and substitutional doping leads to significant changes in the electronic and transport properties of SiNR and GeNRs, projecting them as a suitable contender for interconnect applications. The results presented in the chapter systematically highlight the thermodynamical stability (based on calculation of E_b and E_{FE}), electronic properties (based on calculations of bandgap), and finally R_Q, L_K, and C_Q (as performance parameters for interconnect application). Based on these calculations, it is presented that with the increase in the O-atom concentration at the edge of nanoribbons, their performance for interconnect improves. However, in the case of substitutional doping, the performance depends mainly on the site of the doping. Further, in each case, study pertaining to the impact of the bias voltage has also been carried out. Consideration of the effect of the bias voltage on performance of the interconnect plays a pivotal role in determining the reliability and hence the security and integrity of the interconnect when used in the hardware.

From the future prospective, similar studies can also be conducted through various other metallic impurities and functionalization groups to gauge the performance of SiNR and GeNR interconnects. It is also noteworthy that the reliability or secure interconnect issues can also be probed from other dimensions, such as study of defected SiNRs/GeNRs and use of metal contacts rather than the extension of the same material as electrodes. A systematic probe of such possibilities will lead to both low power and secure nanoscale interconnect for next generation hardware.

REFERENCES

[1] G. E. Moore. Cramming more components onto integrated circuits. Electronics, 38, 8:114–117, 1965.
[2] G. E. Moore. Progress in digital integrated electronics. In International Electron Devices Meeting, IEEE, 1975.
[3] R. H. Dennard, F. Gaensslen, H.-N. Yu, V. L. Rideout, E. Bassous, and A. LeBlanc. Design of ion-implanted MOSFET's with very small physical dimensions. Proceedings of the IEEE, 87:668–678, 1974.
[4] M. M. Waldrop. The chips are down for Moore's law. Nature, 530:144–147, February 2016.
[5] J. A. Hutchby, R. Cavin, V. Zhirnov, J. E. Brewer, and G. Bourianoff. Emerging nanoscale memory and logic devices: A critical assessment. Computer, 41(5):28–32, May 2008.
[6] C. C. Cercignani. The Boltzmann Equation and its Applications. Springer Verlag, New York, 1988.
[7] I. Ferain, C. A. Colinge, and J.-P. Colinge. Multigate transistors as the future of classical metal oxide semiconductor field-effect transistors. Nature, 479:310–316, 2011.
[8] J.-P. Colinge. FinFETs and Other Multi-Gate Transistors. Springer US, New York, 2008.
[9] J. A. del Alamo. Nanometre-scale electronics with III-V compound semiconductors. Nature, 79:317–323, 2011.

[10] X. Huang, W.-C. Lee, C. Kuo, D. Hisamoto, L. Chang, J. Kedzierski, E. Anderson, H. Takeuchi, Y.-K. Choi, K. Asano, V. Subramanian, T.-J. King, J. Bokor, and C. Hu. Sub 50-nm FinFET: PMOS. International Electron Devices Meeting, 1999. Technical Digest (Cat. No. 99CH36318), pp. 67–70, Dec 1999.

[11] Q. Zhang, W. Zhao, and A. Seabaugh. Low-subthreshold-swing tunnel transistors. IEEE Electron Device Letters, 27(4):297–300, April 2006.

[12] A. M. Ionescu and H. Riel. Tunnel field-effect transistors as energy-efficient electronic switches. Nature, 479:329–337, 2011.

[13] A. C. Seabaugh and Q. Zhang. Low-voltage tunnel transistors for beyond CMOS logic. Proceedings of the IEEE, 98:2095–2110, 2010.

[14] N. Singh, A. Agarwal, L. K. Bera, T. Y. Liow, R. Yang, S. C. Rustagi, C. H. Tung, R. Kumar, G. Q. Lo, N. Balasubramanian, and D. L. Kwong. High-performance fully depleted silicon nanowire (diameter=5 nm) gate-all-around CMOS devices. IEEE Electron Device Letters, 27(5):383–386, May 2006.

[15] G. Li, R. Wang, J. Guo, J. Verma, Z. Hu, Y. Yue, F. Faria, Y. Cao, M. Kelly, T. Kosel, H. Xing, and D. Jena. Ultrathin body GaN-on-Insulator quantum well FETs with regrown Ohmic contacts. IEEE Electron Device Letters, 33(5):661–663, May 2012.

[16] Y. C. Yee, V. Subramanian, J. Z. Kedzierski, P. Xuan, T.-J. King, J. Bokor, and C. Hu. Nanoscale ultra-thin-body silicon-on-insulator P-MOSFET with a SiGe/Si heterostructure channel. IEEE Electron Device Letters, 21:161–163, 2000.

[17] S. G. Narendra and A. P. Chandrakasan. Leakage in Nanometer CMOS Technologies. Springer US, New York, 2006.

[18] R. P. Feynman. There's plenty of room at the bottom. Engineering and Science, 23:22–36, 1960.

[19] S. Datta. Electronic Transport in Mesoscopic Systems. Cambridge Studies in Semiconductor Physics and Microelectronic Engineering. Cambridge University Press, Cambridge, 1995.

[20] S. Datta. Quantum Transport: Atom to Transistor. Cambridge University Press, Cambridge, 2005.

[21] S. Datta. Lessons from nanoelectronics. World Scientific, 2012.

[22] M. Chhowalla, D. Jena, and H. Zhang. Two-dimensional semiconductors for transistors. Nature Reviews Materials, 1:16052 (1–15), 2016.

[23] K. S. Novoselov, A. K. Geim, S. V. Morozov, D. Jiang, Y. Zhang, S. V. Dubonos, I. V. Grigorieva, and A. A. Firsov. Electric field effect in atomically thin carbon films. Science, 306(5696):666–669, 2004.

[24] R. R. Nair, P. Blake, A. N. Grigorenko, K. S. Novoselov, T. J. Booth, T. Stauber, N. M. R. Peres, and A. K. Geim. Fine structure constant defines visual transparency of graphene. Science, 320(5881):1308–1308, 2008.

[25] L. Liao, Y.-C. Lin, M. Bao, R. X. Cheng, J. Bai, Y. Liu, Y. Qu, K. L. Wang, Y. Huang, and X. Duan. High-speed graphene transistors with a self-aligned nanowire gate. Nature, 467:305–308, 2010.

[26] P. Avouris, Y. M. Lin, F. Xia, D. B. Farmer, Y. Wu, T. Mueller, K. Jenkins, C. Dimitrakopoulos, and A. Grill. Graphene-based fast electronics and optoelectronics. In 2010 International Electron Devices Meeting, pp. 23.1.1–23.1.4, Dec 2010.

[27] D. Jena. Tunneling transistors based on graphene and 2-D crystals. Proceedings of the IEEE, 101:1585–1602, 2013.

[28] G. Fiori, F. Bonaccorso, G. Iannaccone, T. Palacios, D. Neumaier, A. C. Seabaugh, S. K. Banerjee, and L. Colombo. Electronics based on two-dimensional materials. Nature Nanotechnology, 9:768–79, 2014.

[29] K. S. Novoselov, D. Jiang, F. Schedin, T. J. Booth, V. V. Khotkevich, S. V. Morozov, and A. K. Geim. Two-dimensional atomic crystals. Proceedings of the National Academy of Sciences, 102(30):104510–10453, 2005.

[30] D. A. Abanin and L. S. Levitov. Quantized transport in graphene p–n junctions in a magnetic field. Science, 317(5838):641–643, 2007.

[31] Brian Standley, Wenzhong Bao, Hang Zhang, Jehoshua Bruck, Chun Ning Lau, and Marc Bockrath. Graphene-based atomic-scale switches. Nano Letters, 8(10):3345–3349, 2008. PMID: 18729415.

[32] A. K. Geim and I. V. Grigorieva. Van der waals heterostructures. Nature, 99:419425, 2013.

[33] H. Min, J. E. Hill, N. A. Sinitsyn, B. R. Sahu, L. Kleinman, and A. H. MacDonald. Intrinsic and Rashba spin-orbit interactions in graphene sheets. Physical Review B, 74:165310 (1–5), Oct 2006.

[34] J. Kang, W. Cao, X. Xie, D. Sarkar, W. Liu, and K. Banerjee. Graphene and beyond- graphene 2D crystals for next-generation green electronics. Proceedings of SPIE – The International Society for Optical Engineering, 9083:908305–1–7, 2014.

[35] P. Joensen, R. F. Frindt, and S. R. Morrison. Single-layer MoS_2. Materials Research Bulletin, (4):457–461, 1986.

[36] G.-H. Lee, Y.-J. Yu, X. Cui, N. Petrone, C.-H. Lee, M. S. Choi, D.-Y. Lee, C. Lee, W. J. Yoo, K. Watanabe, T. Taniguchi, C. Nuckolls, P. Kim, and J. Hone. Flexible and transparent MoS_2 field-effect transistors on hexagonal boron nitride-graphene heterostructures. ACS Nano, 7(9):7931–7936, 2013.

[37] K. F. Mak, C. Lee, J. Hone, J. Shan, and T. F. Heinz. Atomically thin MoS_2: A new direct- gap semiconductor. Physical Review Letters, 105:136805–1–4, Sep 2010.

[38] D. Pacile, J. C. Meyer, C. O¨ Girit, and A. Zettl. The two-dimensional phase of boron nitride: Few-atomic-layer sheets and suspended membranes. Applied Physics Letters, 92(13):133107–1–3, 2008.

[39] E. Cinquanta, E. Scalise, D. Chiappe, C. Grazianetti, B. van den Broek, M. Houssa, M. Fanciulli, and A. Molle. Getting through the nature of silicene: An $sp^2 - sp^3$ two-dimensional silicon nanosheet. The Journal of Physical Chemistry C, 117(32):16719–16724, 2013.

[40] P. Vogt, P. De Padova, C. Quaresima, J. Avila, E. Frantzeskakis, M. C. Asensio, A. Resta, B. Ealet, and G. Le Lay. Silicene: Compelling experimental evidence for graphene like two-dimensional silicon. Physical Review Letters, 108:155501–1–5, Apr 2012.

[41] R. W. Keyes. The electrical properties of black phosphorus. Physical Review, 92:580–584, Nov 1953.

[42] S. Balendhran, S. Walia, H. Nili, S. Sriram, and M. Bhaskaran. Graphene analogues: Elemental analogues of graphene: Silicene, germanene, stanene, and phosphorene. Small, 116:640–652, Nov 2014.

[43] X. Li, J. T. Mullen, Z. Jin, K. M. Borysenko, M. Buongiorno Nardelli, and K. W. Kim. Intrinsic electrical transport properties of monolayer silicene and MoS_2 from first principles. Physical Review B, 87:115418–1–9, Mar 2013.

[44] L. Tao, E. Cinquanta, D. Chiappe, C. Grazianetti, M. Fanciulli, M. Dubey, A. Molle, and D. Akinwande. Silicene field-effect transistors operating at room temperature. Nature Nanotechnology, 10:227–231, 2015.

[45] M. Houssa, E. Scalise, K. Sankaran, G. Pourtois, V. V. Afanasev, and A. Stesmans. Electronic properties of hydrogenated silicene and germanene. Applied Physics Letters, 98(22):223107, 2011.

[46] L. C. Lew Yan Voon, J. Zhu, and U. Schwingenschlogl. Silicene: Recent theoretical advances. Applied Physics Reviews, 3(4):040802–1–13, 2016.

[47] J. Zhuang, X. Xu, H. Feng, Z. Li, X. Wang, and Y. Du. Honeycomb silicon: A review of silicene. Science Bulletin, 60(18):1551–1562, 2015.

[48] A. Kara, H. Enriquez, A. P. Seitsonen, L. C. Lew Yan Voon, S. Vizzini, B. Aufray, and Oughaddou. A review on silicene-new candidate for electronics. Surface Science Reports, 67(1):1–18, 2012.

[49] A. Molle, C. Grazianetti, L. Tao, D. Taneja, Md. H. Alam, and D. Akinwande. Silicene, silicene derivatives, and their device applications. Chemical Society Reviews, 47:6370–6387, 2018.

[50] Y.-L. Song, Y. Zhang, J.-M. Zhang, and D.-B. Lu. Effects of the edge shape and the width on the structural and electronic properties of silicene nanoribbons. Applied Surface Science, 256(21):6313–6317, 2010.

[51] S. Cahangirov, M. Topsakal, and S. Ciraci. Armchair nanoribbons of silicon and germanium honeycomb structures. Physical Review B, 81:195120–1–6, May 2010.

[52] Y. Ding and Y. Wang. Electronic structures of silicene fluoride and hydride. Applied Physics Letters, 100(8):083102–1–4, 2012.

[53] Q. Pang, Y. Zhang, J.-M. Zhang, V. Ji, and K.-W. Xu. Electronic and magnetic properties of pristine and chemically functionalized germanene nanoribbons. Nanoscale, 3:4330–4338, 2011.

[54] Y. Lee, K. H. Yun, S. Cho, and Y. C. Chung. Electronic properties of transition metal- decorated silicene. Chem. Phys. Chem., 15(18):4095–99, 2014.

[55] Y.-J. Dong, X.-F. Wang, P. Vasilopoulos, M.-X. Zhai, and X.-M. Wu. Half metallicity in aluminum-doped zigzag silicene nanoribbons. Journal of Physics D: Applied Physics, 47(10):105304–1–6, Feb 2014.

[56] L. Ma, J.-M. Zhang, K.-W. Xu, and Ji. Nitrogen and Boron substitutional doped zigzag silicene nanoribbons: Ab initio investigation. Physica E: Low-dimensional Systems and Nanostructures, 60:112–117, 2014.

[57] Q. Pang, L. Li, C.-L. Zhang, X.-M. Wei, and Y.-L. Song. Structural, electronic and magnetic properties of 3d transition metal atom adsorbed germanene: A first-principles study. Materials Chemistry and Physics, 160:96–104, 2015.

[58] C. Xu, H. Li, and K. Banerjee. Modeling, analysis, and design of graphene nanoribbon interconnects. IEEE Transactions on Electron Devices, 56(8):1567–1578, 2009.

[59] L. Banerjee, A. Sengupta, and H. Rahaman. Carrier transport and thermoelectric properties of differently shaped germanene (Ge) and silicene (Si) nanoribbon interconnects. IEEE Transactions on Electron Devices, 66(1):664–669, Jan 2019.

[60] J. Liang, J. Lee, S. Berrada, V. P. Georgiev, R. Pandey, R. Chen, A. Asenov, and A. Todri-Sanial. Atomistic- to circuit-level modeling of doped SWCNT for on-chip interconnects. IEEE Transactions on Nanotechnology, 17(6):1084–1088, Nov 2018.

[61] S. A. Wolf, D. D. Awschalom, R. A. Buhrman, J. M. Daughton, S. von Molnr, M. L. Roukes, A. Y. Chtchelkanova, and D. M. Treger. Spintronics: A spin-based electronics vision for the future. Science, 294:1488–95, 2001.

[62] C.-Cheng Liu, W. Feng, and Y. Yao. Quantum Spin Hall Effect in Silicene and Two- Dimensional Germanium. Physical Review Letters, 107:076802–1–4, 2011.

[63] W. Liu, J. Zheng, P. Zhao, S. Cheng, and C. Guo. Magnetic properties of silicene nanoribbons: A DFT study. AIP Advances, 7(6):065004–1–9, 2017.

[64] S. Yamacli. Investigation and comparison of large-signal characteristics and dynamical parameters of silicene and germanene nanoribbon interconnects. Computational Materials Science, 141:353–359, 2018.

[65] K. Takeda and K. Shiraishi. Theoretical possibility of stage corrugation in Si and Ge analogs of graphite. Physical Review B, 50:14916–14922, Nov 1994.

[66] S. Cahangirov, M. Topsakal, E. Akturk, H. Sahin, and S. Ciraci. Two- and one- dimensional honeycomb structures of silicon and germanium. Physical Review Letters, 102:236804–1–4, Jun 2009.

[68] M. E. Davila, L. Xian, S. Cahangirov, A. Rubio, and G. L. Lay. Germanene: A novel two- dimensional germanium allotrope akin to graphene and silicene. New Journal of Physics, 16(9):095002–1–10, 2014.

[69] M. Derivaz, D. Dentel, R. Stephan, M.-C. Hanf, A. Mehdaoui, P. Sonnet, and C. Pirri. Continuous germanene layer on Al (111). Nano Letters, 15(4):2510–2516, 2015.

[70] J. Zhuang, N. Gao, Z. Li, X. Xu, J. Wang, J. Zhao, S. X. Dou, and Y. Du. Cooperative electron-phonon coupling and buckled structure in germanene on Au(111). ACS Nano, 11(4):35533559, 2017.

[71] L. Li, S.-Z. Lu, J. Pan, Z. Qin, Y.-qi Wang, Y. Wang, G.-yu Cao, S. Du, and H.-J. Gao. Buckled germanene formation on Pt(111). Advanced Materials, 26(28):4820–4824, 2014.

[72] A. Fleurence, R. Friedlein, T. Ozaki, H. Kawai, Y. Wang, and Y. Yamada Takamura. Experimental evidence for epitaxial silicene on diboride thin films. Physical Review Letters, 108:245501–1–5, Jun 2012.

[73] T. Kaloni, Mh. Tahir, and U. Schwingenschlogl. Quasi free-standing silicene in a superlattice with hexagonal boron nitride. Scientific Reports, 3:3192–1–4, 2013.

[74] M. Eugenia, G. Davila, and G. Le Lay. Few layer epitaxial germanene: A novel two- dimensional dirac material. Scientific Reports, 6:20714–1–6, 2016.

[75] J. Zhu and U. Schwingenschlogl. Silicene on MoS_2: Role of the van der waals interaction. 2D Materials, 2(4):045004–1–6, 10 2015.

[76] T. Aizawa, S. Suehara, and S. Otani. Silicene on Zirconium Carbide (111). Journal of Physical Chemistry C, 118:23049–23057, 2014.

[77] T. P. Kaloni, S. Gangopadhyay, N. Singh, B. Jones, and U. Schwingenschlogl. Electronic properties of Mn-decorated silicene on hexagonal boron nitride. Physical Review B, 88:235418–1–4, Dec 2013.

[78] L. Zhang, P. Bampoulis, A. N. Rudenko, Q. Yao, A. van Houselt, B. Poelsema, M. I. Katsnelson, and H. J. W. Zandvliet. Structural and electronic properties of germanene on MoS_2. Physical Review Letters, 116:256804–1–6, Jun 2016.

[79] A. Podsiadly-Paszkowska and M. Krawiec. Dirac fermions in silicene on Pb(111) surface. Physical Chemistry Chemical Physics, 17:2246–2251, 2015.

[80] Y. Fan, X. Liu, J. Wang, H. Ai, and M. Zhao. Silicene and germanene on InSe substrates: Structures and tunable electronic properties. Physical Chemistry Chemical Physics, 20:11369–11377, 2018.

[81] E. Scalise, K. Iordanidou, V. V. Afanasev, A. Stesmans, and M. Houssa. Silicene on non- metallic substrates: Recent theoretical and experimental advances. Nano Research, 11(3):1169–1182, Mar 2018.

[82] J. Zhuang, C. Liu, Z. Zhao, Z. Li, G. Casillas, H. Feng, X. Xu, J. Wang, W. Hao, X. Wang, S. X. Dou, Z. Hu, and Y. Du. Dirac signature in germanene on semiconducting substrate. Advanced Science, 5(7):1800207–1–8, 2018.

[88] Z. Hong-Xia, Q. Ru-Ge, W. Yang-Yang, S. Jun-Jie, and L. Jin. Silicene on substrates: A theoretical perspective. Chinese Physics B, 24(8):87308, 2015.

[89] Y. Ding and J. Ni. Electronic structures of silicon nanoribbons. Applied Physics Letters, 95(8):083115–1–3, 2009.

[90] Y. Ding and Y. Wang. Electronic structures of zigzag silicene nanoribbons with asymmetric $sp^2 - sp^3$ edges. Applied Physics Letters, 102(14):143115–1–5, 2013.

[91] D. Zhang, M. Long, F. Xie, J. Ouyang, H. Xu, and Y. Gao. Hydrogenations and electric field induced magnetic behaviors in armchair silicene nanoribbons. Scientific Reports, 6:23677–1–7, Mar 2016.

[92] N. B. Le, T. D. Huan, and L. M. Woods. Tunable spin-dependent properties of zigzag silicene nanoribbons. Physical Review Applied, 1:054002–1–6, Jun 2014.

[93] N. Gao, W. T. Zheng, and Q. Jiang. Density functional theory calculations for two-dimensional silicene with halogen functionalization. Physical Chemistry Chemical Physics, 14:257–261, 2012.

[94] D.-Q. Fang, S.-L. Zhang, and H. Xu. Tuning the electronic and magnetic properties of zigzag silicene nanoribbons by edge hydrogenation and doping. RSC Advances, 3:24075–24080, 2013.

[95] J.-M. Zhang, W.-T. Song, K.-W. Xu, and V. Ji. The study of the P-doped silicene nanoribbons with first-principles. Computational Materials Science, 95:429–434, 2014.

[96] F. Ersan, O Arslanalp, G. Gokoglu, and E. Akturk. Effects of silver adatoms on the electronic structure of silicene. Applied Surface Science, 311:9–13, 2014.

[97] C. Si, J. Liu, Y. Xu, J. Wu, B.-L. Gu, and W. Duan. Functionalized germanene as a prototype of large-gap two-dimensional topological insulators. Physical Review B, 89:115429-1–5, Mar 2014.

[98] N. Liu, G. Bo, Y. Liu, X. Xu, Y. Du, and S. X. Dou. Recent progress on germanene and functionalized germanene: Preparation, characterizations, applications, and challenges. Small, 1805147(0):1–11, 2019.

[99] S. M. Aghaei, M. M. Monshi, I. Torres, M. Banakermani, and I. Calizo. Lithium functionalized germanene: A promising media for CO_2 capture. Physics Letters A, 382(5):334–338, 2018.

[100] T. Hussain, T. Kaewmaraya, S. Chakraborty, H. Vovusha, V. Amornkitbamrung, and R. Ahuja. Defected and functionalized germanene-based nanosensors under sulfur comprising gas exposure. ACS Sensors, 3(4):867–874, 2018.

[101] S. Singh, K. Garg, A. Sareen, R. Mehla, and I. Kaur. Doped armchair germanene nanoribbon exhibiting negative differential resistance and analysing its nano-FET performance. Organic Electronics, 54:261–269, 2018.

[102] A. K. Shiraz, A. Y. Goharrizi, and S. M. Hamidi. Structural stability and electron density analysis of doped germanene: A first-principles study. Materials Research Express, 6(10):1–15, 2019.

[103] H. H. Cocoletzi and J. E. Castellanos Aguila. DFT studies on the Al, B, and P doping of silicene. Superlattices and Microstructures, 114:242–250, 2018.

[104] M. A. Pamungkas, Kafi Sobirin, and Abdurrouf. Effects of doping Na and Cl atom on electronic structure of silicene: Density functional theory calculation. AIP Conference Proceedings, 1948(1):020001, 2018.

[105] D.-B. Lu, Y.-L. Song, X.-yu Huang, and C. Wang. Optical properties of a single carbon chain-doped silicene nanoribbon. Journal of Electronic Materials, 47(8):4585–4593, Aug 2018.

[106] A. Lopez-Bezanilla. Substitutional doping widens silicene gap. Journal of Physical Chemistry C, 118(32):18788–18792, 2014.

[107] A. Chen and X. Luo. Boron-doped silicene as a promising anode for Li-ion batteries. In APS Meeting Abstracts, pp. V31.012, 2019.

[108] D. Zou, W. Zhao, C. Fang, B. Cui, and D. Liu. The electronic transport properties of zigzag silicene nanoribbon slices with edge hydrogenation and oxidation. Physical Chemistry Chemical Physics, 18:11513–11519, 2016.

[109] V. Sharma, P. Srivastava, and N. K. Jaiswal. Edge-oxidized germanene nanoribbons for nanoscale metal interconnect applications. IEEE Transactions on Electron Devices, 65(9):3893–3900, Sept 2018.

[110] V. Sharma and P. Srivastava. Probing gold-doped germanene nanoribbons for nanoscale interconnects under DFT-NEGF framework. Journal of Electronic Materials, 49(6), April 2020.

Memristors and Their Applications

5

Hitendra Singh Pawar

Contents

5.1 INTRODUCTION

In today's era of highly interconnected hardware and information technology, hardware and software security becomes even more challenging. Hardware security is specifically concerned about achieving security within the underlying electronics. For hardware security, predominantly complementary metal-oxide-semiconductor (CMOS) technology had been developed more, but with new emerging technologies, unique opportunities arise for advancement of the concept of hardware security. With the help of nanoelectronic devices, high speed and low power consuming miniaturized computational systems can be fabricated that are more rugged, unlike CMOS devices. For example, memristive technology is promising for safe handling of secret keys, which is relatively difficult for CMOS technologies. In recent years, it has been recognized that memristors can be used for hardware security applications.

There are three fundamental elements in electronics, namely resistors, capacitors, and inductors, which can be used for defining four fundamental variables of the electrical circuit, namely current (I), charge (q), voltage (V), and magnetic flux (φ). The relation between voltage and current is defined by resistors. Similarly, capacitors define the relation between charge and voltage, and the relation between current and magnetic flux is defined by inductors. However, no element could define the relation of charge with magnetic flux. Thus, in order to have the symmetry in these relationships, there should be an element showing the relation between charge and flux. In 1971, Professor Chua [1] proposed another basic circuit element and called it the 'memristor', which relates charge with flux and is expressed mathematically as $d\varphi = Mdq$.

The definition of the fourth fundamental element of the electrical circuit in terms of mathematical equation is stated in Table 5.1, where M is memristance of memristor and the magnetic flux of the device (φ) is a function of charge (q) passing through the memristor. Since $v = d\varphi/dt$ and $i = dq/dt$, the ratio of $d\varphi$ and dq can be replaced with the voltage and current equations given by $M(q) = \{(d\varphi/dt)/(dq/dt)\}$. Thus, $v(t) = M \times (q(t)) \times i(t)$. Thus, qualitatively, Chua's fourth fundamental circuit element would have an electrical resistance that depends on the amount of electric charge (q) flowing through its terminals. Also, it would exhibit memory of the last applied voltage level, or the charge passed through it. Its resistance value would be retained even when the power supply is disconnected. Chua introduced this new device as a memristor, a combination of the words memory and resistor.

Chua suggested the following symbol for memristor, which has now been accepted as the standard one (Figure 5.1).

TABLE 5.1   Relation between i, v, q, φ

SR. NO.	DEFINITION	UNIT	RELATION
1.	Resistance (R)	Ohms	$dv = Rdi$
2.	Inductance (L)	Henry	$d\varphi = Ldi$
3.	Capacitance (C)	Farads	$dq = Cdv$
4.	Current (i)	Amperes	$dq = idt$
5.	Voltage (v)	Volts	$d\varphi = vdt$
6.	Memristance (M)	Ohms	$D\varphi = Mdq$

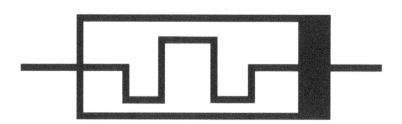

FIGURE 5.1   Symbol for memristor.

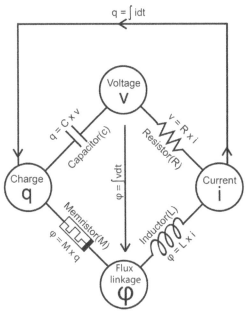

Memristor relationship

FIGURE 5.2   The relation between the four fundamental electrical quantities (i, v, q, and φ) and the fundamental two-terminal elements (resistor (R), capacitor (C), inductor (L), and the memristor (M)). M also denotes the resistance of the memristor element, called memristance.

The proposed fourth fundamental two-terminal element relates the linkage of the flux (φ). The relations between the described electrical quantities and the fundamental one-port elements [2] are presented in Figure 5.2.

5.2 MEMRISTOR MODEL BY HP LABS

In 2008, a team of scientists from Hewlett Packard (HP) Laboratories, led by R.S. Williams, stated that they have fabricated a solid state memristor model whose I–V characteristics resemble the results obtained by Chua's memristor [2]. The device fabricated by HP is formed by using a 50 nm TiO_2 film sandwiched between two electrodes of 5 nm thickness each. The materials used for electrodes are titanium and platinum. In the beginning, one of the two layers of the TiO_2 film has slightly depleted oxygen atoms, creating oxygen vacancies, which act as the charge carriers. The resistance of the oxygen-depleted layer is much lower than that of the non-depleted layer. The oxygen vacancies start drifting on, applying an electric field, which results in the change of boundary between the high and low resistance layers. For a memristor of total length D, the doped region is shown as the width w, as shown in Figure 5.3. Hence, the film resistance totally depends on the amount of charge passing through it. The direction of the drift can be reversed by changing direction of the current. At nanoscale, the HP device shows fast ion conduction and hence is considered a nanoionic device.

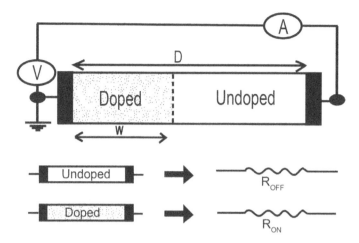

FIGURE 5.3 Memristor resistance levels in undoped and doped condition.

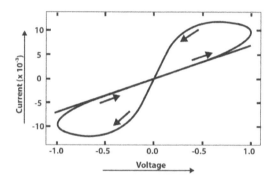

FIGURE 5.4 I–V Characteristics of memristor (hysteresis loop) [2].

5.3 *I–V* CHARACTERISTICS OF MEMRISTOR

Since the doped layer and the depleted layer both show resistivity, memristance is displayed. When the ions can no longer move, because enough charge has passed through the memristor, the device enters into state of hysteresis, as shown in Figure 5.4. Until the current is reversed, it holds the level of charge passed through it for the last time in the form of the memristance. The memristor *I–V* characteristics on application of a sinusoidal wave are shown in Figure 5.4.

5.4 CONDUCTIVE BRIDGE MEMRISTOR

The conductive bridge memristor is composed of a metal insulator-metal sandwich that has a glass-like material that allows metal atoms to move through it under an electric field, thus establishing a conductive path.

This is a nonvolatile and reversible effect, which means that it does not dissipate even when power is turned off, and thus the device can retain its current state for extended periods of time. When the electric field is reversed, the conductive path also gets reversed.

Voltage

FIGURE 5.5   Structure of memristor device with metal/insulator/metal (MIM) configuration.

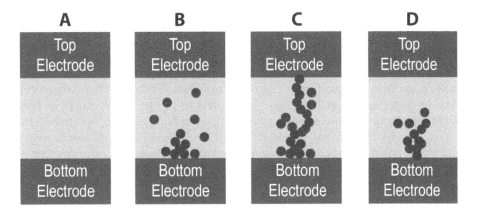

FIGURE 5.6   Resistive switching based on filament formation and rupturing. (a) Native insulator (HRS); (b) electroforming of conductive filament; (c) formed conductive filament depicting a set process, that is, ON-state (transition from HRS to LRS); and (d) ruptured filament during a reset process, that is, OFF-state (transition from LRS to HRS).

The sandwiched metal/insulator/metal (MIM) structure [3] of the memristor is illustrated in Figure 5.5.
In metal-oxide memristors, resistive switching occurs when conductive filaments are formed and ruptured in the active layer. When conductive filaments are formed, the device switches from "OFF" to "ON" state, thus creating a low-resistance state (LRS) and when filaments are ruptured the device switches from "ON" to "OFF" state and hence creating a high-resistance state (HRS), as shown in (Figure 5.6).

5.5 MEMRISTOR MODELING

There are various types of possible memristor models available. They are the linear model, nonlinear model, and exponential model. The linear memristor model describes linear drift in a memristor. But this model has many limitations, one of which is the absence of physical boundary conditions. This limitation is overcome in the nonlinear model, which incorporates a window function to restrict drift between both physical limits ($0 < x < 1$). However, the nonlinear model also has limitations, because it does not match the parameters of real devices due to its linear dependence on the current. This linear dependence is overcome in the exponential model, as it adds nonlinear dependence on the current and also uses an experimental data-based [4] exponential I–V characteristic.

5.6 TYPES OF MEMRISTORS

Several basic types of memristors exist based on different chemical and physical structures [5] and have different principles of operation, which are as follows:

5.6.1 Titanium Dioxide Memristor

It is one of the leading nanostructured elements, having broad applications. The TiO_2-based memristor device structure is shown in Figure 5.7.

5.6.2 Polymeric Memristors

The polymeric memristors are based on unique plastic materials. Single or a group of molecules are able to conduct and switch currents and memorize information using charge accumulation. Polymer-based resistive RAM supplies information as per the low or high conductance when an external electrical field is applied. The conductance states can be read nondestructively. Because electric conductivity is the multiplication of charge volumetric concentration and ionic mobility, changes in either of these, or both, can cause changes in the element resistance states.

5.6.3 Ferroelectric Memristors

Resistance switching has been observed in special sandwich-like structures consisting of ferroelectric thin films when an external signal is applied. These two conductance states (R_{OFF} and R_{ON}) can be used in

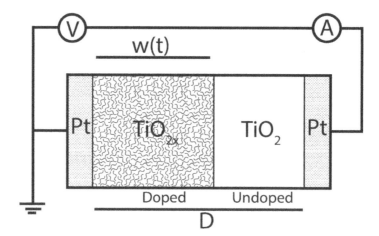

FIGURE 5.7   Device structure of TiO_2 based memristor.

bistable memory components. Resistance-switching behavior in $BiFeO_3$ suggests more independence for flexible devices. Since the polarization transfer has great stability for a chemical alteration and is basically rapid, the interfacial resistance-switching phenomenon is optimistic for the new ferroelectric memristors.

5.6.4 Spintronic Memristors

The spintronic memristor device is based on an oxide thin film placed among two regions with magnetic properties. The first magnetic layer has permanent magnetic characteristics, and the other film is a region with a field barrier. The magnetization of the free region on the left surface of the area divider is associated with the magnetic properties of the first region. The right surface magnetization is in the opposite location to the magnetic field of the first region. The domain barrier moves if the current flows through the structure. This motion depends on the current path. The area wall can be fully transported to the left region, and thus a total anti-parallel magnetization between the regions occurs, leading to low conductance. If the barrier is completely transported to the second region, a parallel magnetization between the first magnetic film and the pinned regions is realized, establishing a high conductance. The state variable describes the length of the parallel magnetized section of the second region. This section has high conductivity, and the anti-parallel layer has lower conductance.

5.7 APPLICATIONS OF MEMRISTORS

Memristors can potentially enhance many areas of computing and IC design. Some of them are:

5.7.1 Nonvolatile Random Access Memory (NVRAM)

Memristors may be used in nonvolatile random access memory (NVRAM) [6], as the memristor-based device memory consumes much less physical area and does not need to draw continuous power. Every single memristor can store one bit of information for digital memory. For this the memristor is forced to its extreme resistance values R_{ON} and R_{OFF}, where each state corresponds to 1 or 0, respectively. The memristor resistance is set by DC voltages. AC signals are used to read stored data without disturbing it. This memory architecture uses a crossbar arrangement [7] that is composed of a grid of vertical and horizontal traces. A memristor connects a horizontal trace to a vertical trace at each intersection. Associative memories map an input pattern to an output pattern based on the comparison of the input pattern with the pattern that is stored in the memory. Memristors can also find use in associative memories [8].

The fundamental cell shown in Figure 5.8 has two memristors, memA and memB. Here memA is set to low resistance and memB is set to high resistance for the logic 1, and vice versa for logic 0.

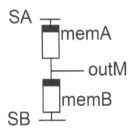

FIGURE 5.8 Fundamental cell using memristors for associative memories [8].

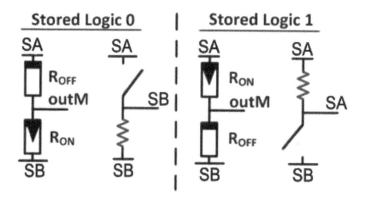

FIGURE 5.9   States of the fundamental cell [8].

		memA	memB	SA	SB	outM
Logic 0		0	1	Searching Logic 0		0 (Match)
				1 (VDD)	0 (GND)	
		0	1	Searching Logic 1		1 (Not Match)
				0 (GND)	1 (VDD)	
Logic 1		1	0	Searching Logic 0		1 (Not Match)
				1 (VDD)	0 (GND)	
		1	0	Searching Logic 1		0 (Match)
				0 (GND)	1 (VDD)	

FIGURE 5.10   Matching logic of the associative cell [8].

The states of the fundamental cell are shown in Figure 5.9, and the corresponding matching logic of the associative cell is depicted in Figure 5.10.

To further expand NVRAM capabilities, memristors could also allow for low power memory and distributed state storage at the nanoscale level.

The topologies of electronic circuits that have characteristics depending on a resistance could be designed using memristors behaving as variable programmable resistances. In this way, a memristor can also be used in an analog manner.

5.7.2 Image Recognition

An important operation used for image recognition and in image processing techniques is Edge detection. It identifies parts of a digital image, where large changes occur. For example the outlines of objects. Memristors can perform edge detection with the use of a memristance grid.

In Figure 5.11a, the intensity of light from the actual photo is applied to points on memristance grid as voltages. The frames shown in Figure 5.11b are the result of the resistance change across the grid with time. Once the system has settled, the recovery of an edge-detected image can be done through the resistance measurement of each grid element.

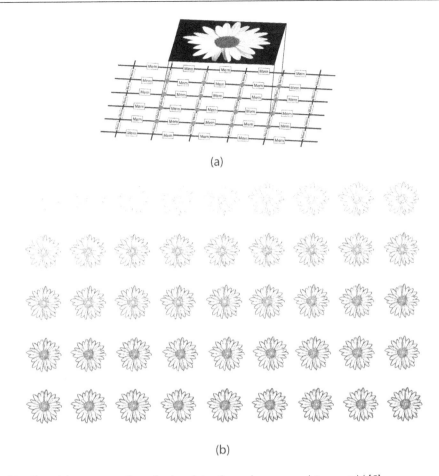

(a)

(b)

FIGURE 5.11   Pictorial representation of edge detection using a memristance grid [6].

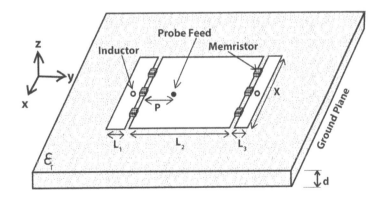

FIGURE 5.12   Geometry of a reconfigurable antenna [9].

5.7.3 Radio Frequency (RF) Antenna

Memristors can also find uses as switches between conductive elements in a reconfigurable (RF) antenna. The wide difference between the frequencies of the two signals can be used to separate the programming mode from the resistive mode. In Figure 5.12, memristors are used as switches between the copper elements of the antenna. The resonance frequency changes when the elements of the antenna are joined or split electrically.

5.8 APPLICATIONS OF MEMRISTORS IN HARDWARE/SOFTWARE SECURITY

Metal-oxide memristor devices have the following unique characteristics, depending on the material used, that can be used for security applications:

1. Nonvolatile: Memristors are nonvolatile, as even when the power is turned OFF, they can retain their memristance value.
2. Bidirectional: As evident from the theoretical I–V curve, which is symmetrical in nature, some of the bipolar memristors show similar I–V characteristics regardless of the polarity of the voltage or the current applied.
3. Nonlinear: Because of the time-dependent behavior of memristors, their I–V characteristics are extremely nonlinear.
4. Forming process: In many cases, for initializing a memristor to the low resistance state (LRS), a separate forming step (V_f) is required, before which the memristor acts as a linear resistor [10].
5. Memristance drift: In some metal-oxide memristors, when an input voltage ($+v_i$ or $-v_i$) is applied, there is a change in the memristance due to the movement of dopants. This process is called memristance drift [11]. It depends on the amplitude, polarity, and duration of the applied voltage.
6. Process variation: The variations in the process can induce changes in the concentration of the dopant and the dimensions of the memristor, hence affecting its memristance.
7. Radiation hardness: Due to the inherent properties of the material, some memristor-based devices exhibit radiation hardness characteristics [11].
8. Temperature stability: In a TiO_2 memristor, the temperature coefficient of resistance is very small ($< -3.82 \times 10^{-3}$/K) for TiO_2, so the low resistance state and high resistance state are extremely stable. But due to the change in mobility of the dopant atom, the switching speed of the memristor can vary with variation in temperature.

These unique features of memristors can be useful for the following security-based nanoelectronic devices.

5.8.1 Nanoelectronic Physical Unclonable Functions (NanoPUFs)

Physical unclonable functions (PUFs) are a type of hardware-based security primitives. The variations in their fabrication process can be used to derive their cryptographic 'keys'. Physical unclonable disorders that occur during the fabrication process of the integrated circuit (IC) can be used to generate unique outputs (responses) when inputs (challenges) are applied. PUFs that are based on crossbar architectures with integrated memristors can be leveraged effectively for hardware-based security applications. This is because they have a low-cost and simple fabrication process and a very small footprint and are compatible for CMOS circuit integration. They also exhibit process-induced variations in their I–V characteristics due to their memory mechanism and mixed electronic-ionic transport. PUFs map a challenge to a response. PUFs have been used in the following application areas:

1. Execution of secure software in a processor.
2. Device authentication.

3. Storage encryption.
4. Trusted configuration of FPGAs.

The inherent process variations in memristors can be used to create PUFs called NanoPUFs [12], which have three main sections:

i. A memristors-based crossbar.
ii. The challenge (Input) circuit: They enable the application of an input to the crossbar. A memristor is selected by the challenge circuit through the column and row decoders. A particular column is selected by the column decoder and then connected to the load resistance (R_{load}). A voltage of V_{dd} magnitude is applied to a particular row by the row decoder. At the same time, the rest of the columns and rows stand floating.
iii. The response circuit: This gathers the response for a challenge from the crossbar. The response circuit consists of a current comparator and load resistance (R_{load}). The comparator compares I_{out} (the current which is flowing out of the column) with I_{ref} (the reference current). The response bit is logic 0 when I_{out} is less than I_{ref}; else the logic is 1.

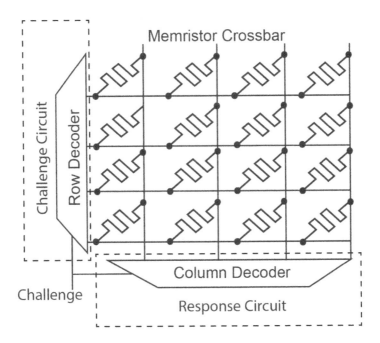

FIGURE 5.13 Memristor-based NanoPUF [12].

Advantages: As memristors are more dense as compared to SRAM cells, for the same amount of area, NanoPUFs based on memristors generate more response bits. They also consume less power.

5.8.2 Nanoelectronic Public Physical Unclonable Functions (NanoPPUFs)

While the simulation models of PUFs are hidden from an attacker, the simulation models of a PPUF are public. However, if an attacker tries to simulate the PPUF, the simulation time would be very large – could even be several years – as compared to PUF primitive, whose simulation time could be just a few nanoseconds.

To further enhance security, a 2-layer security protocol can be implemented by NanoPPUFs [13], like key exchange, authentication, time stamping, and bit commitment. Memristor-based nanoelectronic devices are best to implement NanoPPUFs. A NanoPPUF uses memristor properties like bidirectionality, process variations, and the complex simulation behavior of memristors and its crossbar models and polyominoes shapes, which are geometrical structures formed by joining more than one independent blocks.

The NanoPPUF has five main parts, which are:

1. Crossbar.
2. Challenge and characterization circuit.
3. Refresh and characterization circuit.
4. Controller circuit.
5. Response circuit.

Every cross-point in the crossbar is composed of a memristor.

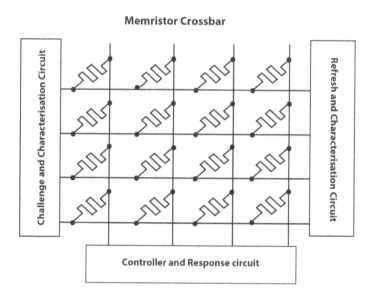

FIGURE 5.14   A 4 × 4 memristor-based NanoPPUF.

5.8.3 Memristor-Based True Random Number Generators (NanoTRNGs)

Random number generators (RNGs) can be used to create session keys that are necessary in establishing secure communication channels; hence, they are considered to be important security primitives.

Some types of memristors can be utilized for generating random numbers [14]. In random number generator devices, the electrons that are trapped in the insulation layer will randomly create an impact on the current that flows in the filament channel. When a high voltage (3 V) is applied to the memristor-based RNG device, the current that flows in the filament would be very large to be affected by the trapped electrons. On the other hand, when a low voltage (1.2 V) is applied, the filament shrinks down in width and the current that flows in the filament is greatly impacted by the trapped electrons. As the number of electrons that are trapped in the insulation layer is random, the nature of the output current also will be random.

Advantages: As NanoTRNGs have a high density, consume less power, and have more randomness in their physical properties, they are superior to their CMOS counterparts.

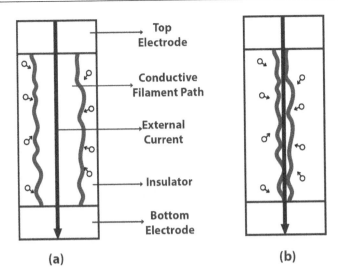

FIGURE 5.15 A memristor-based RNG device. (a) Formation of a conductive filament when a high voltage is applied. The electrons trapped in the insulation layer have insignificant impact on the current that flows through the filament. (b) Impact on conductive filament when a low voltage is applied. The electrons trapped in the insulation layer have a large impact on the current flowing through the filament.

5.8.4 Memristor-Based Unique Signatures (MUS)

Some nanoelectronic devices based on memristors can generate unique signatures for the hardware by utilizing two distinct features of memristors [15], which are:

1. Inherent process variations that are nonuniform and irreproducible during fabrication.
2. Forming process to make memristors functional.

For random bit generation, a memristor-based unique signature (MUS) utilizes a pair of nonpolar memristors that are connected in series. Here the generation of a bit is a function of the position of the filament with low resistance [15]. Thus, a random word can be generated through multiple instances of random bit generators. Being nonvolatile, these signatures can be used for the purpose of hardware identification. This type of embedded hardware identification can be used to block electronics fraud or detect refurbished parts.

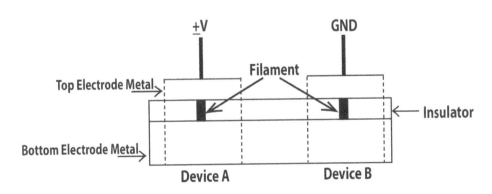

FIGURE 5.16 Memristor-based unique signatures (MUS) [15].

5.8.5 Tamper Detection Circuits Based on Memristors

Tamper detection means the identification of unauthorized use of any hardware. Memories that are tamper-evident make sure that the stored data remains confidential and also disclose if there is any read/write attempt. The properties in memristor-based devices such as device formation, nonvolatility, and run-time drift can be utilized to detect tampering, particularly manufacture-time and run-time tamper detection. Compared to CMOS devices, which require system-level techniques to enable tamper detection, resulting in high power and performance overhead, memristor-based devices use inherent properties of the device and hence need less power and performance overhead for enabling detection.

5.8.5.1 Manufacture-Time Tamper Detection

The device-forming step in memristors is used to differentiate between an unformed device and a formed one. For instance, a circuit comprising a set of memristors could be used for data encryption in a secure microprocessor. Any unauthorized or authorized user would first have to form these devices before obtaining any important information. The tamper detection of a circuit can be done using techniques that can also be used to verify the authenticity of the new IC received from an unreliable fabrication unit. In this technique, first a known value is written onto the memristors; then the value is read back; and then the complimentary value of the known value is written onto the memristors. After this, the next value is read back, and the results are compared. If both the read values are the same, then this would indicate that the formation step did not occur, and it would not be possible to write onto the memristors. On the other hand, if the read values are different, it implies that the forming step in memristors has occurred, which indicates a likely tampering of the circuit.

5.8.5.2 Run-Time Tamper Detection

Unauthorized memory reads in memories based on memristors can be traced by monitoring the related memristance drift. An attacker, to hide an unauthorized read, can restore the device memristance to its original value. For this, the attacker applies a read pulse of same magnitude and width but of opposite polarity, to unread the device. The memory read operation is changed so that the attacker cannot change the memristance value back to the original. For this, two consecutive read pulses (as shown in Figure 5.17), are used by the modified memory read operation [16], one of which is public and also known

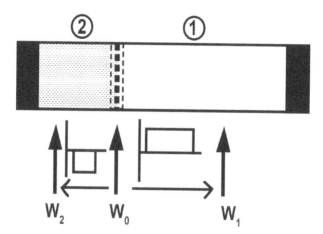

FIGURE 5.17   Memristor based run-time tamper detection device [16]. The dashed region represents high-resistive region, while the white region represents low-resistive region. The dotted line depicts the position of the domain wall determining the current resistance value.

to the attacker, while the other read pulse is private and is known only to the authorized user. Thus, an attacker cannot restore the memristance value to its original, which was created by the second pulse with its nonpublic parameters, even though he can revert back the memristance drift created by the first pulse. This memristance drift can be detected by an authorized user and thus help in determining possible tampering.

5.8.6 Nanoelectronics-Based Forensics (NanoForensics)

Digital forensics involve the recovery of data from hardware. For forensic applications, digital logic gates based on memristors utilize the run-time drift property of memristors.

When logic 0 is applied to an input, there will be no change in memristance value. But when logic 1 is applied, there is a drift in the memristance value of that particular memristor. The number of logic 1s applied to the input determines the amount of drift. Consider an memristor-based threshold logic (MTL) gate (Figure 5.18), realizing an AND function. The location of the gate within the circuit determines the number of 1s received by its input. Thus, the memristance drift at the inputs of different gates will differ.

The number of 1s that are received by that input is determined by measuring the memristance drift of that particular memristor and, in the same way, can be determined for all the memristors in the circuit. These measurements can help make a forensic analysis. If no memristance drift was observed in any of the memristors, then it can be concluded that the hardware was not used at all.

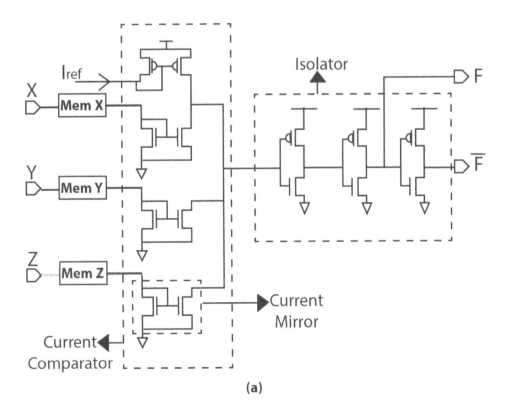

(a)

FIGURE 5.18 Forensic analysis in memristor-based nanoelectronic architecture. A three-input memristor-based threshold logic gate (MTL) [17].

5.9 LIMITATIONS AND FUTURE SCOPE OF MEMRISTORS

The main challenges regarding the use of memristors are that they have rather low speed, and at high frequency, major changes can occur in the memristor's characteristics. Also, because this is a new circuit element, designers need to learn how to build circuits using memristors. Besides this, a standard memristor model is needed that can be used for the purpose of fabrication and simulation.

The use of memristors in NVRAM can greatly revolutionize memory storage systems. In the field of neural network systems, the use of memristors could reveal crucial information about how the human brain works.

5.10 CONCLUSION

Research is going on to explore different aspects of the memristor, its other applications, and efforts to overcome its limitations. Since the need for an ultimate source for nonvolatile memory storage system is increasing every day, the application of memristors in NVRAM may bring another era in computer systems.

REFERENCES

[1] L. O. Chua, "Memristors – The missing circuit element," IEEE Transactions on Circuit Theory, vol. CT-18, no. 5, pp. 507–519, Sep. 1971.

[2] D. B. Strukov, G. S. Snider, D. R. Stewart & R. S. Williams, "The missing memristor found," Nature, vol. 453, pp. 80–83, 2008.

[3] R. Waser & M. Aono, "Nanoionics-based resistive switching memories," Nature Materials, vol. 6, pp. 833–840, 2007.

[4] Alon Ascoli, Fernando Corinto, Vanessa Senger & Ronald Tetzlaff, "Memristor model comparison," IEEE Circuits and Systems Magazine, pp. 89–105, May 2013, doi:10.1109/MCAS.2013.2256272.

[5] Valeri Mladenov, Advanced Memristor Modeling, Memristor Circuits and Networks Memristor Modeling, Memristor Devices, Circuits and Networks; MDPI: Basel, Switzerland, pp. 4–840, 2019.

[6] T. Prodromakis & C. Toumazou, "A review on memristive devices and applications," in Proceeding of 17th IEEE International Conference on Electronics, Circuits, and Systems (ICECS 2010), pp. 934–937, 2010.

[7] Kuk-Hwan Kim, Siddharth Gaba, Dana Wheeler, Jose M. Cruz-Albrecht, Tahir Hussain, Narayan Srinivasa & Wei Lu, "A functional hybrid memristor crossbar-array/CMOS system for data storage and neuromorphic applications," Nano Letters, American Chemical Society, pp. 389–395, 2012.

[8] Y. Yuanfan, "Matching in memristor based auto-associative memory with application to pattern recognition," Proceedings of 12th International Conference on Signal Processing (ICSP), pp. 1463–1468, 2014.

[9] M.D. Gregory & D.H. Werner, "Application of the memristor in reconfigurable electromagnetic devices," IEEE Antennas and Propagation Magazine, vol. 57, no. 1, pp. 239–248, 2015.

[10] Q. Xia et al., "Memristor-CMOS hybrid integrated circuits for reconfigurable logic," Nano Letters, vol. 9, no. 10, pp. 3640–3645, 2009.

[11] R. Williams, "How we found the missing memristor," IEEE Spectrum, vol. 45, no. 12, pp. 28–35, 2008.

[12] G. Rose, N. McDonald, L.-K. Yan & B. Wysocki, "A write-time based memristive PUF for hardware security applications," in Proceedings of IEEE/ACM International Conference on Computer-Aided Design, 2013, pp. 830–833.

[13] J. Rajendran, G. Rose, R. Karri, & M. Potkonjak, "Nano-PPUF: A memristor-based security primitive," in Proceedings of IEEE Computer Society Annual Symposium on VLSI, 2012, pp. 84–87.

[14] C. Huang, W. Shen, Y. Tseng, C. King & Y. C. Lin, "A Contact-resistive random-access-memory-based true random number generator," IEEE Electron Device Letters, vol. 33, no. 8, pp. 1108–1110, 2012.
[15] N. McDonald, "Al/CuxO/Cu memristive devices: Fabrication, characterization, and modeling," M.S. Thesis, College of Nanoscale Science and Engineering, Albany, NY, USA, 2012, vol. 1517153.
[16] X. Wang & Y. Chen, "Spintronic memristor devices and application," in Proceedings of IEEE/ACM, Design, Automation and Test in Europe, Conference Exhibition, 2010, pp. 667–672.
[17] J. Rajendran, H. Manem, R. Karri & G. Rose, "An energy-efficient memristive threshold logic circuit," IEEE Transactions on Computers, vol. 61, pp. 474–487, 2012.

Memristor
A Novel Device for Better Hardware/Software Security

6

Ms. Chhaya Verma and Jeetendra Singh

Contents

DOI: 10.1201/9781003126645-6

6.1 INTRODUCTION

The memristors are the fourth fundamental and nonvolatile electrical two-terminal devices. These behave as nonlinear resistors with memory, hence named memristors [1]. Memristors are axiomatically defined by a relationship between electric charge and magnetic flux [1, 2]. It is known to be flux controlled or charge controlled, depending on whether this relationship can be expressed as a single-valued function of the flux-linkage (φ) or charge (q). The voltage developed across charge-controlled memristors are given by

$$v(t) = M(q(t))i(t) \tag{6.1}$$

Where $M(q)$ is memristance, measured in ohms.

$$M(q) = \phi(q)/dq \tag{6.2}$$

Similarly, the charge-controlled memristors are given by

$$i(t) = W(\phi(t))v(t) \tag{6.3}$$

Where $W(\varphi)$ is menductance, measured in mhos.

$$W(\phi) = dq(\phi)/d\phi \tag{6.4}$$

The memristors can behave as simple resistors at any instance for high-frequency range, while memristance and inductance rely on the past of memristors [1, 2].

6.1.1 Device Structure and Basic Model

In 2008, R. Stanley Williams, with several colleagues at HP Laboratories, fabricated a memristor using titanium oxide and platinum electrodes. Williams's memristor consists of a transition metal oxide (TiO_2) separated by two platinum metal electrodes, as shown in Figure 6.1 [2]. Some oxygen vacancies are created on one side of the transition metal oxide so that this part of the layer will become conductive.

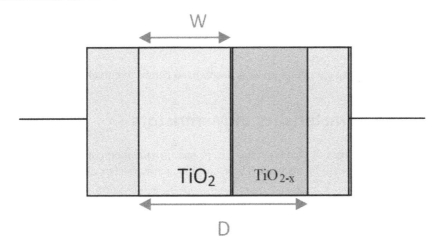

FIGURE 6.1 Device structure of memristor.

Titanium dioxide is an electrical insulator with high resistance, but altering the amount of oxygen in titanium oxide reduces its resistance. The TiO_{2-x} layer is said to be 'doped', as these oxygen deficiency effects are similar to an added impurity atom in the semiconductor. In general, a memristor can be fabricated by a variety of insulating materials sandwiched between metal layers, including chalcogenides [3, 4], metal oxides [5, 6], perovskites [7, 8], and organic films [9, 10].

6.1.2 Memristors Operations

The operation of memristors can be understood by hydraulic analogies. In this analogy, memristors are compared to a fictitious pipe with a mobile sand filter in it that regulates the flow of water. When water flows through a bed of sand and gravel, sediment slowly clogs the pores of the filter and thereby restricts the flow of water. Reversing the flow flushes out the sediment, and water flows smoothly. It is important to note that in both cases it is the direction of flow that controls the state of the device, as the resistance of the sand filter is the same in both directions.

Similarly, the memristor resistive state is controlled by the applied voltage polarity in a bipolar memristor and voltage magnitude in a unipolar memristor [11]. Memristors are said to have two resistance sets; a high resistance state (HRS) referred to as R_{OFF} and a low resistance state (LRS) referred to as R_{ON}. In SET operation, a voltage bias of suitable polarity and magnitude, V_{SET}, must be applied to switch it from HRS to LRS. The RESET operation is done by switching LRS to HRS by applying a lower voltage, V_{RESET}. The polarity of V_{SET} and V_{RESET} is the same in unipolar and opposite in bipolar, [12]. There are several simulated models of the memristors that are designed adopting the principle of device physics behavior and the link between the memristance and the flux $\varphi(t)$ of the device, $M(\varphi(t))$, can be noted as in equation 6.5, by Y.N. Joglekar et al. [13].

$$M\left(\phi\left(t\right)\right) = \frac{HRS}{D}\sqrt{D^2 - 2\eta\frac{LRS}{HRS}\phi\left(t\right)\mu} \qquad (6.5)$$

η defines the polarity of the applied bias +1 for positive voltage and −1 for negative applied bias, and the dopants' mobility is μ. The movement rates the domain wall estimated by S. Kvatinsky et al. [14] as in equation 6.6, when the current (I) is flowing through the memristor.

$$\frac{dw}{dt} \frac{R_{ON}}{D} \left(1 - x^{2p}\right) \tag{6.6}$$

Here, P is accounted as tuning parameter and x is normalized conductive thickness.

6.1.3 Various Characteristics of Memristors

Memristors bear different types of characteristics. Here metal-oxide memristors have noteworthy characteristics that make them useful for security. These specific characteristics of the memristor rely on the utilized material.

6.1.3.1 Nonvolatility

Memristors are stable and remember their state even if the device loses power; that is, it retains its memristance value even the power is OFF [15].

6.1.3.2 Nonlinearity

The current-voltage characteristics of memristors are extremely nonlinear, since these characteristics are resistance state-dependent and time dependent [11], as shown in equation 6.5.

6.1.3.3 Bidirectionality

The obtained current-voltage characteristic shown in Figure 6.2 is symmetric, and the curve is similar in opposite quadrants; hence, it is concluded that memristor has bidirectional characteristics [16].

6.1.3.4 Formation Process

Most of the memristor needs to be initialized to exhibit the switching characteristics. This is also known as conductive filament formation, which is achieved by applying a high voltage across it; before this, it behaves as a linear resistor [17, 18].

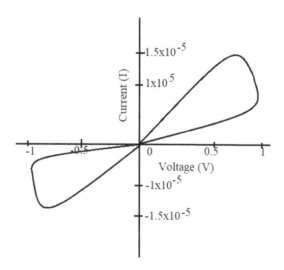

FIGURE 6.2   Pinched *I–V* characteristics of memristors.

6.1.3.5 Memristance Drift

In some metal-oxide memristors, a process of change in memristance due to the movement of dopants has occurred, which is known as memristance drift [2]. This movement of dopants occurs by varying the applied input voltage (positive or negative). The drifting amount depends on the magnitude, polarity, and applied voltage duration [11].

6.1.3.6 Process Variations

The process variation in memristor fabrication leads to change in its dopant concentration and dimensions, which ultimately affects the value of memristance. Moreover, the length variation has a highly nonlinear effect on memristance value [19].

6.1.3.7 Radiation-Hardness

Memristors that are made from a material that has the inherent property of radiation hardness are immune to radiation [2].

6.1.3.8 Temperature Stability

Due to the very small temperature coefficient of resistance of TiO_2 (less than $-3.82 \times 10^{-3}/K$), the LRS and HRS values of Williams's memristor are highly stable. However, temperature variation leads to a change in mobility of dopant atoms, which varies the switching speed [19]. All these characteristics have different advantages and disadvantages in memory and logic circuits designed using metal-oxide memristors. Moreover, nonvolatility, nonlinearity, and radiation-hardness are beneficial in the context of security.

6.2 MEMRISTOR-BASED PUF AND PPUF

6.2.1 Physically Unclonable Function (PUF)

Today, every single work environment, either public, private, business, or even national defense, has gone digitalized, and digital security has become a necessity in the current world. And the security of this digital world mostly relies on the hardware-based root of trustworthy security functions such as storage and confidential key generation. The current transitions to the Internet of things, autonomous intelligent systems, and embedded systems are highly prone to physical attack, which increases the importance of hardware solutions for security. The two main hardware primitives which can solve this issue are physical unclonable functions (PUF) and true random number generators (TRNGs). TRNGs are utilized for generating masks, session keys, padding bits, initialization values, and other cryptographic data [20].

A PUF is a hardware-dependent security primitive that provides a specific unique identification for each integrated circuit implementation, and this specificity in PUFs is due to large variations in the manufacturing process. Earlier, security keys were stored in the form of nonvolatile memory (EEPROM, flash memory, etc.) that was used for the cryptography of data. The safe storage of these keys is quite complex and is not economic. Furthermore, these are fixed keys stored on nonvolatile memory that are vulnerable to attacks; once attackers unethically get access to the keys, there is no recovery. Despite incorporating a single cryptographic key, PUFs implement challenge–response authentication, which is intentionally designed in such a manner that the responses depend on both the applied challenge and the physical variations present in PUF; hence, they are said to have a unique fingerprint, which strengthens the security of the integrated circuit.

The security in software execution is primarily provided by PUFs [21], which is necessary for authentic configuration of field-programmable gate arrays (FPGAs) [21], trusted device working, and encrypted storage [22]. Several types of PUFs architecture have been designed, and several new architectures are under research. A few of the currently used architectures are the arbiter PUF [23, 24], butterfly PUF [25], flip-flop PUF [26], SRAM PUF [27], and ring oscillator PUF [23, 28, 29].

6.2.2 Memristor-Based PUF (NanoPUF/M-PUF) Architecture

The inherent characteristics of process variation in memristors are the main reason behind designing memristor-based PUFs. These memristor-based PUFs are also referred to as NanoPUFs [30, 31, 52]. The NanoPUF/M-PUF consists of three major parts:

1. Memristor crossbar.
2. Challenge circuit and Row decoder.
3. Response circuit and column decoder with a current comparator.

The input to the crossbar is enabled by a challenge circuit through the row and column decoders. A voltage of magnitude V_{dd} is applied to the row decoder, which is connected to a particular column resistance, also called the load resistance (R_{load}). The remaining columns and rows are in floating mode. Although the magnitude of pulse voltage is V_{dd}, the response circuit collects the response from the column decoder. It consists of R_{load} and a current comparator that compares the reference current (I_{ref}) with the current coming out of the column (I_{out}). If the I_{out} is greater than I_{ref}, then the response bit is logic 1; otherwise, it is 0 [11]. The direct path and sneak path are two major paths that are available in the crossbar. The current flowing from a row (input) to a column (output) is the function of the resistance of the device at the cross-point of that row and column; this flow path is called a direct path. The current flows as a function of device resistance from one input to another output in the crossbar in a sneak path. A 4 × 4 memristor-based PUF circuit is shown in Figure 6.3.

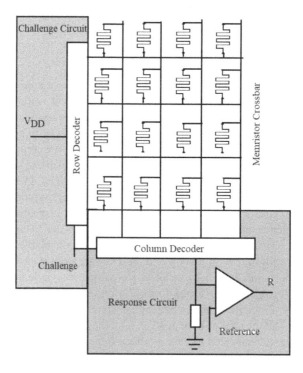

FIGURE 6.3 A 4 × 4 memristor-PUF/NanoPUF.

6.2.3 Operation of Memristor-Dependent PUF

For the same challenge (input), the response is different for the different integrated circuits. Likewise, in the same PUF circuit, each memristor has its identical switching time due to process variation, as for an applied pulse of fixed duration to a selected memristor in the crossbar, some memristors turn ON, while the remaining are OFF. The response is considered as 1 if the memristor is turned ON. Otherwise, the response bit is 0. A variable CRP that researches varies only in input pulse duration, while others vary both in magnitude and duration of input pulse [30, 31, 53].

6.2.4 Security Analysis

Uniformity, uniqueness, and bit-aliasing are the three matrices that can be utilized to investigate the ability of NanoPUFs in generating unique IDs [32, 54].

1. The responses achieved from the two distinct crossbars give the hamming distance by which uniqueness is measured after applying the same challenge. The ideal value of the uniqueness is 50%.
2. The uniformity determined the randomness of the response, and it is ascertained by proportions of 1s and 0s in the response. Uniformity will also have a 50% ideal value.
3. Bit aliasing gives affinity to achieve 1s and 0s in the response bit. Its ideal value also should be 50%. The results of these PUFs can be foreseeable because distinct PUFs may generate similar responses due to the bit-aliasing. The qualities of these results are listed in Table 6.1; these responses are generated by a NanoPUF by G. Rose et al. [30].

When the performance of same sized M-PUFs was compared with CMOS-based PUFs (SRAM-based PUFs), it was found that M-PUFs generate more response bits, as memristors are much denser than SRAM cells. Moreover, they consume lower power than SRAM-based PUFs [55–58].

There are some issues with memristor NanoPUFs that need to be solved and are under research. These issues are stability and entropy. Various challenges at different temperatures and voltage need a more stable response, as these are lacking in M-PUFs, whereas in the case of entropy, the various responses generated by PUF should be a linear function of the nanoelectronics devices used, and this should be an exponential function of the elements utilized in the PUFs' circuit.

6.2.5 Memristor-Based PPUF (M-PPUF/ Memristor NanoPPUF)

Public physical unclonable functions (PPUFs) are a form of PUF also considered as a public key that is used to exchange information in a remote location. In PPUFs, the challenge or simulation models are available publically to the attacker, but the simulation time is too long (in years), whereas the simulation time for the recipient, that is, PUF primitive, is in nanoseconds [33]. A PPUF is used to implement two classes' security protocols like a key exchange, authentication, stamping, and bit commitment. The characteristics of memristors that are the ideal choice for designing NanoPPUFs are bidirectionality, process variation sensitive, and memristor-crossbar models and their simulation complexity, with a higher chip density of M-PPUF compared to the same-sized CMOS PPUF. A memristor NanoPPUF protocol for time-bounded authentication proposed by J. Rajendran is shown in Figure 6.4, which outlines the

TABLE 6.1   Qualities of Unique Responses Generated by M-PUF [24]

METRIC	UNIQUENESS	UNIFORMITY	BIT-ALIASING
Values	49.85%	49.99%	49.99%

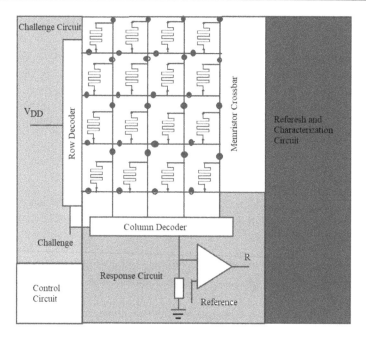

FIGURE 6.4 A 4 × 4 memristor-based NanoPPUF architecture.

two-party security protocols [34, 59]. He has used the concept of polyominos as challenge constraints. A polyomino can be determined as a geometric plane figure made by combining the number of discrete blocks. In this NanoPPUF's crossbar, each memristor is considered as a block. These polyominoes can be of different sizes. For sound security, it is wise to have large-sized polyominoes so that the simulation time can be greater, since for the PPUF primitive (recipient), the simulation time will increase in a few nanoseconds, but for the attacker, the simulation time will increase in years.

6.2.6 The Architecture of Memristor NanoPPUF

The memristor NanoPPUF consists of five major parts [34, 60], which are shown in Figure 6.4.

1. A memristor crossbar.
2. A refresh and characterization circuit.
3. The challenge and characterization circuit.
4. The controller circuit.
5. The response circuit.

In the crossbar, there are cross-points and tap points. Each cross-point consists of a memristor. Tap points denoted by dots (in Figure 6.4) are connected to a voltage sensor measuring voltages to give the boundary conditions. The polyomino can be realized by selecting the tap points in a crossbar.

6.2.7 Operation of Memristor NanoPPUF

The working of a NanoPPUF is described in four phases: characterization phase, refresh phase, challenge–response phase, and measurement phase. These phases are described in the following sections.

6.2.7.1 Characterization Phase

In this phase, the characteristics such as length (L), width (W), and thickness (D) of the device are determined, then its HRS, LRS, and time to switch on (t_{ON}) are determined. It is a one-time process for a NanoPPUF. Determining the HRS and LRS value for calculating HRS value, the first device is switched OFF, then the negative pulse is applied for a long duration, followed by applying positive pulse for a short duration of magnitude V_{dd} [35, 61]. Then current through the load resistor is measured. And for calculating LRS, the device is switched ON, and a positive voltage is applied for a long duration. Then for a short duration negative voltage ($-V_{dd}$) is applied. Finally, the load current is measured. The value of HRS or LRS is given as

$$HRS / (or)LRS = \frac{V_{dd}}{I_{Load}} - R_{Load} \tag{6.7}$$

For the determination of switch-on time (t_{ON}), the device is switched OFF for long-duration by applying negative voltage. Then a positive pulse voltage is applied until the device is switched ON. The switch of t_{ON} is the duration of the applied positive pulse voltage. The measurement of LRS, HRS, and t_{ON} determine the values of L, W, and D from the following equations:

$$M_{OFF} = \frac{HRS \times D}{L \times W} \tag{6.8}$$

$$M_{ON} = \frac{LRS \times D}{L \times W} \tag{6.9}$$

$$\frac{dw}{dt} = \frac{M_{OFF} - M_{ON}}{t_{ON}} \tag{6.10}$$

This step should be done for each memristor in the crossbar. Finally, at the end of this, the phase simulation mode for M-PPUF is designed.

6.2.7.2 Refresh Phase

Before starting the protocol, all memristors need to be refreshed in a known state; that is, a negative voltage of specified duration is applied in all memristors and switched to HRS. In this phase both a challenge-characteristics circuit and refresh-characteristics circuit are used.

6.2.7.3 Challenge–Response Phase

The challenge is applied in this phase using a challenge-characteristics circuit, and the response is measured by using a response circuit. The fluctuations, which are produced by voltage and temperature, are compensated for by the controller circuits.

6.2.7.3 Measurement Phase

The boundaries voltages of polyomino nodes picked by the verifier are sensed in this phase utilizing voltage sensors.

6.2.8 Protocols

The utilized protocol can be simply understood by a two-party security protocol proposed by Rajendran; it is a user authentication protocol using NanoPPUF/M-PPUF, shown in Figure 6.5. Here, Alice and Bob

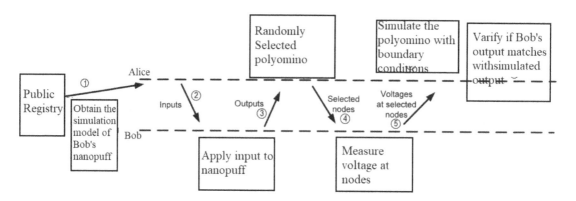

FIGURE 6.5   Time-bounded authentication protocol using NanoPPUF between two parties.

are friends, where Molly is an adversary. Alice sends a challenge vector C to Bob to authenticate before sending a piece of secret information, while Molly is willing to act as Bob to get that information. The challenge vector C sent by Alice is from challenge set X, where the number of bits in X and C is the same. Both Bob and Alice are using NanoPPUFs to authenticate each other; the Alice NanoPPUF is denoted by PPUFA and Bob's is PPUFB. There, PPUF is designed using memristors whose bidirectionality and process variation characteristics produce an exponentially large number of polyominoes, and the response generated by these PPUF to the applied challenge is random and unique. Bob sends a unique response R to Alice; the PPUFA randomly selects a polyomino after receiving the outputs from Bob to validate the challenge and response on PPUFB. Meanwhile, Molly is also trying to generate the response to challenge C, but it is not possible to generate the response R simulated by PPUFB. Molly has a near-zero probability of randomly predicting the selected polyomino. So any message exchanged between Alice and Bob is safe; for Molly, it will take years to simulate.

Due to unidirectionality, CMOS XOR-based PPUFs need a large number of XOR gates for the desired security, which results in a large latency (in several seconds). This issue is resolved in memristor-based PPUFs due to its bidirectionality, and simulation time is in few nanoseconds. Moreover, the speed and size of the memristors cover less area and hence make the circuit more compact, which operates fast compared to CMOS-based PPUFs.

6.3 MEMRISTOR-DEPENDENT TRUE RANDOM NUMBER GENERATORS (NANOTRNGS)

6.3.1 True Random Number Generators

A true random number generator (TRNG) is one of the hardware security primitives that is used in cryptographic applications to generate security keys, initial vectors, nonce, masking values, and padding. The TRNG generators keep track of system activities such as the number of times a hard drive in a computer has been accessed; it counts the number of clicks from the mouse or the thermal noise coming from the circuits, which creates an array of distinctive numbers. The random numbers are taken from this array of unique numbers, but it is safe and more random if values far from initials or final are selected. These random numbers are vital for the digital security of connection and transactions. A security key generated using random numbers is unique for a user, and the public keys made from these security keys using algorithm cryptography are larger in number than the original security key. Hence the original security key remains safe.

Many researchers have designed the TRNG focusing on the jitter entropy. These jitter entropies are produced by oscillators created by the noise of the semiconductor component, since the amount of jitter in a period is vital for the determination of randomness of output bits. Currently, ring oscillators (ROs) and differential ROs are mostly used to design any random number generator. In 2007, B. Sunar et al. [36] designed a secure true random number generator by using CMOS; the design is shown in Figure 6.6. To gain sufficient entropy, Sunar increased the number of ROs in his design. But a large number of ROs acquires more area and consumes more power; hence, it was recommended to reduce the number of stages of ROs and to rather work on how to increase the entropy of a single-stage RO.

Following these suggestions, in 2016, Hashim et al. [37] proposed a modified RO TRNG by changing the PMOS of the inverter RO design of Sunar with a memristor, shown in Figure 6.7. Several stages of RO are designed using a single-stage RO, as shown in Figure 6.8, which provides high entropy. This memristor-based design is found to be more promising because it provides more random oscillation frequency and higher entropy; meanwhile, due to fewer stages of ROs, it acquires less area and consumes minimal power.

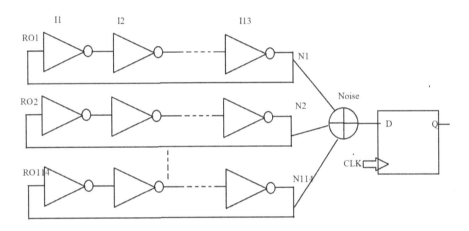

FIGURE 6.6   A random number generator in Sunar's design.

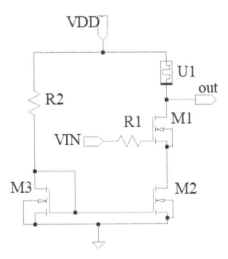

FIGURE 6.7   A single-stage ring oscillator.

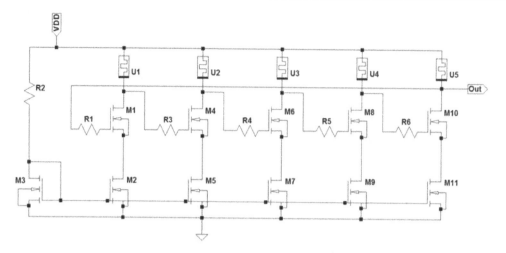

FIGURE 6.8   A five-stage ring oscillator utilizing memristors.

6.4 MEMRISTOR-BASED TAMPER DETECTION CIRCUIT

The unauthorized access to target hardware or its unauthorized usage is termed as tampering, and the ability of the device to sense such unauthorized activities is defined as tamper detection. A device that has a tamper detection circuit along with a defense circuit to respond to such practices keeps the stored data safe and confidential. Memory devices designed using memristors are temper proof due to their nonvolatility and memristance drifting characteristics.

6.4.1 The Architecture of Tamper Detection Circuit

A memristor-based memory device designed in a crossbar pattern is similar to that shown in Figure 6.3. In these memristors, the HRS represents logic 0, and LRS represents logic 1, respectively. The two operations write and read are governed as follows:

Write operation: Apply V_{RESET} (logic 0) and V_{SET} (logic 1) to corresponding rows, and apply 0 V to the corresponding columns.

Read operation: A small amplitude positive pulse is applied as a read voltage to the corresponding row. The output current and reference current are compared with each other; logic 1 is read if the output current is larger than the reference current, and logic 0 is read if the output current is smaller than the reference current. In the memristor-based devices, if memristance drift is exhibited during the read operation, generally two steps of the read operation are utilized. In a bipolar memristor, the same magnitude and duration positive pulse followed by the negative pulse is the ideal read pattern, which creates a zero net change in memristance.

Tampering of the device can be done two times; first, it is possible that the manufacturer is not trustworthy and that the delivered device is used or tampered with by the manufacturer. Second, the device may be tampered with by an outside attacker during the run time [38]. To prevent such activities, the two types of temper detection are (1) manufacture-time tamper detection and (2) run-time tamper detection.

6.4.2 Manufacture-Time Tamper Detection

The memristor-based device can easily be checked whether it has been used earlier or not by a device formation step. To confirm the trustworthiness of newly integrated circuits purchased, a known value

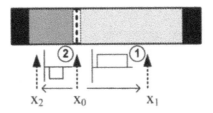

FIGURE 6.9 The idea of dual read pulses to detect run-time tampering in a memristor device; the dotted line represents the current resistance state of the memristor.

and then the complement of the known value is written in the memristor, and these values are then read back. The compared results suggest its trustworthiness. The compared results would be the same if there is no formation occurring in the memristor, and this will facilitate writing in the memristor [39]. The compared read output will be different if the memristor is already formed, which means that the circuit has been tampered with. It is like making a mirror image of information and overlapping it to check whether they match or not; if overlapped, then the device is not formed, but if not, then it might have formed, that is, used.

6.4.3 Run-Time Tamper Detection

The main idea to detect an unauthorized read operation in memristor-based memories is to observe the unauthorized drift in memristance. To hide an unauthorized read, the attackers would try to reinstate the memristance of the memristor back its actual state by 'un-reading'; this can be achieved by just applying an opposite polarity read with a pulse of the same magnitude and duration [40]. To prevent such restoring of memristance, the read operation is corrected, as shown in Figure 6.9. In this modified read operation, two consecutive read pulses are used. Out of these two consecutive pulses, the first is public (known to the attacker as well) and the second is private (only known to the authorized user) [41, 62, 63, 64]. In such a case, the attacker will not be able to restore the memristance after reading, as the second pulse is unknown. Since the memristance value due to an unauthorized read will differ from the original value before reading, the authorized user can monitor this difference and become aware of this act of tampering.

6.5 MEMRISTOR-BASED FORENSICS

In this digital world, the evidence for any crime could also be seen in the form of 0s and 1s. This digital evident may get tampered with by the culprit. To retrieve these shreds of evidence, forensics research is done on those devices. Evidence devices made from CMOSs are sometimes in such a condition that retrieval of data is either impossible or too time-consuming. Memristor-based memory of the digital device is relatively easier and less time consuming compared to a CMOS-based device [42].

6.5.1 The Architecture of Memristor-Based Forensics

The run-time drift property of memristors can be used in memristor-dependent digital logic gates. The drift in the memristance will provide the sets of the inputs applied in the past to the memristor with the help of forensics. Memristor-dependent threshold gates were proposed by Rajendran in 2012 [43]. In this memristor-dependent threshold logic (MTL), memristors are utilized as weights for the input of the gates. A three-input

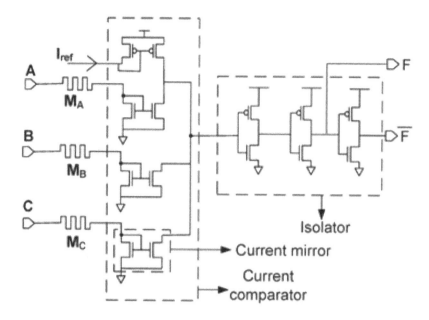

FIGURE 6.10   Three input memristor dependent threshold logic gate (MTL) used in forensic analysis.

threshold gate using memristors is shown in Figure 6.10. The memristors are denoted as M_A, M_B, and M_C, which is used to weigh the current coming from input A, B, and C, respectively. The current mirrors are used to isolate the currents of distinct inputs. The sum of weighted currents is compared to reference the current using a comparator. Logic is 1 if the sum of weighted currents is greater than I_{ref}; else, the logic is 0. Logic 1 is denoted by a positive voltage and logic 0 is denoted by zero voltage. A programming circuit can be used to program the MTL circuit, just like the one used in memristor-based memory [44]. To address the memristors in the design, one can use either one write circuit or an individual write circuit for each group of memristors.

6.5.2 Operation of Memristor-Based Forensics

The emristance value changes only when logic '1 is applied and has no change when logic 0 is applied. Therefore, the number of logic 1s of a particular input determines the amount of drift in the memristance. The memristance drift leads to change in memristance value; if the final memristance value of the memristor is known, then the numbers of 1s applied in inputs can be estimated. For example, if the memristance value of memristors is 2 M, and assuming the magnitude and width of input pulses to be 1.1 V and 2 ns, respectively, then the value of M_A, M_B, and M_C will be 2.12 M, 2.25 M, and 2.25 M, respectively, and these values can be used to evaluate about millions of 1s applied to the inputs. This process of estimation is referred to as forensic analysis [45].

6.6 MEMRISTOR-BASED CRYPTOGRAPHY

The special attractive feature of nonvolatile main memories (NVMM) is also a cause of threat to security, as the ability to save data even while power-off may avail attackers the chance to read its information. The prevention of such an attack can be done by storing the data in encrypted form using algorithms (e.g., AES) and decrypt it while reading. But this cryptographic technique is not attack proof, as an attacker can suggest the algorithms and tamper with the data [46]. There is also a possibility of side-channel attacks, as it may retrieve the security key by observing the side-channel contents like timing and power. Along with the security issue, this cryptographic technique also results in huge power consumption and size overhead. To resolve this issue, researchers have suggested encryption utilizing nanoelectronic devices. The sneak path encryption (SPE) technique is mostly referred for creating secure NVMM (SNVMM).

6.6.1 Architecture Memristor-Based Cryptography

There are four main part of this architecture:

a. Processor memory.
b. Cache memory.
c. Sneak path encryption control unit (SPECU).
d. SNVMM/NVMM.

Two-level memory architecture was proposed by S. Kannan et al. in 2014 [47]. The processor and cache memory operate on unencrypted data. The cache memory is divided into two levels, Level 1 and 2. The SPECU is just below the Level 2 cache before NVMM. The data stored in NVMM is encrypted, and the sneak path encryption of this data is controlled by SPECU [48]. The SPECU architecture and modified NVMM are shown in Figure 6.11. The SNVMM mainly consists of a crossbar, row decoder, column decoder, and transistor control unit. The crossbar is similar to that of Figure 6.3; the only difference is that each cross-point in the crossbar has an additional transistor along with the memristor. These transistors are used to regulate voltage, causing the memristor to change its memristance value [49]. This transistor control unit enables or disables the selected memristor across which it is connected. The selected memristor is referred to as the point of encryption (PoE).

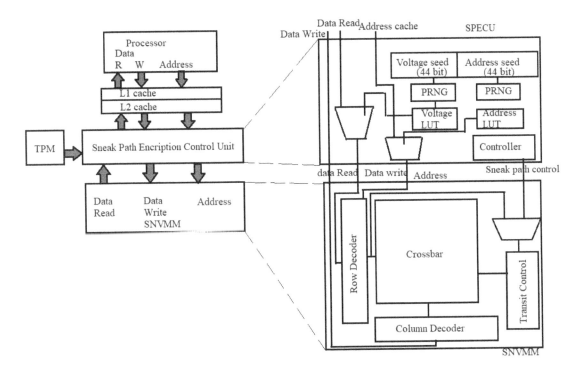

FIGURE 6.11   Sneak path encryption using SPECU architecture and NVMM modifications.

6.6.2 Operation Memristor-Based Cryptography

The read and write operation in NVMM is controlled by SPECU. A selected set of the transistor is switched on by the transistor control unit in NVMM to create a sneak path that can be used by SPECU for encryption and decryption. The sneak paths are disabled by the transistor during read and write operations. The amplitude and duration of write pulses of SNVMM changes to enable sneak paths for encryption and decryption, which are ascertained by lookup tables and PRNGs. There are two PRNGs in SPECU: one is used to generate a random number for the first half of the key that controls the magnitude and width (duration) of write pulse, and the other is used to produce random numbers for the second half, which is used to select the PoEs. The two lookup tables are used to map the random number to the amplitude and duration of the write pulse and position of PoEs [50]. The memristor-based crypto-architectures consume low power, and performance overhead is also less, and it is strong enough to deal with side-channel attacks that are based on power analysis.

6.7 MEMRISTOR-BASED NEUROMORPHIC SECURITY SYSTEM

The fifth-generation computing systems, defense appliances like drones, automated vehicles, and home appliances, etc., are based on artificial intelligence, which uses neural networks for machine learning, also called deep learning. These neural networks process input data by multiplying them with layers of weighted connections and are fast and efficient. The hardware, which is designed to accelerate neural networks or machine learning algorithms, is called a neuromorphic computing system. The designing and manufacturing of the neuromorphic computing system is complex and advance. Companies like IBM, Micron, Intel, etc., are working on research and development of these brain-like chips, that is, neuromorphic computing chips. IBM's TrueNorth (2014), Intel's Loihi (2017), and Micron's Automata Processor are the latest CMOS-based neuromorphic integrated circuits.

Since a CMOS-based neuromorphic system is the most recent, many researchers are currently working on it. Yang et al. (2016) [51] in a paper have discussed the possible threat to the security of the CMOS-based neuromorphic system. They have considered a defense AI-drone that somehow got captured by an adversary who used reverse engineering to understand the hardware implementation of the system. Foreseeing many possible ways this could be a security threat, they have proposed a secure memristor-based neuro-morphic computing system (MNCS) design. The most attractive property mentioned that is drawing the researchers' attention is that memristors are highly dense and can be stacked in a 3D structure, which makes it extremely difficult for physical attacks [52].

6.8 MEMRISTOR-BASED CHAOTIC CIRCUIT FOR SECURE COMMUNICATION

In the past two decades, chaos-based communication has been an indispensable research topic, as this system can spread a spectrum that provides multiple access and is expected to have anti-jamming capabilities. The realization of a chaotic secure communication system on a chip is simple. The synchronization of chaotic systems, first introduced by Fujisaka and Yamada [53], plays a key role in the implementation of secure communication. In this communication, a chaotic signal is used as a carrier, which is transmitted after modulation. At the receiver's end, chaos synchronization is used to recover the message signal.

Memristor-based chaotic oscillators of high frequency have enormous potential for application in the field of secure communications, as the memristor is a nanoelectronic device.

6.9 CONCLUSION

The study of the memristor has become essential for the current scenario because of its attractive characteristics. This chapter provides a complete understanding of advanced security systems employed with a memristor and set a strong background for the readers and is also useful to develop a new aspect of them using the memristor, which will possibly dominate in next-generation technology. The memristor-based M-PUF, with its architecture and functional behavior, is discussed in various phases. The utilization of the memristor in a true random number generator, for tamper detection in a confidential device, and for forensics inspection was briefed. Moreover, more security purposes such as cryptography, neuromorphic security, and memristor-based chaotic circuits are also explored with their architectures and functionality. This chapter aims to give a better understanding of the memristor and its applications in the field of hardware/software security.

REFERENCES

[1] L. Chua, "Memristor: The missing circuit element," IEEE Transactions on Circuit Theory, vol. 18, no. 5, pp. 507–519, 1971.
[2] R. Williams et al., "How we found the missing memristor," IEEE Spectrum, vol. 45, no. 12, pp. 28–35, 2008.
[3] A. Oblea, A. Timilsina, D. Moore, and K. Campbell, "Silver chalcogenide based memristor devices," in Proceedings under International Joint Conference on Neural Networks, 2010, pp. 1–3.
[4] R. Waser and M. Aono, "Nanoionics-based resistive switching memories," Nature Materials, vol. 6, pp. 833–840, 2007.
[5] Priya Bansal and B. Raj, "Memristor: A versatile nonlinear model for dopant drift and boundary issues," JCTN, American Scientific Publishers, vol. 14 (5), pp. 2319–2325, May 2017; B. Briggs et al., "Influence of copper on the switching properties of hafnium oxide-based resistive memory," MRS Online Proceedings Library, vol. 1337, 2011.
[6] A. Sawa, T. Fujii, M. Kawasaki, and Y. Tokura, "Interface resistance switching at a few nanometer thick perovskite manganite active layers," Applied Physics Letters, vol. 88, no. 23, pp. 232112-1-232112-3, 2006.
[7] Anil Kumar Bhardwaj, Sumeet Gupta, B. Raj, and Amandeep Singh, "Impact of double gate geometry on the performance of carbon nanotube field effect transistor structures for low power digital design," Computational and Theoretical Nanoscience, ASP, vol. 16, pp. 1813–1820, 2019.
[8] J. C. Scott and L. D. Bozano, "Nonvolatile memory elements based on organic materials," Advanced Materials, vol. 19, pp. 1452–1463, 2007.
[9] Neeraj Jain and B. Raj, "Thermal stability analysis and performance exploration of Asymmetrical Dual-k underlap Spacer (ADKUS) SOI FinFET for security and privacy applications," Indian Journal of Pure & Applied Physics (IJPAP), vol. 57, pp. 352–360, May 2019.
[10] Jeetendra Singh and Balwinder Raj, "Comparative analysis of memristor model for memories design," Journal of Semiconductor, IOP Science, vol. 39, no. 7, pp. 1–12, 2018.
[11] Candy Goyal, Jagpal Singh Ubhi and B. Raj, "Low Leakage Zero Ground Noise Nanoscale Full Adder using Source Biasing Technique," Journal of Nanoelectronics and Optoelectronics, American Scientific Publishers, vol. 14, pp. 360–370, March 2019.
[12] Y. N. Joglekar and S. J. Wolf, "The elusive memristor: Properties of basic electrical circuits," European Journal of Physics, vol. 30, p. 661, 2009.
[13] S. Kvatinsky, E. Friedman, A. Kolodny, and U. Weiser, "TEAM: ThrEshold Adaptive Memristor model," IEEE Transactions on Circuits Systems I: Regular Papers, vol. 60, no. 1, pp. 211–221, 2013.

[14] Jeetendra Singh and Balwinder Raj, "Design and investigation of 7T2M NVSRAM with enhanced stability and temperature impact on store/restore energy," IEEE Transaction on VLSI Systems, vol. 27, no. 6, pp. 1322–1328, 2019.

[15] Jeetendra Singh and Balwinder Raj, "Enhanced nonlinear memristor model encapsulating stochastic dopant drift," Journal of Nanoelectronics and Optoelectronics, American Scientific Publishers, vol. 14, pp. 1–6, 2019.

[16] Jeetendra Singh and Balwinder Raj, "Modeling of mean barrier height levying various image forces of metal insulator metal structure to enhance the performance of conductive filament based memristor model," IEEE Transaction on Nanotechnology, vol. 17, no. 2, pp. 268–275, January 15, 2018.

[17] Jeetendra Singh and Balwinder Raj, "Tunnel current model of asymmetric MIM structure levying various image forces to analyze the characteristics of filamentary memristor," Applied Physics A, Springer, vol. 125, no. 3, pp. 203–213, 2019.

[18] Jeetendra Singh and Balwinder Raj, "Temperature dependent analytical modeling and simulations of nanoscale memristor," Engineering Science and Technology, an International Journal, Elsevier, vol. 21, no. 5, pp. 862–868, 2018.

[19] W. Schindler, "Evaluation criteria for physical random number generators," Cryptographic Engineering, 25–54, 2009.

[20] Neeraj Jain and B. Raj, "Dual-k spacer region variation at the drain side of asymmetric SOI FinFET structure: Performance analysis towards the analog/RF design applications," Journal of Nanoelectronics and Optoelectronics, American Scientific Publishers, vol. 14, pp. 349–359, March 2019.

[21] J. Guajardo, S. Kumar, G.-J. Schrijen, and P. Tuyls, "Physical unclonable functions and public-key crypto for FPGA IP protection," in Proceedings of International Conference on Field Programmable Logic and Application, 2007, pp. 189–195.

[22] Neeraj Jain and B. Raj, "Analysis and performance exploration of high-k SOI FinFETs over the conventional low-k SOI FinFET toward analog/RF design," Journal of Semiconductors (JoS), IOP Science, vol. 39, no. 12, pp. 124002–1–7, December 2018.

[23] U. Chatterjee, S. R. Chakraborty, J. Mathew and D. K. Pradhan, "Memristor based Arbiter PUF: Cryptanalysis Threat and its Mitigation," in 29th International Conference of VLSI Design, 2016.

[24] Candy Goyal, Jagpal Singh Ubhi, and B. Raj, "A reliable leakage reduction technique for approximate full adder with reduced ground bounce noise," Journal of Mathematical Problems in Engineering, Hindawi, vol. 2018, Article ID 3501041, 16 pp., October 15, 2018.

[25] Girish Wadhwa and B. Raj, "Label free detection of biomolecules using charge-plasma-based gate underlap dielectric modulated junctionless TFET," Journal of Electronic Materials (JEMS), Springer, vol. 47, no. 8, pp 4683–4693, August 2018.

[26] D. E. Holcomb, W. P. Burleson, and K. Fu, "Power-up SRAM state an identifying fingerprint and source of true random numbers," IEEE Transactions on Computers, vol. 58, no. 9, pp. 1198–1210, 2009.

[27] F. Kotydek, R. Lorencz, and J. Bucek, "Improved ring oscillator PUF on FPGA and its properties," Microprocessors and Microsystems, pp. 1–9, 2016.

[28] Aakash Jain, Sanjeev Sharma, and B. Raj, "Analysis of Triple Metal Surrounding Gate (TM-SG) III-V nanowire MOSFET for photosensing application," Opto-electronics Journal, Elsevier, vol. 26, no. 2, pp. 141–148, May 2018.

[29] G. Rose, N. McDonald, L.-K. Yan, and B. Wysocki, "A write-time based memristive PUF for hardware security applications," in Proceedings of IEEE/ACM International Conference on Computer-Aided Design, 2013, pp. 830–833.

[30] Neeraj Jain and B. Raj, "Parasitic capacitance and resistance model development and optimization of raised source/drain SOI FinFET structure for analog circuit applications," Journal of Nanoelectronics and Optoelectronics, ASP, USA, vol. 13, pp. 531–539, April 2018 (SCI).

[31] Shradhya Singh, S. K. Vishvakarma, and Balwinder Raj, "Analytical modeling of split-gate junction-less transistor for a biosensor application," Sensing and Bio-sensing, Elsevier, vol. 18, Pages 31–36, April 2018.

[32] G. Saiphani Kumar, Amandeep Singh, and Balwinder Raj, "Design and analysis of gate all around CNTFET based SRAM cell design," Journal of Computational Electronics, Springer, vol. 17, no. 1, pp 138–145, March 2018. (SCI) IF=1.52.

[33] J. Rajendran, G. Rose, R. Karri, and M. Potkonjak, "Nano-PPUF: A memristor-based security primitive," in Proceedings of IEEE Computer Society Annual Symposium on VLSI, 2012, pp. 84–87.

[34] Jeetendra Singh and Balwinder Raj, "An accurate and generic window function for non-linear memristor model," Computational Electronics, Springer, vol. 18, no. 2, pp. 640–647, 2019.

[35] B. Sunar, W. J. Martin, and D. R. Stinson, "A provably secure true random number generator with built-in tolerance to active attacks," IEEE Transactions on Computers, vol. 56, pp. 109–119, 2007.

[36] Nor Hashim et at., "Implementing memristor in ring oscillators based random number generator," IEEE Student Conference on Research and Development (SCOReD), 2016.

[37] Gurinder Pal Singh, B. S. Sohi, and B. Raj, "Material properties analysis of Graphene Base Transistor (GBT) for VLSI analog circuits," Indian Journal of Pure & Applied Physics (IJPAP), vol. 55, pp. 896–902, December 2017 (SCI).

[38] Neeraj Jain and B. Raj, "Impact of underlap spacer region variation on Electrostatic and analog/RF performance of symmetrical high-k SOI FinFET at 20 nm channel length," Journal of Semiconductors (JoS), IOP Science, vol. 38, no. 12, pp. 122002, Dec 2017; H. Manem, J. Rajendran, and G. S. Rose, "Design considerations for multilevel CMOS/Nano memristive memory," Journal on Emerging Technologies in Computing Systems, vol. 8, no. 1, pp. 6:1–6:22, 2012.

[39] X. Wang and Y. Chen, "Spintronic memristor devices and application," in Proceedings of IEEE/ACM Design, Automation & Test in Europe Conference & Exhibition, 2010, pp. 667–672.

[40] Jeetendra Singh, Annuradha, Balwinder Raj, and Mamta Khosla, "Design and performance analysis of nano-scale memristor-based nonvolatile static random access memory," Sensor Letters, American Scientific Publishers, vol. 16, no. 10, pp. 798–805, 2018.

[41] Neeraj Jain and B. Raj, "Device and circuit co-design perspective comprehensive approach on FinFET technology – A review," Journal of Electron Devices, vol. 23, no. 1, pp. 1890–1901, 2016.

[42] Gurmohan Singh, R. K. Sarin, and B. Raj, "A novel robust exclusive-OR function implementation in QCA nanotechnology with energy dissipation analysis," Journal of Computational Electronics, Springer, vol. 15, no. 2, pp. 455–465, June 2016.

[43] H. Manem, J. Rajendran, and G. Rose, "Stochastic gradient descent inspired training technique for a CMOS/ nano memristive trainable threshold gate array," IEEE Transactions on Circuits Systems I: Regular Papers, vol. 59, no. 5, pp. 1051–1060, 2012.

[44] B Raj, A. K. Saxena, and S. Dasgupta, "Nanoscale FinFET based SRAM cell design: Analysis of performance metric, process variation, underlapped FinFET and temperature effect," IEEE Circuits and System Magazine, vol. 11, no. 2, pp. 38–50, 2011.

[45] Aakash Jain, Sanjeev Sharma, and B. Raj, "Design and analysis of high sensitivity photosensor using cylindrical surrounding gate MOSFET for low power sensor applications," Engineering Science and Technology, an International Journal, Elsevier's, vol. 19, no. 4, pp. 1864–1870, December 2016.

[46] G. Khedkar and D. Kudithipudi, "RRAM motifs for mitigating differential power analysis attacks (DPA)," in Proceedings of IEEE Computer Society Annual Symposium on VLSI, 2012, pp. 88–93.

[47] B. Cambou, "Match-In-Place: A Novel Way to Perform Secure and Fast Users Authentication, 2014." [Online]. Available: www.crocustechnology.com/pdf/Crocus_MIP_White_ Paper_v6.

[48] The Embedded Security Challenge, 2014. [Online]. Available: http://isis.poly.edu/esc/.

[49] Chaofei Yang et al., "Security of neuromorphic computing: Thwarting learning attacks using memristor's obsolescence effect," ICCAD'16, November 07–10, 2016.

[50] S. Saini et al., "Secure communication using memristor based chaotic circuit," International Conference on Parallel, Distributed and Grid Computing, 2014.

[51] H. Fujisaka and T. Yamada, "Stability theory of synchronized motion in coupled- oscillator systems," Progress of Theoretical Physics, vol. 69, no. 1, pp. 32–47, 1983.

[52] Karmjit Singh and B. Raj, "Comparison of temperature dependent performance and analysis of SWCNT bundle and copper as VLSI interconnects," Research Cell: An International Journal of Engineering Sciences, vol. 17, no. 1, pp. 583–595, January 2016.

[53] Sunil Kumar and B. Raj, "Analysis of ION and ambipolar current for dual-material gate-drain overlapped DG-TFET," Journal of Nanoelectronics and Optoelectronics, American Scientific Publishers, USA, vol. 11, pp. 323–333, June 2016.

[54] Sunil Kumar, Sumit Kumar, Karamveer, Keshav Kumar Gupta, and B. Raj, "Analysis of double gate dual material TFET device for low power SRAM Cell design," Quantum Matter, American Scientific Publishers, vol. 5, pp. 762–766, December 2016.

[55] Raghav Aggarwal, B. Raj, Lakhwinder Singh Teji, and Ritesh Kumar, "Analysis of leakage current and power dissipation in 5T SRAM Cell," Quantum Matter, American Scientific Publishers, vol. 5, pp. 792–797, December 2016.

[56] S. Khandelwal, B. Raj, and R. D. Gupta, "Design and optimization of 6T FinFET SRAM in 45 nm technology," Quantum Matter, American Scientific Publishers, vol. 5, pp. 552–556, August 2016.

[57] Sunil Kumar and B. Raj, "Compact channel potential analytical Modeling of DG-TFET based on Evanescent-mode approach," Journal of Computational Electronics, Springer, vol. 14, no. 2, pp. 820–827, July 2015.

[58] Karmjit Singh and B. Raj, "Temperature dependent modeling and performance evaluation of multi-walled CNT and single-walled CNT as global interconnects," Journal of Electronic Materials, Springer, vol. 44, no. 12, pp 4825–4835, December 2015.

[59] S. Khandelwal, B. Raj, and R. D. Gupta, "FinFET based 6T SRAM cell design: Analysis of performance metric, process variation and temperature effect," Journal of Computational and Theoretical Nanoscience, ASP, USA, vol. 12, pp. 2500–2506, 2015.

[60] Sumit Singh, Shekhar Yadav, Jagdeep Rahul, Anurag Srivastava, and Balwinder Raj, Impact of HfO2 in graded channel dual insulator double gate MOSFET," Journal of Computational and Theoretical Nanoscience, American Scientific Publishers, vol. 12, no. 6, pp. 950–953, April 2015.

[61] Vijay Kumar Sharma, Manisha Pattanaik, and Balwinder Raj, "PVT variations aware low leakage INDEP approach for nanoscale CMOS Circuits," Microelectronics Reliability, Elsevier, vol. 54, pp. 90–99, 2014.

[62] V. K. Sharma, M. Pattanaik, and Balwinder Raj, "INDEP approach for leakage reduction in nanoscale CMOS circuits," International Journal of Electronics, Taylor & Francis, vol. 102, no. 2, pp. 200–215, 2014.

[63] Ashvini Kumar Dogre, Manisha Pattanaik, Balwinder Raj, and Arti Naik, "A novel switched capacitor technique for NBTI tolerant low power 6T-SRAM cell design," Journal of VLSI and Signal Processing (IOSR-JVSP), pp. 68–75, vol. 4, no. 2, 2014.

[64] Hardik Vaghela, Mamta Khosla, and Balwinder Raj, "Ambipolar effect free double gate PN diode based tunnel FET," World Academy of Sciences, Engineering and Technology, International Journal of Electronics and Communication Engineering, vol. 1, no. 11, 2014.

A State-of-the-Art Study on Physical Unclonable Functions for Hardware Intrinsic Security

7

Vivek Harshey, S. K. Bansal, and Devendra Chack

Contents

DOI: 10.1201/9781003126645-7

7.1 HISTORICAL BACKGROUND AND PRELIMINARY CONCEPTS

Gustav Simmons et al. [1] first conceptualized and used the properties of unclonable structures for paper authentication by using object identification. Their concept was to allow the paper's authentication by third parties where public-key cryptography was combined with derived fingerprints. However, the first persons to elaborate on the idea of fingerprinting the unclonable structures or random variations called PUF and to subsequently define the terminology were Pappu et al. [2].

Some earlier work in the direction of authentication utilized random arrangement of the physical nature of the material used for the manufacturing of currency, for example, small optical fiber, which could be instrumental in authenticating the same (i.e., currency) [3]. However, the first practical microelectronics circuit that appeared in 2002 [4] triggered researchers' interest.

As we witness the increasing complexity of computing devices, it poses a grave challenge for designing a robust security system. The current state-of-the-art electronic devices may incorporate roughly a few billion transistors. Providing security to such complex systems could easily become intractable. Therefore, the need arises to ensure that various subcomponents of modern design and their frequent interactions work precisely in the same manner as they are intended to do.

This validation of modern devices' security becomes a gray area and, therefore, prone to security attacks. People of *mala fide* intention may take advantage of such weak links. For example, someone can get access to an Internet modem and can save monthly subscription money. Also, adversaries can hack into your bank account details or credit card details and dent your wealth or cripple financial savings. Criminals with evil motives may cut into the smart health care system. They may increase or decrease the dosage of drugs beyond a permissible limit, therefore endangering the life or attacking intelligent vehicles' security and pose a threat to life.

The physically unclonable function that harnesses the random inherent disorder or entropy of the real world may thwart such sinister goals. Such systems are also useful for generating random numbers, and the intrinsic randomness also helps in designing more robust systems than the traditional ones. Random process variations in the fabrication of integrated circuit chips are the most sought-after physical randomness extracted in chip-unique signatures and lead to PUF architecture [5].

PUF instances are provided with challenges, inputs, and natural random variations that generate virtually unique *responses* as outputs against these inputs. This uniqueness of the challenge–response pair is analogous to the unique fingerprints of us humans. The real world's intrinsic noise or entropy works as a seed in generating a secure key for authentication scenarios. The creation of this key is instantaneous, as the system is powered up and the challenge–response pair is generated. Thanks to the intrinsic random nature of PUFs, the secure essential functions are, in principle, considered to be unclonable and hence provide credible strength against security attacks [6].

7.2 THE NECESSITY FOR HARDWARE PROTECTIONS

The physical layer or the hardware layer is the first defense mechanism against mounting security attacks (Figure 7.1). This layer defines the interface and mechanism for putting data bits on a network medium. The hardware layer is also fundamental to the security of higher-level functions and is the first target of hardware attacks. Hence, understanding the attacker's modus operandi at this layer may help successfully mitigate the severity of the attack and prevent hackers from accessing secure information at higher-level layers.

7.2.1 Physical Attacks on IoT Devices

As per the latest trend, billions of computing devices are connected to the Internet and give impetus to technology. These IoT devices are pervasive and found everywhere around us, in smartphones, smartwatches, human bodies, baby monitors, body implants, in our home automation, autonomous cars, and

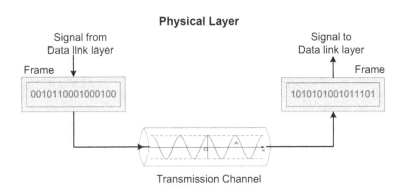

FIGURE 7.1 PUF is utilized in the physical layer.

medical implants. IoT devices' ubiquitous existence facilitates an adversary to succeed in securing the possession of a chip and, therefore, mount a systematic, targeted attack or reverse engineer its design and extract confidential information or plot a virus.

7.2.2 Hardware Trojans

Some other factors are also responsible for making the security of hardware prone to attack. Today's semiconductor chips' production is carried out at various locations, geographically separated from the outsourcing of each step's economic viability, and keeps the overall fabrication cost reasonable [7]. Because of this multinational distributed business model, we have access to cheap smartphones, PCs, tablets, and various other electronic devices. The foundry that fabricates the chips may counterfeit the ICs and overproduce the chips [8]. The pirated information, for example, the trade secret of initial design and layouts of the chips, may be sold to interested parties, making the chips susceptible to further security attacks.

The distributed nature of the integrated circuits supply chain offers competitors or enemies ample chances to manipulate through original design and add harmful features (Figure 7.2). Some recent reports justify these suspicions, for example, a parasitic antenna to leak information or a kill switch to disable the chip remotely. The complexity of IC manufacturing's resource chains makes it easy for a hacker to implant hardware Trojans in any step of manufacturing based on the adopted method. [6]

7.2.3 Hardware-Assisted Authentication System

Secure hardware tokens are examples of hardware-assisted authentication used for a range of security-sensitive applications, for example, smart cards, cell phones, and PDAs. Also, hardware-assisted client authentication is used by various financial institutions across the globe to carry out bona fide transaction utility bill payments, loan repayment, money transfers, and balance checking. These hardware tokens are portable devices and easy to carry and help confirm beneficiaries' or clients' identities. These hardware-assisted authentication systems also find digital signature usage to establish a financial deal's authenticity, anonymous transactions, private data mining, etc. But their security protocol is not robust and is prone to attack [6].

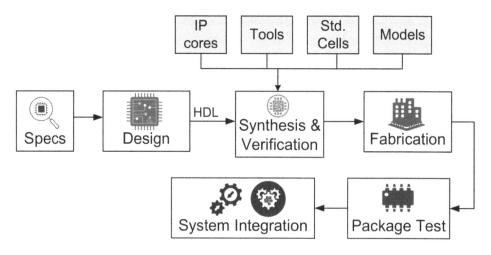

FIGURE 7.2   IC manufacturing supply chain.

7.3 A SUMMARY OF ATTACKS

Before moving on to the next section about fundamental principles and the working of PUF circuits and their properties, it is imperative first to understand the various kinds of attacks and threats that need to be defended against. Prepared with this information about PUFs, it will be clearer how to enforce specific security measures and why it is essential to implement particular measures. There are various forms of attack to which an IoT device, a hardware chip, or a computer system may be exposed, and some of which will be covered in this section. The insights into these attacks' tactics help us outline the steps crucial to secure hardware devices and network systems.

7.3.1 Replay Attacks

A resourceful intruder can quickly attack a communication channel link to obtain unauthorized access to classified information and therefore be able to influence transmitted data. A legitimate transmission of the data is repeated or delayed maliciously or fraudulently in a playback attack (replay attacks) without the customers' knowledge [6, 11].

7.3.2 Programming Attacks and Machine Learning Attacks

These attacks target the computing device's software and maliciously modify it or install programs that exploit a vulnerability in design and cause a system failure. An adversary may gain remote entry to a computer device when it is connected to public networks. Machine learning attacks are those techniques that attempt to fool hardware models or computing devices by supplying deceptive inputs. The most common reason for a programming attack is to cause a malfunction in IoT devices or hardware systems.

7.3.3 Hardware Incidents

These attacks, in theory, can occur at any point throughout the life span of IC chips. When chips are transferred to a foundry for production, attacks can occur there (see Figure 7.3). The noteworthy aspect of hardware attacks related to the security of embedded systems and portable hardware is an extremely

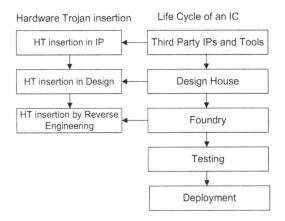

FIGURE 7.3 Hardware Trojan attack during the IC lifecycle.

hostile environment in which the hardware is used. These attacks aim at maliciously altering or interfering with a computer device's physical implementation.

Such attacks consist of snooping by gaining access to protected memory and can only be performed by directly accessing the design files or hardware. A small component in the overall system could be modified by a malicious individual aiming for espionage or sabotage. These attacks can be especially devastating in industries such as the military, which is security-critical.

Based on the life cycle of a hardware device, it can also be prone to attacks post-fabrication.

1. Invasive attacks.
2. Noninvasive physical attacks.
3. Semi-invasive attacks.

An adversary who may damage the chip will access hardware and manipulate the internal components that can also carry out an intrusive or invasive attack. For this to occur, he should move the accessed hardware or IoT device to an expensive lab where specialized pieces of equipment are available. However, the very nature of the attack, that is, moving the chip to a technical lab, makes it less feasible for IoT devices. Since many tools are available in public places, these devices cannot be brought into the laboratory, because bringing these publicly accessible devices to the specialized laboratory is out of the question.

In contrast, the main motive behind the noninvasive attack is to uncover secret information without accessing the internal components directly. For such an attack to happen effectively, the dominant approach is to utilize machine learning techniques. These are sophisticated algorithms that can guess the response for a given challenge with a high success rate [12, 13, 14]. Another type of noninvasive attack is side-channel attacks (SCA), which rely on the efficacy of forensic techniques to extract information about the embedded hardware's physical implementation (Figure 7.4) [15].

In between the extremes of these two approaches lies the third attack method, which is a semi-invasive technique where, without damaging the layers of the chip, a skillful adversary gets access to the chip surface. Electromagnetic probing, laser fault injection, contactless optical probing, and photonics emission are some of the reported semi-invasive PUF attack techniques. [6]

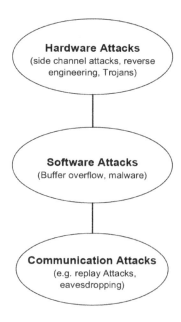

FIGURE 7.4 Various forms of attack on embedded devices [6].

7.3.4 PUF-Based Threats on Hardware Devices

The PUF itself can be a tool for attackers if an ingenious enemy can replicate the challenge–response pair and hence mimic the presented challenge's accurate response. This replication of behavior is possible by anticipating CRPs with a machine learning attack with very high accuracy or cloning the PUF behavior. Such imaginative attack techniques pose a significant challenge toward designing security systems and require creative solutions to mitigate the risk.

7.4 PHYSICALLY UNCLONABLE FUNCTIONS

In manufacturing methods of physical devices, a random element presents certain constraints, making it difficult to produce tightly controlled devices. These limitations give rise to small variations in the sizes and properties of the fabricated items. These effects become more pronounced in deep submicron technology nodes, that is, below 90 nm [14, 15]. Fabricating two PUF systems with identical physical parameters is impossible due to intrinsic random fluctuations. The main advantages of PUFs are that they have easy-to-extract digital fingerprints, are inexpensive, and do not require nonvolatile memories such as EEPROM or a nonvolatile static random access memory (nvSRAM).

The fundamental characteristic of PUFs is that there is not any storage for keeping the secret key. Instead, the PUF extracts the key from the features and attributes of the device. Temperature, humidity, and air pressure affect characteristics for each PUF.

A physically unclonable function (PUF) takes advantage of intrinsic fluctuations in device parameters. It maps them to a device's unique physical signature, which can be considered a fingerprint of that device. There are different ways to realize a PUF, and silicon-based integrated circuits are one of them. For example, a ring oscillator implemented on a chip produces a different frequency for that circuit and provides a unique fingerprint to that chip. PUFs do not need any external source or battery resource; however, RAMs such as SRAMs require some external energy source to store the secret key.

The strength of PUFs is characterized by the complexity of reproducing their precise physical structure and hence differ from brute force approaches. Such security-related functionality specific to a PUF circuit can be considered a hardware-level security algorithm and can be termed a cryptographic primitive. The various cryptographic primitives are classified based on their usage and functionality and are the basic building blocks to design secure systems. Some popular cryptographic primitives are shown in Figure 7.5.

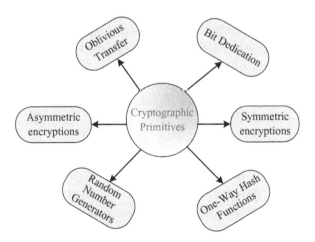

FIGURE 7.5 Conceptual representation of cryptographic primitives [6].

7.5 PROPERTIES OF PUFS

Many properties are desirable for a PUF design. The PUF's performance is judged on these properties, and it also tests its consistency and suitability for a specific use. The uniformity in the randomness of the PUF responses is critical for its utility in various applications. Some of the desired properties of the PUF are enumerated in the following subsections.

7.5.1 Unclonability

This property is the primary desirable property of the PUF and states that it should be impossible to manufacture a PUF circuit replica. Also, mathematically, it should be intractable to model its behavior. This makes it very difficult to clone its functionality, so the responses cannot be simulated [16].

7.5.2 Reproducibility

Reproducibility is described in terms of spreading the intra-distance responses of the whole PUF category, that is, taking into account unexpected challenges on arbitrary PUF cases. It essentially implies that a particular PUF must produce a similar response with a high probability of the same challenge. Expressed in terms of hamming distance means that the two reactions for similar inputs should be very close. The number of bit positions in which the two binary strings differ is termed as hamming distance.

7.5.3 Reliability

This measures the capacity of the PUF to produce a steady output for a given bitstream, irrespective of any changes in environmental parameters (Figure 7.6), viz. voltage supply, surrounding temperatures, etc. This property can be measured by evaluating intra-chip hamming distance.

7.5.4 Exclusivity

A PUF circuit or device should generate exclusive or distinct identification, and this property is defined as the exclusivity or uniqueness of PUF. When two chips are implemented with the same PUF structure, each instance of PUF compared with other PUF should have a unique signature. Talking simply, this means,

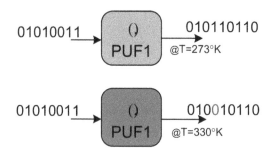

FIGURE 7.6 An example of a reliability evaluation of a PUF design.

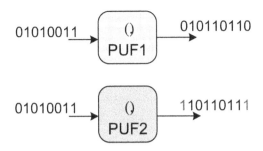

FIGURE 7.7   An example of uniqueness evaluation of a PUF design.

for example, if two RO PUFs are implemented on two different chips, they should produce distinctive responses.

For Figure 7.7, two RO PUF circuits on two different chips generate exclusive responses for the same input pattern. A value of uniqueness close to 50% is considered ideal. When a bit string (011001) is presented to each PUF circuit, they yield a different output. Here two outcomes differ in two bits, and therefore their Hamming distance is 2. This means approximately one-fourth of bits are in uncertainty.

7.5.5 Withstand Manipulation

This property signifies that if an attacker gains access to the PUF devices and invasive attack methods tamper with the PUF construction, then the response of PUF is distorted and therefore destroys the PUF functionality, rendering it useless. This property characterizes the ability of a PUF design to withstand tampering and accordingly change its behavior. This property can also be measured in terms of hamming distance mathematically [6].

7.6 CLASSIFICATION OF PUF

The PUF circuit classification can be based on the source of randomness, vulnerability to the presented challenge, manufacturing method, and novelty of the design. Some of these PUF categories are presented in this section.

7.6.1 Electronic PUFs

Among the various designs of PUF reported in the literature, the prevailing class of PUF design is intrinsic PUF or electronic PUF. Due to its easy process of fabrication for embedded hardware, it can be integrated with existing processing steps. This design is based on the implicit randomness in the fabrication of electronics hardware. We may further classify them into PUFs based on delay properties, or delay-based PUFs, and bistable PUFs. Those semiconductor circuits that have two stable states in circuit operations can serve as the SRAM cell. The second category of circuits exploits unique variations of delay differences in adjacent matching transistors or symmetric wires and logic gates on the chip. Ring oscillator PUF and arbiter PUF are two famous delay-based PUF architectures and can also be implemented in field-programmable gate array (FPGA), compared to custom hardware ASICs.

However, well-planned machine learning attacks can successfully break into the defenses of an arbiter PUF. We may make them resistant to such attacks by adapting them to the XOR arbiter PUF design.

However, they have also shown vulnerabilities against a well-planned machine learning attack in recent studies [17, 18]. It is typically understood as a latch and emits a one or a zero, based on the first step. According to initial inputs, an arbiter PUF can also produce a one or a zero, depending on which adjustment comes first. When a circuit of the identical configuration mask is manufactured on various chips, the implemented logic function is different for every chip because of random variation in delay.

7.6.2 Weak and Strong PUFs

PUFs can be shaped as a black-box challenge–response system. It means that similar to a function, there is an input for the PUF, which is called challenge C, and based on the function $f(C)$, response R will be calculated. In this black-box, $f(.)$ explains the relationship between challenge and response. The domain of $f(.)$ and the number of challenges that each PUF can process are the leading difference between strong and weak PUFs [19].

In cryptographic functions, key generation can be satisfactorily achieved by weak PUF design, which has a limited number of CRPs. Instead, a more stringent requirement of authentication is fulfilled by an exponential number of such unique challenge–response pairs (CRPs), which is typical of strong PUFs. A key generation algorithm utilizes a seed or a key generated by the PUF's response to a fixed challenge. In cryptographic applications that require a secure key, the PUF's response can be used as the key [6].

Weak PUFs, as well as strong PUFs, should exhibit uniqueness and credibility to secure their properties [20]. An example of the weak PUF is the SRAM PUF, and optical PUFs and arbiter PUFs fall under the strong PUF category. The ring oscillator PUF and arbiter PUF designs enjoy extensive challenge–response domains. On the other hand, the SRAM PUF possesses limited challenge–response pairs. The key generation often requires the use of a fuzzy extractor, which combines error correction with hash-based entropy amplification.

Unfortunately, most stable implementations of PUFs are vulnerable to attack, since by observing a relatively limited number of CRPs, an adversary can construct a machine-learning model, making it possible to predict the performance response of the PUF to a future challenge. The possibility of the creation of such a model is considered a breach of security.

In physically unclonable functions with limited challenge–response pairs, the chip must never be left in the safe. In the case of dynamic challenge–response use, the challenge–response pairs determined during enrollment must be shielded as secrets but not stored on the chip. In both cases, where necessary, the unknown can be estimated from the PUF and then erased from volatile memory. This erasing of transient memory significantly decreases the vulnerability of the key to attackers and represents a substantial improvement in hardware security [21].

7.6.3 Reconfigurable PUFs

The classical implementation of PUF is fixed and cannot be adapted or configured dynamically. If someone knows the challenge–response routine of the initial PUF design, then an attack can be launched, and therefore the security provided by the PUF is compromised. The solution is suggested in [22], and a reconfigurable PUF (rPUF) is built in the FPGA logic such that a programmed function adjusts the input–output relationship of the PUF unpredictably and uncontrollably. This way, the PUF reconfiguration is hard to restore, even by an intrusive attack. There are two ways in which we can reconfigure the PUF system – physical reconfiguration and logical reconfiguration.

To better utilize the resources in terms of usage and improve their efficiency, a better design is proposed [23]. However, such methods are unproductive in terms of resource utilization. The entropy sources, in terms of logic and routing resources, can now be effectively used. The entropy extracted from a logic block is maximized by way of modern FPGAs' partial reconfiguration ability. This complex reconfiguration without altering the overall structure of a PUF circuit is still considered a desirable feature for crucial applications such as revocation and regeneration.

7.6.4 Erasable PUFs

The erasable PUF design is motivated by a reconfigurable PUF design. In an RPUF, each CRP is not refreshed independently, and such a system is not a strong PUF [24]. The erasable PUF's CRP can be independently revived or reconfigured, which makes it a strong PUF design and robust against reuse model attack. It also widens its range of uses and hence is more versatile than an exclusive reconfigurable system. A SHIC-PUF-based first erasable PUF design was proposed in 2011 and appreciated [25]. However, this design is still plagued by many flaws, and the development of an inexpensive and functional erasable PUF continues to be a significant field of study.

7.6.5 Novel Architectures

In addition to traditional architecture, recently, novel PUF architectures have been proposed, including transient effect ring oscillator (TERO) PUFs (Figure 7.8). The basic building block of the TERO PUF [26] is the TERO cell, which is characterized by a stable state and a transient oscillating state. It can be thought of as a RO PUF. An RSlatch is utilized in TERO PUFs, and the circuit starts oscillation, setting the initial signal to one for a short period.

The public PUF (PPUF) is an alternative development of arbiter PUFs that utilizes XORs. PPUF systems have good area effectiveness, side-channel attacks immunity, and low power utilization and are potential candidates for general key protocol implementations for the IoT [28]. PPUFs and arbiter PUFs based on hybrid ring oscillators are also recently proposed PUF architectures suitable for use in IoT applications. They have reportedly low power consumption, making them a powerful candidate for battery-powered IoT devices. PPUFs are yet another evolution of PUF arbiters. PUFs based on the current array are also designed to withstand machine learning attacks and have been reported in the literature [29, 30].

TABLE 7.1 Desirable PUF Properties in Various PUF Architectures

COMPARISON OF PROPERTIES OF VARIOUS PUF	OPTICAL PUF	ARBITER PUF	RING OSCILLATOR PUF	SRAM PUF	LATCH PUF	BUTTERFLY PUF
Uniqueness	✓	✓	✓	✓	✓	✓
Reproducibility	✓	✓	✓	✓	✓	✓
Physically Unclonable	✓	✓	✓	✓	✓	✓
Unpredictability	✓	–	–	✓	✓	✓

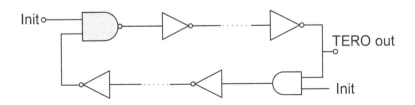

FIGURE 7.8   TERO PUF architecture [27].

7.6.6 Nanotechnology-Based PUF

Today, almost all designs of PUF concentrate on the utilization of random fluctuations of process parameters inherent to CMOS technology. Substantial improvement has been made recently in emerging nanoelectronics devices [32], particularly for nano-elements like CNFETs and resistive RAM, spin-transfer-torque magnetic random-access memory (STT-MRAM), etc. These innovations are evolving for memory construction, using other storage mechanisms, such as magnetism and memresistance, instead of maintaining the load on the capacitor, such as flash memory, dynamic RAM, and static RAM [31, 33].

7.7 A DETAILED STUDY OF SOME POPULAR PUFS

After understanding the various attack scenarios and gone through a handful of PUF designs, we are ready to dig deeper. Here we provide a detailed overview of some popular PUF designs.

7.7.1 Arbiter PUFs

A VLSI (very large scale integration) design-based APUF [4] was introduced in 2002 (Figure 7.9). The PUF design exploits fabrication randomness resulting in random time delays and interconnects for the gate transistors. This PUF design is uncomplicated and dense and possesses an extensive CRP space. However, the APUF design is vulnerable to multiple simulation attacks, unlike its counterpart, because of its linear additive structure, the optical PUF. Here an intruder uses documented CRP findings to construct an incorrect CRP behavior model [34]. While the new APUF variants [35, 36] can withstand numerous modeling threats, the PUF security race is unlikely to end. In other words, there might be other modeling methods of attack that would degrade or bypass the protection of the new proposed APUF variants.

Lim et al. [37] implemented PUFs that used a distinctive structure and, additionally, an arbiter to differentiate the gap in the delay among the directions. The specific design of the PUF system makes it possible for these PUFs to be more resilient to surrounding conditions. The same has also been implemented in FPGA. The APUF is the primary type of time-delay-based PUF but requires careful circuit layout. For example, manual tuning of delay wires is typically needed when APUF is implemented on field-programmable gate array (FPGA) platforms to minimize response bias.

The arbiter PUF families take advantage of the small delay discrepancies between the two symmetrically constructed chip paths to produce Boolean outcomes. A single arbiter PUF consists of several cascaded stages and an arbiter to produce an output at the end; see Figure 7.10. This construction comprises

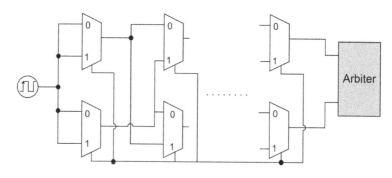

FIGURE 7.9 Arbiter PUF architecture [38].

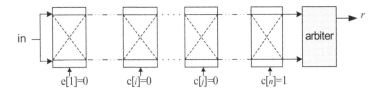

FIGURE 7.10   Arbiter PUF construction details.

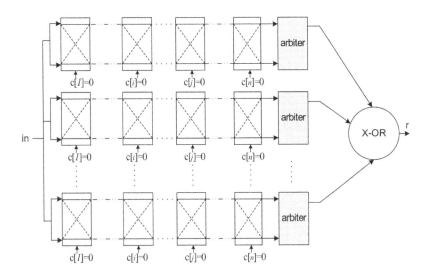

FIGURE 7.11   XOR arbiter PUF schematic.

two signal inputs, two signal outputs, and a single input challenge. The arbiter PUF's first stage inputs are linked to a standard activation signal.

Here, two signals scatter down two similar paths to the end of the chain, providing an electrical pulse as an activating signal. The signal propagates within a stage via the crossed paths if the input to the challenge is set to one. Then it uses direct routes. Though the cumulative delays for natural paths and crossed paths are identical, the propagation gap on one of the tracks can be longer or shorter due to chip limitations. Therefore, various problems result in varying delays in propagation at the last stage outputs.

An arbiter PUF's primary physical protection presumption is that an attacker cannot calculate the internal delays inside the Arbiter PUF stages without breaking the PUF itself, i.e., modifying its challenge–response conduct. The intruder can only attempt challenges from an infinite space in this scenario and observe the respective answers.

For the authentication scenario, attractive candidates are the arbiter PUF circuits because of their comparatively broad CRP space. However, arbiter PUFs are not safe from machine learning (ML) attacks and demonstrated vulnerability to such attacks. The arbiter PUF, based on the XOR logic function, was implemented as a defense to inhibit the efficacy of ML attacks. The XOR arbitrator PUF has k parallel arbitrator chains, each with an N stage and an arbitrator. See Figure 7.11. XOR-ing the responses of all individual arbiter chains produces the combined binary response. Two independent linear arbiter PUFs can be XOR-mixed to enforce improved statistic properties of an arbiter PUF. [30]

7.7.2 Ring-Oscillator PUF

An alternative delay-dependent PUF design that exploits the inherent temporal spacing on the chip is a ring oscillator PUF or RO PUF [43] (Figure 7.12). An odd number of serially connected inverter gates

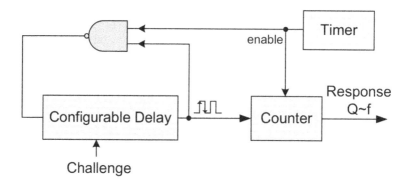

FIGURE 7.12   Construction of a simple ring oscillators PUF as proposed by Gassend et al.

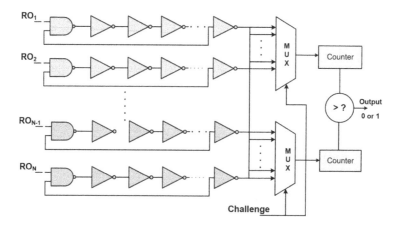

FIGURE 7.13   An alternative design of ring oscillator PUF [6].

makes the ring oscillator circuit, and to further enable or disable the circuit optionally, an AND gate is also included. RO PUF consists of an individual ring oscillator with the same number of gates, where all ring oscillators are connected to multiplexers; see Figure 7.13.

The vulnerability to environmental factors is one of the shortcomings of ring oscillator PUFs. For example, if the RO pairs have adjacent frequencies, the flipping of a response bit can occur due to variations in temperature (see Figure 7.13).

7.7.3 Optical PUFs

An optical PUF, which also has been called a one-way physical function, consists of a translucent substance that is doped with particles that emit light. A spontaneous and unusual speckle pattern can appear as a laser beam shines upon the material. Placing the particles dispersing light is an unregulated mechanism, and the contact between the particles and the laser is very complicated. It is also tough to reproduce the optical PUF so that there would be an identical spatter configuration.

Pappu et al. [2] used a bubble-filled transparent epoxy wafer and shined a laser beam through it, resulting in a pattern of interference reaction. Here the aim was to establish a one-way mechanism by practical means instead of numeral theory. A two-dimensional spatter shape was created under lighting at a given viewpoint from intricate interference within an inhomogeneous transparent plastic token, captured by a charging-coupled device camera (Figure 7.14).

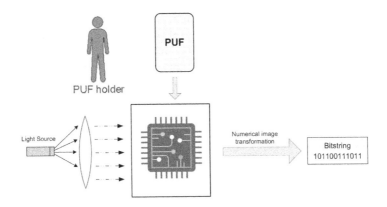

FIGURE 7.14   Conceptual design of an optical PUF.

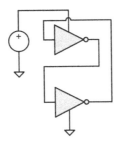

FIGURE 7.15   SRAM cell.

The angle is treated like the optical PUF's question (input) with the one-way physical feature, while the speckle configuration is treated as the solution (output). A CRP may be placed and then used to determine an optical PUF's validity.

Since the 2D speckle pattern relies on random inhomogeneities, copying or counterfeiting the optical PUF is impossible. Validation can be performed remotely until the CRP is registered only by the responsible party (server) in a secure database. So the optical PUF has an extensive area for CRP. If the hamming distance between responses is below a predetermined threshold, and thus the incorporated optical PUF entity's validity is identified, the received response is compared to the lodged one; otherwise, validity is denied. Each CRP is used only once to prevent replay-based attacks [24].

7.7.4 SRAM PUFs

A CMOS SRAM cell is a six-transistor unit made up of two cross-coupling inverters and two control transistors connecting the word-line signal to the data bit lines. The transistors that form the cross-coupling inverters are made incredibly weak to allow them to be powered easily to 1 or 0 during a write phase. An SRAM cell is a cross-coupled connection of two inverters and may be implemented in CMOS technology (Figure 7.15). Due to this cross-coupled connection, there is a metastable state in the SRAM cell characteristics. The positive feedback of this circuit may push this metastable state to any of the two stable states (1 or 0) by amplifying the random noise present at the time of powering up. An SRAM cell's startup activity depends on the disparity of its transistor threshold voltages. Even the slightest variations would drive the SRAM cell to one of the two stable states.

This randomness of powering up behavior will produce different stable states (1 or 0) for every available SRAM cell independently of each other. An individual SRAM cell will latch to the same condition for every power-up with relatively high probability. Provided that every SRAM cell has its preferred form any time it is driven, the SRAM response yields a unique and random sequence of zeros and ones. This pattern is like the fingerprint of a chip because it is unique to a certain SRAM.

Since the SRAM PUF can be attached directly to the regular circuitry implemented on the same hardware, these PUFs are instantly installed in cryptographic applications as a hardware component, presenting themselves as a preferred candidate for security solutions. One may regard many memory locations within an SRAM memory block as a task for SRAM-PUFs. A further advantage of these PUF is that a stream of 0 or 1 is produced at the output when presented with a challenge. However, to obtain a bitstream from other PUF implementations, a lengthy analog to digital conversion process must be followed.

7.7.5 Memory-Based Intrinsic PUFs

At the time of manufacturing, the APUF circuit is not embedded in a hardware device and therefore is not considered an intrinsic PUF. Because of this, to design an APUF, we require additional layout and design steps. However, an SRAM PUF is deemed to be an intrinsic PUF and was first introduced in 2007.

The SRAM cell address within the schematic is treated as a challenge, and the response is derived from its power-up state (Figure 7.16). Due to their pervasiveness in commercial electronic products, DRAM PUFs, SRAM PUFs, and flash memory-based PUFs are more convenient in practice [39, 40]. However, the primary limitation of a memory-based PUF is that its challenge–response pair size is limited and, therefore, cannot be used in public CRP-based authentication applications. Simultaneously, its application in the generation of an on-demand volatile cryptographic key is a direct consequence of its preferred freedom among response bits.

7.7.6 Other Memory-Based PUFs

The definition of the butterfly PUF was suggested in the literature [41]. In the FPGA matrix, structures are constructed such that they mimic the behavior of the startup process of an SRAM cell. Using the latches accessible in the FPGA matrix, a cross-coupled configuration is utilized. This design causes an excitation

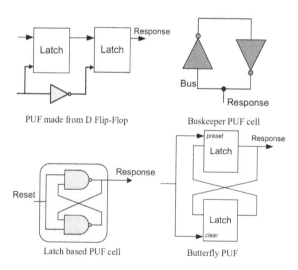

FIGURE 7.16 Bistable memory elements based on various PUF designs.

signal to create an unstable state, settling down to a stable condition due to positive feedback. Flip-flops-based PUF designs have also been suggested in the literature [42].

7.8 APPLICATIONS OF PUF

Nowadays, PUF circuits are ubiquitous and widely used for various security applications. Here, some of the famous and well-known applications of PUF designs are considered.

7.8.1 Real-World Applications

Highlights for manufacturers and customers are the introduction of IoT technologies in autonomous cars, smart homes, etc., applications, the maintenance of protection and trust, privacy protection, and the fight against counterfeits. Various proven cryptographic algorithms have been proposed to resolve these issues, both of which depend on the assumption that the key must be kept secure. Several distinctive applications are considered in the following subsections [43].

7.8.2 Symmetric Ciphers

This is a pseudorandom function designed on hardware, which internally utilizes PUF building blocks, is successfully used by the researchers, and is termed as PUF-PRFs [44]. The abstraction used in cryptography to model symmetric ciphers is defined by the pseudorandom method. The PUF-based block cipher implemented allows for simultaneous defense against algorithmic and physical attackers against memory attacks.

7.8.3 Identification and Authentication

If there is a PUF with N challenges on the IC, the IC classification can be done on the basis of this PUF. This recognition mechanism comprises presenting the input challenges to the PUF first and then loading the answers in a secure archive, then rechallenging the PUF at authentication time and comparing recent responses to those previously collected. But there is a limited lifetime of this authentication protocol, which is based on challenge–response matching. The reason for this deficiency is that during the enrolment phase, only a small number of challenges can be measured and stored in the memory. Using more advanced key-based encrypted verification protocols (e.g., zero-knowledge protocols) will overcome this restriction. Such a system may be used to create a genuinely unclonable RFID tag. [47]

7.8.4 Secure Key Storage

Most cryptographic algorithms used in computer devices (e.g., encryption, authentication, etc.) necessitate that a key is generated and stored beforehand and acts like the root of trust. Typically, the key generation part is performed externally, and the device's memory is used for storage.

PUF technology offers another tactic to storage and generation of the key in which a root key is constructed by combining the responses of the PUF circuit. This method eliminates the necessity of on-chip memory storage of the generated key and delivers enhanced security to counter the attacks. Further, it also obviates the requirement of management of the keys by the manufacturers [6].

7.8.5 Key Storage and IP Protection

Additional processing steps are required to generate a private key from the response of the PUF. In particular, the PUF response needs to be converted into a private key. The method consists of applying a fuzzy extractor or practical data algorithm to a resounding PUF answer, resulting in robust and secure access. The hardware implementation of the aid data-based key extractor is defined in more detail in [45]. Symmetric and asymmetric key primitives have been used to implement hidden key storage in related IP security protocols and FPGA [9], [46].

7.8.6 Spurious Proof Design

Counterfeiting of integrated circuits is a known problem in IC chip design and incurs severe economic losses to IC manufacturers. A PUF, with its unique spurious proof design, can be used to curb the menace of counterfeiting IC chips and saves lots of money. The method of integrated circuit chip can be locked by utilizing a PUF circuit in an embedded chip. At the post-manufacturing stage, each PUF is characterized and verified by the IC chip manufacturers. Also, to activate authenticated chips, a passkey can only be generated by the design house.

7.8.7 Inexpensive Verification

The comparatively basic architecture of the PUF makes it an appealing alternative for low-cost authentication schemes, in which a PUF's unique challenge/response behavior is used as a physical object identity. It is particularly useful for resilient systems such as the Internet of things, which cannot afford ideal security solutions [6], [10].

7.8.8 Secure Hardware Tokens

Thanks to their complex challenge/response behaviors, PUFs can also be used in hardware-assisted cryptographic protocols. It can be achieved by making them intrinsically more resistant to manipulation than other tokens that rely on digitally stored information (e.g., smart cards and secure memory devices) [48]. Physical alteration or modification of the physical entity embedding $f(.)$ of PUF transforms it into $f'(.)$ of PUF so that f' is not equal with a high probability to f.

7.8.9 Safe Detection

PUF circuits are sensitive to fluctuations in environmental parameters such as temperature or supply voltage, which essentially means that a PUF's reaction relies on both the constraints imposed and the atmospheric parameters; this makes it possible to use the PUF as a monitor to quantify changes in environmental conditions.

An ideal PUF must have several characteristics [49], including unclonability, unpredictability, and deceptive proof. Although unclonability is about believing that a PUF's action is neither physically nor mathematically clonable, unpredictability is about the attacker's failure to anticipate the PUF's reaction by analyzing a series of challenge–response pairs (CRPs). Furthermore, tamper-evidence means that if the intruder performs a semi- or full-invasive attack against a PUF, the PUF's challenge–response behavior is altered with a high probability of degradation of the PUF and the loss of the hidden key or fingerprints.

7.9 PUF EXTENSIONS

The utility of the PUF circuit is not restricted to only the fields, as mentioned earlier. Here, we discuss the extended applications of PUF in allied areas.

7.9.1 POKs: Physically Obfuscated Keys

Gassend [50] introduced the idea of a physically obfuscated key or POK, and Bringer et al. [51] extended it to physically obfuscated algorithms. The only condition for a POK is that a key is permanently held in a 'physical' manner instead of a digital form, which makes it impossible for an attacker to discover the key from a poking attack. Besides, an intrusive attack on the computer that holds the key could break the key and render future use unnecessary, thereby presenting proof of wrongdoing. It is evident that POKs and PUFs are very relative terms, and Gassend [50] has already pointed out that POKs can be constructed from (tamper-evident) PUFs' versa.

7.9.2 CPUFs: Controlled PUFs

A managed PUF or CPUF, in conjunction with other (cryptographic) primitives, as proposed by Gassend et al. [52], is a mode of operation for a PUF. A PUF is said to be managed if it can only be reached by an algorithm that is inextricably connected to the PUF. Attempts should break the connection between the PUF and the access algorithm, leading to destruction.

7.9.3 RPUFs: Reconfigurable PUFs

Kursawe et al. implemented reconfigurable PUFs, or RPUFs [53]. The concept behind an RPUF is to expand a PUF's standard challenge–response actions with a further procedure called reconfiguration. This reconfiguration consequence is that the PUF's partial or total challenge–response activity is altered arbitrarily and preferably irreversibly, resulting in a new PUF.

7.10 CHALLENGES AND OPPORTUNITIES

Despite being referred to as 'unclonable functions,' various successful modeling and replicating incidents against PUFs have occurred [54]. Machine learning methods can model virtual ring oscillator PUFs and arbiter PUFs precisely beyond those designs' experimental strength. In this case, the machine learning models presume that the adversary can increase availability to several thousand challenge–response pairs. An attacker can obtain this information by various means, like robbing the list created for training the models during enrollment or equivalently measuring a system under control. Models will then forecast the answer to fresh inputs. It is necessary to remember, though, that a good attack on a PUF damages just the particular PUF instantiation. The intruder must restart the procedure of learning the solution to the next PUF instantiation.

These simulation attacks are not sufficient for PUFs that do not have an extensive challenge–response mechanism. The fuzzy extractors usually used in such PUFs are nevertheless subject to side-channel and modeling attacks. Researchers have shown the ability to mechanically clone SRAM PUFs by using near-infrared emissions to describe the reaction of one SRAM PUF, accompanied by targeted ion beam (FIB) circuit improvements to cause the same PUF reaction in the other schematic. Propagation delays in ring oscillators may often be changed with FIB circuit modifications. Still, delay-based PUFs are likely vulnerable to cloning attacks if the adversary can better classify the system to be cloned [55].

Infrastructure problems still occur. A big corporation could sell millions of ICs a year. If each is fitted with a PUF for authentication and anti-counterfeiting, fabrication houses need a verification system capable of serving millions of devices. There are also issues with utilities. A big corporation could sell billions of ICs in a year. Suppose ICs manufactured by each company is armed with a PUF for validation and anti-cloning. In that case, vendors will need a validation scheme capable of catering to many millions of products.

It is another task to ensure the reliability of the PUF production over time. PUF outputs are expected to change over time because of aging effects such as adverse bias temperature fluctuations or electromigration, but almost no research has been conducted in this field [56]. If minimal aging effects are considered, the result can be reverted with fuzzy extractors and makes it feasible to identify and reduce more significant aging-induced shifts.

Lastly, resources exist to create uniform security strategies around PUFs. Wide acceptance of PUFs may be encouraged by setting industry standards for verification and PUF-based key generation. These specifications will accelerate universal interfaces and services to allow users to validate the reliability of their devices and facilitate technology adoption. It will help reduce the dissemination of fake electronics in industry, sensitive infrastructure, and military networks.

The PUF response is sensitive to thermal disturbances, changing environments, and aging. As such, direct responses automatically cannot be utilized as a cipher. Usually, the PUF's leading architecture consists of two components: A sound sketch module and an entropy accumulator used to transform raw, noisy reactions to cipher. These two elements are generally termed a fuzzy extractor [57].

In addition, sketching and recovery operations are carried out by encoding and decoding the error correction code (ECC), respectively. Most ECC decoding algorithms depend on hard decision decoding, which treats any answer bit with the same probability of error [58].

As per the requirement of protection, the generated private key must comprise at least 128 bits of entropy to satisfy the safety specifications. Because ECC deciphering is computationally more expensive than encrypting, therefore resource-constrained PUFs can contain the encrypting logic. A well-resourced processor does computationally difficult deciphering logic. This configuration is called an inverted/reusable fuzzy extractor. [24]

7.11 GUARDING TACTICS

There are difficulties in incorporating PUFs into security systems, particularly when device resources are limited, mainly due to unreliability. Also, numerous attacks endanger the stability of the PUF and ask for preventive measures. Invasive and semi-invasive side-channel attacks involve straight admission to the PUF, assuming that an attacker can take the computer to a facility and use advanced tools to strike the machine. Adequate physical rotation techniques for IoT devices (e.g., PCB and epoxy adhesive) and incongruity recognition (e.g., sensing irregular system movement through low-cost gyroscope sensors) may reduce the threat level but still cannot provide complete security.

Remember, though, that these types of attacks are typically costly, and cost-efficiency can be used as an obstacle to low-cost IoT systems. Noninvasive threats, on the other hand, will act with relatively necessary equipment outside the lab without violating the physical security of the system. Passive noninvasive attacks add a variety of problems to the PUF and track external variables such as power consumption to build a machine-based PUF learning model. Many challenges are required to do this, ranging from thousands to even hundreds of thousands [59]. To counter this, the PUF can be built in such a way that it either

TABLE 7.2   Strengths and Weaknesses of Different PUF Designs

PUF ARCHITECTURE	WEAKNESS	STRENGTH
Ring Oscillator PUF	Environmental vulnerability	Easy to design and fabricate
Arbiter PUF	Delay path must be identical	(Partly)
		Resistance against machine learning attack
SRAM PUF	Low number of CRPs	Good statistical properties
TERO PUF	Need further investigations	Environmental toughness
Public PUF	Need further investigations	No secret info

accepts only very particular challenges, making it very challenging to get all of them, or only accepts a limited amount of challenges every second, slowing down the assault by many orders of magnitude [60].

The best defensive technique is to make the PUF more resilient to external threats, effectively reducing aggressive attacks to the same level as passive attacks. There are two options: first, CRPs can be kept secret by encryption. Alternatively, we may select a (healthy) PUF configuration with several CRPs that are big enough never to regenerate one [30].

In a more complex attack example, the adversary uses machine learning techniques to conduct a prediction attack and simulate challenge–response pairs, using previously captured challenge–response pairs as feedback to train the machine learning model. Again, thousands of CRPs are expected to carry out successful attacks. One approach to addressing these attacks was to improve the robustness of the PUF architecture by the spatial nonlinearity. Unfortunately, in the end, this technique has proven unsuccessful. Another solution was to use cryptographic methods to use machine-learning to tackle simulation attacks. This strategy was more successful but ignored the design of the PUF, opening the field for side-channel attacks. A physical defense can be used to prevent aggressive and semi-invasive side-channel threats. [30]

Higher robustness can be accomplished by making the PUF software robust to specific environmental changes, and the software of the PUF used decreases the possibility of noninvasive side-channel attacks. Choosing the right PUF architecture is, therefore, essential. Finally, the PUF must be paired with an effective authentication protocol to incorporate an automated solution.

7.12 CONCLUSION

This chapter explained the brief history of PUF technology, its working, and various types of PUF circuits and architecture. We also discussed the multiple sources of security threats and how to mitigate them. The chapter also emphasized the applications of PUFs in various domains that have been pointed out, along with challenges faced by the technology. PUFs circuits can be a useful security tool in designing embedded systems as well as for IoT systems. Commercially, IoT products started utilizing PUFs, for example, SRAM PUFs and butterfly PUFs, for security enhancement. Despite covering so much ground, there are many challenges, and therefore active research is going on in this field.

New techniques and state-of-the-art approaches to design more secure PUFs are being proposed regularly, and the field is a growing research area [61, 62]. Some of the solutions offered recently indicate the consistent evolution of novel PUF design, which points in the direction of better focus on security necessities. ASIC-based PUF interface, which is gaining fame, are costly due to their custom made architecture and therefore are not suitable for low-cost IoT solutions. However, an alternate PUF design strategy targeting IoT devices can be an answer to this problem. PUF based security solutions are being increasingly adopted in commercial IoT applications. The credit of such widespread acceptability is due to the intrinsic nature of the SRAM PUF, catering to every prevalent electronic product. The number of IoT devices will cross 20 billion marks in 2020, as per a study conducted by Gartner [63]. Seeing this massive number of IoT devices, the PUF-based authentication protocol will emerge as viable security solutions for these devices.

REFERENCES

1. Simmons, Gustavus J. "Identification of data, devices, documents, and individuals." In *Proceedings. 25th Annual 1991 IEEE International Carnahan Conference on Security Technology*, pp. 197–218. IEEE, 1991.
2. Pappu, R. S. *Physical One-Way Functions*. PhD thesis, Massachusetts Institute of Technology, Cambridge, March 2001.
3. Simmons, Gustavus J. "A system for verifying user identity and authorization at the point-of sale or access." *Cryptologia* 8, no. 1 (1984): 1–21.
4. Gassend, Blaise, Dwaine Clarke, Marten Van Dijk, and Srinivas Devadas. "Silicon physical random functions." In *Proceedings of the 9th ACM Conference on Computer and Communications Security*, pp. 148–160. 2002.
5. Lao, Yingjie, Bo Yuan, Chris H. Kim, and Keshab K. Parhi. "Reliable PUF-based local authentication with self-correction." *IEEE Transactions on Computer-Aided Design of Integrated Circuits and Systems* 36, no. 2 (2016): 201–213.
6. Halak, Basel. *Physically Unclonable Functions: From Basic Design Principles to Advanced Hardware Security Applications*. Springer, Cham, Switzerland, 2018.
7. Koushanfar, Farinaz. "Hardware metering: A survey." In *Introduction to Hardware Security and Trust*, pp. 103–122. Springer, New York, NY, 2012.
8. Rostami, Masoud, Farinaz Koushanfar, and Ramesh Karri. "A primer on hardware security: Models, methods, and metrics." *Proceedings of the IEEE* 102, no. 8 (2014): 1283–1295.
9. Maes, R. *Physically Unclonable Functions: Constructions. Properties and Applications*. Springer, Berlin, 2013.
10. Coombs, Rob. "Securing the future of authentication with ARM TrustZone-based trusted execution environment and fast identity online (FIDO)." *ARM White Paper* (2015).
11. Tajik, Shahin, Enrico Dietz, Sven Frohmann, Jean-Pierre Seifert, Dmitry Nedospasov, Clemens Helfmeier, Christian Boit, and Helmar Dittrich. "Physical characterization of arbiter PUFs." In *International Workshop on Cryptographic Hardware and Embedded Systems*, pp. 493–509. Springer, Berlin, Heidelberg, 2014.
12. Mahmoud, Ahmed, Ulrich Rührmair, Mehrdad Majzoobi, and Farinaz Koushanfar. "Combined Modeling and Side Channel Attacks on Strong PUFs." *IACR Cryptol. ePrint Arch.* 2013 (2013): 632.
13. Rührmair, Ulrich, Xiaolin Xu, Jan Sölter, Ahmed Mahmoud, Mehrdad Majzoobi, Farinaz Koushanfar, and Wayne Burleson. "Efficient power and timing side channels for physical unclonable functions." In *International Workshop on Cryptographic Hardware and Embedded Systems*, pp. 476–492. Springer, Berlin, Heidelberg, 2014.
14. Halak, Basel, Santosh Shedabale, Hiran Ramakrishnan, Alex Yakovlev, and Gordon Russell. "The impact of variability on the reliability of long on-chip interconnect in the presence of crosstalk." In *Proceedings of the 2008 International Workshop on System Level Interconnect Prediction*, pp. 65–72. 2008.
15. Alexander, Craig, Gareth Roy, and Asen Asenov. "Random-dopant-induced drain current variation in nano-MOSFETs: A three-dimensional self-consistent Monte Carlo simulation study using 'ab initio' ionized impurity scattering." *IEEE Transactions on Electron Devices* 55, no. 11 (2008): 3251–3258.
16. Helfmeier, Clemens, Christian Boit, Dmitry Nedospasov, and Jean-Pierre Seifert. "Cloning physically unclonable functions." In *2013 IEEE International Symposium on Hardware-Oriented Security and Trust (HOST)*, pp. 1–6. IEEE, 2013.
17. Ganji, Fatemeh. *On the Learnability of Physically Unclonable Functions*. Springer, Cham, Switzerland, 2018.
18. Becker, Georg T. "The gap between promise and reality: On the insecurity of XOR arbiter PUFs." In *International Workshop on Cryptographic Hardware and Embedded Systems*, pp. 535–555. Springer, Berlin, Heidelberg, 2015.
19. Herder, Charles, Meng-Day Yu, Farinaz Koushanfar, and Srinivas Devadas. "Physical unclonable functions and applications: A tutorial." *Proceedings of the IEEE* 102, no. 8 (2014): 1126–1141.

20. Santiago, Leandro, Vinay C. Patil, Charles B. Prado, Tiago A. O. Alves, Leandro A. J. Marzulo, Felipe M. G. França, and Sandip Kundu. "Realizing strong PUF from weak PUF via neural computing." In *2017 IEEE International Symposium on Defect and Fault Tolerance in VLSI and Nanotechnology Systems (DFT)*, pp. 1–6. IEEE, 2017.
21. Bauer, Todd, and Jason Hamlet. "Physical unclonable functions: A primer." *IEEE Security & Privacy* 12, no. 6 (2014): 97–101.
22. Kursawe, Klaus, Ahmad-Reza Sadeghi, Dries Schellekens, Boris Skoric, and Pim Tuyls. "Reconfigurable physical unclonable functions-enabling technology for tamper-resistant storage." In *2009 IEEE International Workshop on Hardware-Oriented Security and Trust*, pp. 22–29. IEEE, 2009.
23. Gehrer, Stefan, and Georg Sigl. "Reconfigurable PUFS for FPGA-based SOCS." In *2014 International Symposium on Integrated Circuits (ISIC)*, pp. 140–143. IEEE, 2014.
24. Gao, Yansong, Said F. Al-Sarawi, and Derek Abbott. "Physical unclonable functions." *Nature Electronics* 3, no. 2 (2020): 81–91.
25. Rührmair, Ulrich, Christian Jaeger, and Michael Algasinger. "An attack on PUF-based session key exchange and a hardware-based countermeasure: Erasable PUFs." In *International Conference on Financial Cryptography and Data Security*, pp. 190–204. Springer, Berlin, Heidelberg, 2011.
26. Bossuet, Lilian, Xuan Thuy Ngo, Zouha Cherif, and Viktor Fischer. "A PUF based on a transient effect ring oscillator and insensitive to locking phenomenon." *IEEE Transactions on Emerging Topics in Computing* 2, no. 1 (2013): 30–36.
27. Marchand, Cédric, Lilian Bossuet, Ugo Mureddu, Nathalie Bochard, Abdelkarim Cherkaoui, and Viktor Fischer. "Implementation and characterization of a physical unclonable function for IoT: A case study with the TERO-PUF." *IEEE Transactions on Computer-Aided Design of Integrated Circuits and Systems* 37, no. 1 (2017): 97–109.
28. Xu, Teng, James B. Wendt, and Miodrag Potkonjak. "Security of IoT systems: Design challenges and opportunities." In *2014 IEEE/ACM International Conference on Computer-Aided Design (ICCAD)*, pp. 417–423. IEEE, 2014.
29. Xi, Xiaodan, Haoyu Zhuang, Nan Sun, and Michael Orshansky. "Strong subthreshold current array PUF with 2 65 challenge-response pairs resilient to machine learning attacks in 130 nm CMOS." In *2017 Symposium on VLSI Circuits*, pp. C268–C269. IEEE, 2017.
30. Babaei, Armin, and Gregor Schiele. "Physical unclonable functions in the Internet of Things: State of the art and open challenges." *Sensors* 19, no. 14 (2019): 3208.
31. Tarik, Farhan Bin, Azadeh Famili, Yingjie Lao, and Judson D. Ryckman. "Robust optical physical unclonable function using disordered photonic integrated circuits." *Nanophotonics* 9, no. 9 (2020): 2817–2828.
32. Yu, Shimeng, and Pai-Yu Chen. "Emerging memory technologies: Recent trends and prospects." *IEEE Solid-State Circuits Magazine* 8, no. 2 (2016): 43–56.
33. Wong, Philip H.-S., and Sayeef Salahuddin. "Memory leads the way to better computing." *Nature nanotechnology* 10, no. 3 (2015): 191–194.
34. Becker, Georg T. "The gap between promise and reality: On the insecurity of XOR arbiter PUFs." In *International Workshop on Cryptographic Hardware and Embedded Systems*, pp. 535–555. Springer, Berlin, Heidelberg, 2015.
35. Nguyen, Phuong Ha, Durga Prasad Sahoo, Chenglu Jin, Kaleel Mahmood, Ulrich Rührmair, and Marten van Dijk. "The interpose PUF: Secure PUF design against state-of-the-art machine learning attacks." *IACR Transactions on Cryptographic Hardware and Embedded Systems* (2019): 243–290.
36. Sahoo, Durga Prasad, Debdeep Mukhopadhyay, Rajat Subhra Chakraborty, and Phuong Ha Nguyen. "A multiplexer-based arbiter PUF composition with enhanced reliability and security." *IEEE Transactions on Computers* 67, no. 3 (2017): 403–417.
37. Lim, Daihyun, Jae W. Lee, Blaise Gassend, G. Edward Suh, Marten Van Dijk, and Srinivas Devadas. "Extracting secret keys from integrated circuits." *IEEE Transactions on Very Large-Scale Integration (VLSI) Systems* 13, no. 10 (2005): 1200–1205.
38. Idriss, Tarek, Haytham Idriss, and Magdy Bayoumi. "A PUF-based paradigm for IoT security." In *2016 IEEE 3rd World Forum on Internet of Things (WF-IoT)*, pp. 700–705. IEEE, 2016.
39. Wang, Yinglei, Wing-kei Yu, Shuo Wu, Greg Malysa, G. Edward Suh, and Edwin C. Kan. "Flash memory for ubiquitous hardware security functions: True random number generation and device fingerprints." In *2012 IEEE Symposium on Security and Privacy*, pp. 33–47. IEEE, 2012.
40. Orosa, Lois, Yaohua Wang, Ivan Puddu, Mohammad Sadrosadati, Kaveh Razavi, Juan Gómez-Luna, Hasan Hassan et al. "Dataplant: Enhancing system security with low-cost in-DRAM value generation primitives." *arXiv preprint arXiv:1902.07344* (2019).

41. Kumar, Sandeep S., Jorge Guajardo, Roel Maes, Geert-Jan Schrijen, and PimTuyls. "The butterfly PUF protecting IP on every FPGA." In *2008 IEEE International Workshop on Hardware-Oriented Security and Trust*, pp. 67–70. IEEE, 2008.

42. Maes, Roel, PimTuyls, and Ingrid Verbauwhede. "Intrinsic PUFs from flip-flops on reconfigurable devices." In *3rd Benelux Workshop on Information and System Security (WISSec 2008)*, vol. 17, p. 2008. 2008.

43. Jajodia, Sushil, and Henk C. A. van Tilborg, eds. *Encyclopedia of Cryptography and Security: A-K*. Springer US, New York, 2011.

44. Armknecht, Frederik, Roel Maes, Ahmad-Reza Sadeghi, Berk Sunar, and Pim Tuyls. "Memory leakage-resilient encryption based on physically unclonable functions." In *Towards Hardware-Intrinsic Security*, pp. 135–164. Springer, Berlin, Heidelberg, 2010.

45. Bösch, Christoph, Jorge Guajardo, Ahmad-Reza Sadeghi, Jamshid Shokrollahi, and Pim Tuyls. "Efficient helper data key extractor on FPGAs." In *International Workshop on Cryptographic Hardware and Embedded Systems*, pp. 181–197. Springer, Berlin, Heidelberg, 2008.

46. Guajardo, Jorge, Sandeep S. Kumar, Geert-Jan Schrijen, and Pim Tuyls. "FPGA intrinsic PUFs and their use for IP protection." In *International Workshop on Cryptographic Hardware and Embedded Systems*, pp. 63–80. Springer, Berlin, Heidelberg, 2007.

47. Jajodia, Sushil, and Henk C. A. van Tilborg, eds. *Encyclopedia of Cryptography and Security: A-K*. Springer US, New York, 2011.

48. Acquisti, Alessandro, Sean W. Smith, and Ahmad-Reza Sadeghi, eds. *Trust and Trustworthy Computing: Third International Conference, TRUST 2010, Berlin, Germany, June 21–23, 2010, Proceedings*. Vol. 6101. Springer, Berlin, Heidelberg, 2010.

49. Maes, Roel. "Physically unclonable functions: Properties." In *Physically Unclonable Functions*, pp. 49–80. Springer, Berlin, Heidelberg, 2013.

50. Gassend, Blaise Laurent Patrick. *Physical Random Functions*. PhD diss., Massachusetts Institute of Technology, 2003.

51. Bringer, Julien, Hervé Chabanne, and Thomas Icart. "On physical obfuscation of cryptographic algorithms." In *International Conference on Cryptology in India*, pp. 88–103. Springer, Berlin, Heidelberg, 2009.

52. Gassend, Blaise, Dwaine Clarke, Marten Van Dijk, and Srinivas Devadas. "Controlled physical random functions." In *18th Annual Computer Security Applications Conference, 2002. Proceedings*, pp. 149–160. IEEE, 2002.

53. Kursawe, Klaus, Ahmad-Reza Sadeghi, Dries Schellekens, Boris Skoric, and PimTuyls. "Reconfigurable physical unclonable functions-enabling technology for tamper-resistant storage." In *2009 IEEE International Workshop on Hardware-Oriented Security and Trust*, pp. 22–29. IEEE, 2009.

54. Helfmeier, Clemens, Christian Boit, Dmitry Nedospasov, and Jean-Pierre Seifert. "Cloning physically unclonable functions." In *2013 IEEE International Symposium on Hardware-Oriented Security and Trust (HOST)*, pp. 1–6. IEEE, 2013.

55. Schlangen, Rudolf, Rainer Leihkauf, Ted Lundquist, Peter Egger, and Christian Boit. "RF performance increase allowing IC timing adjustments by use of backside FIB processing." In *2009 16th IEEE International Symposium on the Physical and Failure Analysis of Integrated Circuits*, pp. 33–36. IEEE, 2009.

56. Maiti, Abhranil, Logan McDougall, and Patrick Schaumont. "The impact of aging on an FPGA-based physical unclonable function." In *2011 21st International Conference on Field Programmable Logic and Applications*, pp. 151–156. IEEE, 2011.

57. Delvaux, Jeroen, Dawu Gu, Dries Schellekens, and Ingrid Verbauwhede. "Helper data algorithms for PUF-based key generation: Overview and analysis." *IEEE Transactions on Computer-Aided Design of Integrated Circuits and Systems* 34, no. 6 (2014): 889–902.

58. Bösch, Christoph, Jorge Guajardo, Ahmad-Reza Sadeghi, Jamshid Shokrollahi, and Pim Tuyls. "Efficient helper data key extractor on FPGAs." In *International Workshop on Cryptographic Hardware and Embedded Systems*, pp. 181–197. Springer, Berlin, Heidelberg, 2008.

59. Becker, Georg T., and Raghavan Kumar. "Active and passive side-channel attacks on delay based PUF designs." *IACR Cryptol. ePrint Arch.* 2014 (2014): 287.

60. Delvaux, Jeroen, Roel Peeters, Dawu Gu, and Ingrid Verbauwhede. "A survey on lightweight entity authentication with strong PUFs." *ACM Computing Surveys (CSUR)* 48, no. 2 (2015): 1–42.

61. Nguyen, Phuong Ha, Durga Prasad Sahoo, Chenglu Jin, Kaleel Mahmood, Ulrich Rührmair, and Marten van Dijk. "The interpose PUF: Secure PUF design against state-of-the-art machine learning attacks." *IACR Transactions on Cryptographic Hardware and Embedded Systems* (2019): 243–290.

62. Jin, Chenglu, Charles Herder, Ling Ren, Phuong Ha Nguyen, Benjamin Fuller, Srinivas Devadas, and Marten van Dijk. "FPGA implementation of a cryptographically-secure puf based on learning parity with noise." *Cryptography* 1, no. 3 (2017): 23.

63. *Which-50*. https://which-50.com/iot-connected-devices-reach-20-4-billion-2020-says-gartner/ (2017).

Review of the Medical Internet of Things-Based RFID Security Protocols

8

Nagarjuna Telagam and Nehru Kandasamy

Contents

8.1 INTRODUCTION

RFID technology includes memory functions and radio frequency interface with the address. It has a security module that enables users' data exchange with reader facility into attachable tags by using magnetic fields [1]. The development in future generation devices mostly depends on the Internet of things and RFID technology. RFID is considered a core subject for the Internet of things and has many applications such as automatic toll payments, theft detection, vehicle tracking, infant protection, medical shops identification, patient monitoring and tracking, medicine management, blood transfusion, and nursing house identification, etc. [2]. The privacy issues that need to be addressed in medical hospitals and health care colleges/institutions have already deployed RFID systems. The algorithms, such as crucial encryption in symmetric and asymmetric schemes, are proposed, and they use hash functions [3],[4]. For the maintenance of such algorithms, high-security back-end servers are required. This server contains information about tagged objects, session keys, and tags used. Such knowledge leads to the leakage of many confidential messages, and these tags also leak private information. The reader tags are more vulnerable to attacks when the server is malicious; the commands will execute, which leads to different authentication results of the particular user tag. The protocol is to identify the leakage in the data. This protocol also ensures data integrity, which is caused by malicious server attacks [5].

The IoT plays an essential role in physical devices containing sensors, RFID, etc. The embedded technology is the core of IoT, which helps communicate with all other external environments [6]. The ITU (International Telecommunication Union) pointed out that there are five leading core technologies to realize the Internet. They are sensor technology, RFID technology, intelligent embedded technology, nanotechnology, and an ultra-light authentication protocol proposed using RFID tags. It depends on bit-wise operations, and it suffers from vulnerability [7],[8].

The RFID technology has been replaced by bar codes identification technology in recent times. RFID mainly has three components, that is, readers, tags, and a server [9]. RFID tags have an inbuilt transponder that processes the information with an antenna's help and communication with the readers. This RFID server communicates with an inbuilt antenna. Once the signal is received, the data information is authenticated for identification and further information processing. The reader emits an interrogation signal through an antenna, which helps to identify multiple tags. The signal range will also play a crucial role in determining, further processing, and retrieving information [10]. The interfacing between labels and the reader is always insecure, but the interfacing between the RFID server and reader interface is still secure and fixed. These insecure communication channels have many privacy problems. Hacking or frequent attacks will always happen in this insecure channel. Spoofing, eavesdropping, de-synchronization, and counterfeiting will be the most frequent attacks. The authentication protocols help to solve the drawbacks in RFID technology [11].

The RFID authentication protocols are proposed based on symmetric-key cryptography. Still, this suffers from storage space and scalability problems [12], leading to the elliptic curves cryptography topic, where the solution for these issues is explained briefly. These elliptic curve cryptography-based authentication protocols will prevent cloning attacks and also offer good privacy [13]. The new modified elliptic curve cryptography-based RFID authentication protocols suffer from tag masquerade attack, location tracking, cloning of tag attack, and server spoofing attacks [14].

The security in medication for patients is an issue nowadays. Many researchers propose RFID grouping proof protocols. Some of the proposed protocols suffer from denial of service (DoS) attacks, relay attacks, and impersonation attacks [15]. The inpatient safety-based RFID security system is intended to increase medication safety for patients in hospitals. This protocol can't find the DoS attacks, and corresponding hospital staff can modify the medication details. New RFID novel tamper-resistant prescription access control protocols [16] are proposed. Unfortunately, this novel protocol is also suffering from traceability and impersonation attacks. Then RFID-based mutual authentication protocols [17] for health care environments are offered, and these also suffer from traceability attacks.

Public key cryptography is developing in recent times, and elliptic curve cryptography is receiving tremendous attention. This technology has a small key size, takes less space, and has a high speed of execution. The first RFID authentication protocol was proposed using ECC in 2006 [18]. But how this protocol suffers from privacy laws is explained [19]. The authors in [20] proposed a secure ECC authentication protocol that withstands various attacks [20]. Still, it suffers from fundamental compromise problems, and hackers can get information about the private tag key [21]. The authors introduced the new, improved protocol in [22] to overcome the fundamental compromise problem. This enhanced protocol also suffers from forwarding traceability and backward traceability. A lightweight ECC authentication protocol with the identification verifier and strong security properties is proposed [23].

This chapter discusses the RFID technology authentication protocols for medical data security, security attacks, architectures, specifications, and the impact of threats on medical IoT devices. This chapter concludes with many research directions for researchers in medical IoT RFID authentication protocols.

Section 8.1 discusses a brief introduction about RFID technology, RFID components, RFID security protocols for IoT, security attacks, and consequences in recent years. Section 8.2 offers a brief introduction about the blocks and functions of medical IoT architecture. Section 8.3 discusses medical device evaluation and vulnerabilities in medical IoT. Section 8.4 discusses device vulnerability analysis, section 8.5 discusses RFID authentication protocols for securing medical data in medical IoT devices, and section 8.6 concludes the chapter with future scope.

8.2 MEDICAL INTERNET OF THINGS

IoT technologies are an extensive area in which medical IoT is one technique that supports only the health care domain to provide the best medical facility for increasing population in-country or population in hospitals that require long-term treatment [24]. The mobile applications in smartphones will play an essential role in the next generation, where appointment bookings and medical staff video conferencing takes place; software applications have an impact on IoT. The medical IoT devices have the capability of programmability.

The medical IoT system mainly monitors devices and tracks the patient's condition remotely. It will record health measurements in every period and sends the data to the back-end system. The collected data in the back-end system will generate reports to the doctors as well as clinicians. In this way, doctors can detect health problem issues and react to them accordingly [25]. The medical device, which acts as a monitoring device, mostly depends on software applications in the user cellular phone or smartwatches. The health records of the patient can be monitored strictly using forms. The hackers can analyze the data information with big data, cloud services, and databases.

Figure 8.1 shows the architecture and protocols corresponding to the energy-efficient methods and offers different applications such as image transmission of medical data and signal monitoring. The Federal Trade Commission refers to an IoT that can connect devices worldwide, and it can simultaneously transceive the data [26]. In the meantime, the medical Internet-connected devices will increase rapidly, and concerned government organizations will deploy these devices in individual houses and hospitals. The device's data will be legally obligated to device manufactures, and they will follow the rules of the Health Insurance Portability and Accountability Act (HIPAA) of 1996. HIPAA's privacy of information and security rules required entities for electronic patient health information (ePHI) [27]. The individual user information, payment details, address information, and social security numbers are collected by electronic patient health information to ensure data integrity, confidentiality, and protection against impermissible users or organizations. The encryption of user data is most convenient for medical IoT communication devices; the packets of bulk information that are sent can be intercepted by hackers or network observers. Suppose the exact text authenticated data is compressed and sent over the network. The hackers can decrypt this compressed message by using algorithms; somewhere, the hackers will identify the design flaws. They scrutinize the data and interpret the information [28]. The medical IoT devices' data are studied and analyzed, such as blood pressure monitoring and heartbeat

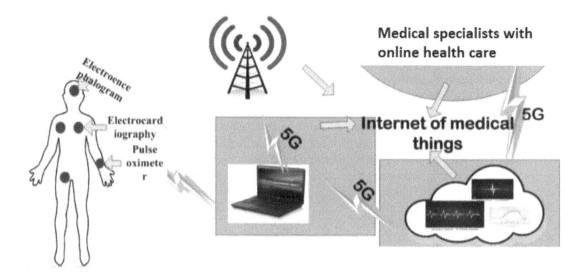

FIGURE 8.1   The architecture of M IoT.

FIGURE 8.2   Medical IoT-connected devices can deter illnesses and manage diseases.

monitoring. The network traffic is also monitored, and health information identification will infer ePHI from encrypted data communications. Figure 8.2 shows the medical Internet of things connected through doctors and patients [29].

The vulnerability in precise text transmission techniques in health privacy information is observed in multiple devices, which leads to leakage of patient data. The data is sent over cookies, uniform resource locators, with the help of specific IP addresses. The crucial information is leaked from the network observers. Network address translation also shields the individual IP addresses, and medical devices are tethered leads to the identification of medical data. The article has presented a user interface for medical IoT device users to continually check or monitor the information that will travel through the Internet to identify the data's misuse. The authority will issue a warning to the user if the data is leaked with the association of access points, as shown in Figure 8.3.

The vulnerabilities and challenges in IoT devices and the need for consumer visibility of medical data are addressed with a unified web-based dashboard system [30].

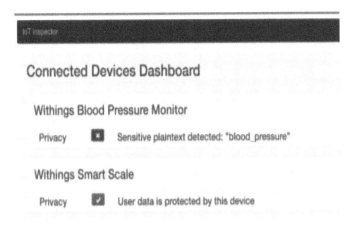

FIGURE 8.3 The user interface display.

8.3 MEDICAL IOT DEVICE EVALUATION AND VULNERABILITIES ANALYSIS

The privacy vulnerabilities are broadly classified into three phases, that is, data collection, clear text identification, and metadata analysis.

8.3.1 Data Collection

IoT data collection is the process of using sensors to track the conditions of physical things. In the created test environment, different devices are connected to network and their corresponding traffic is captured live. It helps to separate traffic on devices and filter the traffic if any unwanted or illegal traffic is observed. Wireshark is demonstrated in the Coursera course, and it is used to capture all the ethernet traffic or packets traveling from access points to each IoT device. The hackers can save these packets in PCAP files for analysis in offline mode. The data is collected in many instruments, along with user registration and full details with health information.

8.3.2 Cleartext Identification

The analysis of captured packets is broadly explained in this section. The separation of data packets depends on the protocol. These protocols mainly focus on IITTP packets and TCP packets and ignore the SSL packets, which are encrypted. After separating the data packets, the separation of the payload is done in sequence. The payload consists of application data that helps analyze the open system interconnection (OSI)-based application layer. This classification can be analyzed using three different schemes [31]: i.e., the naïve ASCII approach, Shannon entropy test, and method comparison.

8.3.3 Metadata Analysis

Patient health information details such as blood sugar are analyzed, and the data is measured at regular periods. Wireshark software helps to determine patient behavior. With IP addresses, hackers can track the data. Figure 8.4 shows the medical IoT device's vulnerability [32].

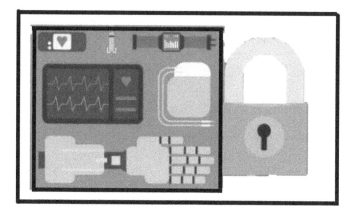

FIGURE 8.4   Securing medical IoT devices?

8.4 MEDICAL DEVICE VULNERABILITY ANALYSIS

The vulnerability is observed in HTTP (hypertext transfer protocol) GET protocol requests, packet header data, and Internet protocol conversion tables in devices. The most secure, which is not vulnerable to any attack so far, is the 1 by One digital wireless body fat scale.

8.4.1 Hypertension Monitor Devices

The data leaks are observed in cleartext within the tools, and the second-order metadata is captured sensitively in the blood pressure monitor IoT devices. The network observers will detect the information sections of all the queries or issues generated by the users and track them from traffic-based IP addresses such as static.withings.com.

8.4.2 Diabetes Mellitus Monitor Devices

These devices usually used TLSv1.2 on port 443 to send the encrypted information. The packets are not labeled; the network observers can quickly know the destination Internet protocol addresses. This evidence suggests that the user's information and device manufacturers' data is available to hackers.

8.5 RFID AUTHENTICATION PROTOCOLS FOR SECURING MEDICAL DATA IN MEDICAL IOT DEVICES

For the past few years, RFID protocols have been proposed; some of the protocols are HB, HB+, HB++ [33], EMAP, M2AP, and LMP+ [34]; the lightweight protocols also introduced are mostly based on PUFs and LFSRs [35]. The article [36] proved that protocols in [35] are suffering from attacks such as message injection and faced many vulnerabilities. The article [38] explains attacks such as desynchronization and confidential disclosure against [37] RFID protocols. In the year 2014, the paper [39] introduced a protocol to overcome the attacks shown in [38], and another article [40] showed the protocol proposed in [39] has a vulnerability in tracking and tagging problems and also suffers from desynchronization attacks.

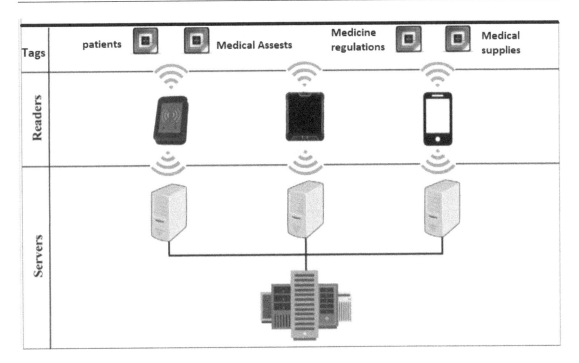

FIGURE 8.5   The architecture of RFID tag readers in the medical Internet of things.

The authors in article [41] proposed the new authentication protocol for the IoT system, which depends on RFID technology and suffers from reader compromised attacks, which is discussed in the article [42]. Figure 8.5 shows the architecture of RFID tag readers.

The denial of service attack executed on the IoT is presented in [43]; the compromise problem in the tag is observed, that is, the information is confidentially leaked in the proposed protocol, which shows support against various attacks; this is explained in [44]. The TMIS is suggested, which entirely depends on cryptographic algorithms between the user RFID tag and information storage server in [45].

Most of these protocols depend on time-stamp and are not supported by electronics product codes class-1 generation 2 standards. Another article [46] proposed a protocol for mobile health applications for bilinear pairing devices, which also is vulnerable to a tracking system attack, and this protocol depends on ECC. The multi-server environment can be applied to privacy issues like mobility, security, and scalability problems. A new secure protocol was proposed [47] and showed anonymity for tags, reader, and forward-backward untraceable features based on the cloud server. After analyzing the protocols since the year 2014, some of the requirements not presented for RFID authentication protocols are mutual authentication, entity intractability, message integrity, entity privacy-preserving, immunity against disclosure, and robustness against replay attacks.

8.5.1 Chunhua Jin Scheme

This paper [48] structured a shared verification convention for RFID dependent on ECC. Regarding security, this algorithm can accomplish privacy, untraceability, standard validation, labels' namelessness, accessibility, and forward security.

Every RFID tag or reader has mainly three participants, that is, the trusted tag issuer, trusted reader, and charged tag. Mostly the trusted reader connects to the database and public keys of legal tags. The list of symbols is shown in Table 8.1; Figure 8.6 shows the Chunhua et al. [48] proposed scheme.

TABLE 8.1 List of Symbols Used in Chunhua Jin Scheme

SYMBOL	INTERPRETATION
n and q	Both are large prime numbers
P	A generator with order n
$F(q)$	A finite field
E	An elliptic curve defined over a finite field $F(q)$ by the equation $y^2 = x^3 + ax + b$, where a, b belongs to $F(q)$
ID_{Ti}	The identity of the ith tag, where ID_{Ti} belongs to $0,1$
S_R, P_R	Private/public key of the tag reader, where $P_R = S_R P$, S_R belongs to Zn
S_{Ti}, P_{Ti}	Private/public key of the tag reader, where $P_{Ti} = S_{Ti} P$, S_{Ti} belongs to Zn
H_1, H_2	Secure hash functions

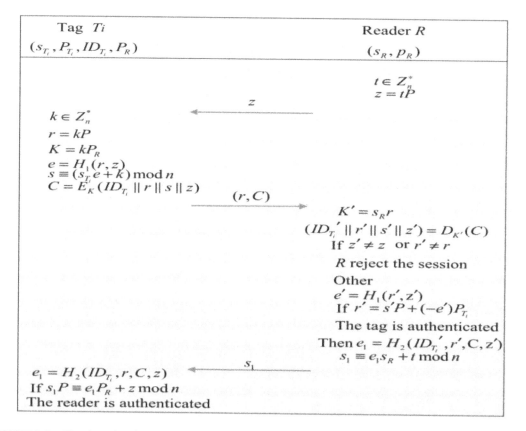

FIGURE 8.6 Chunhua Jin scheme.

8.5.2 Zhang and Qi's Scheme

Prescription mistakes are hazardous, even deadly, since they could make genuine patients suffer. To decrease medicine mistakes, robotized persistent prescription algorithms utilizing the RFID technology are used in numerous emergency clinics. In the last decade, multiple security protocols have been proposed to guarantee the protection for security of medical data. This paper [49] proposes an RFID

TABLE 8.2   Symbols Used in Zhang and Qi's Scheme

SYMBOL	INTERPRETATION
G	An additive group on elliptical curve
P	A generator of G
X_i	The identifier of ith tag
Y	Server private key
Y	$Y = yP$, server public key
H	Hash function

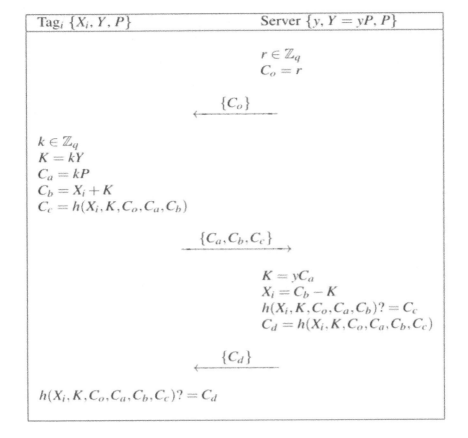

FIGURE 8.7   Zhang and Qi's scheme.

improved authentication protocol to improve persistent medical mistakes by utilizing ECC. The list of symbols is shown in Table 8.2. Figure 8.7 shows the Zhang et al [49] proposed scheme.

8.5.3 Zhao's Method

The vast majority of RFID authentication protocols depend on hash function or symmetric cryptography. The ECC has been utilized in the structure of the RFID protocol validation. Liao et al. [14] proposed another RFID protocol using ECC and guaranteed their algorithm will withstand all the attacks. This paper [50] shows that their convention experiences the critical trade-off issue; for example, an enemy

TABLE 8.3 Symbols Used in Zhao's Scheme

SYMBOL	INTERPRETATION
q, n	Large prime numbers
$F(q)$, E	Finite field and elliptic curve
P	Generator point with order n
X_s, $(P_s = X_s P)$, X_t, $(Z_t = Xt\,P)$	Private/public key pair of the server and tag

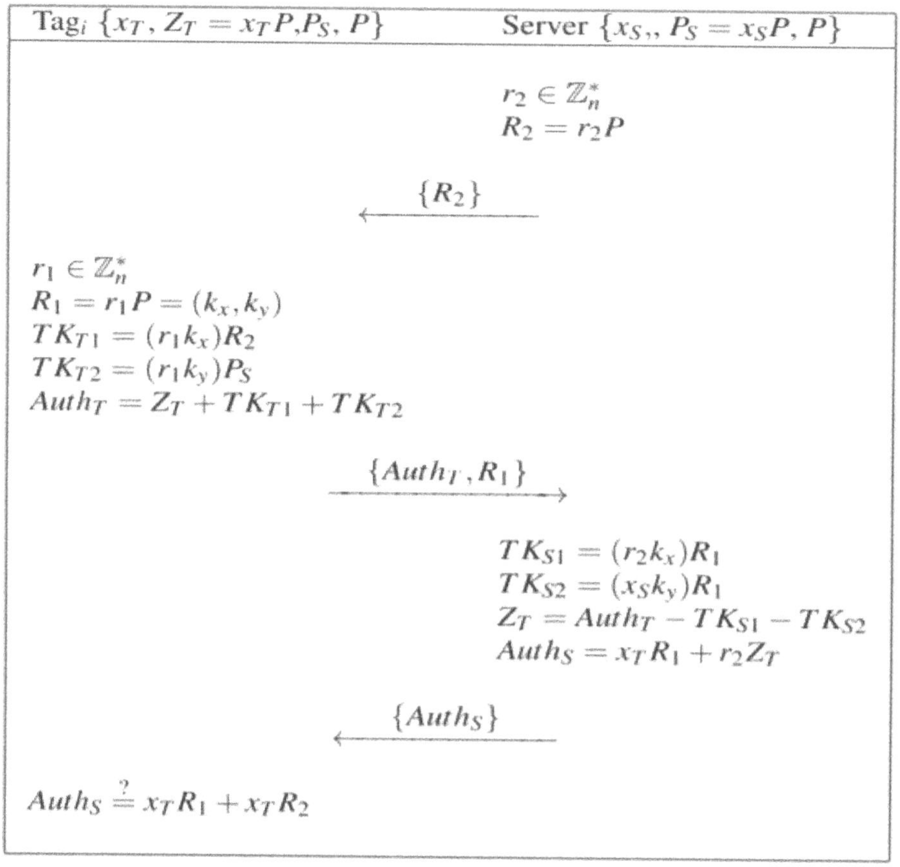

Tag$_i$ $\{x_T, Z_T = x_T P, P_S, P\}$ Server $\{x_S,, P_S = x_S P, P\}$

$$r_2 \in \mathbb{Z}_n^*$$
$$R_2 = r_2 P$$

$\{R_2\}$

$$r_1 \in \mathbb{Z}_n^*$$
$$R_1 = r_1 P = (k_x, k_y)$$
$$TK_{T1} = (r_1 k_x) R_2$$
$$TK_{T2} = (r_1 k_y) P_S$$
$$Auth_T = Z_T + TK_{T1} + TK_{T2}$$

$\{Auth_T, R_1\}$

$$TK_{S1} = (r_2 k_x) R_1$$
$$TK_{S2} = (x_S k_y) R_1$$
$$Z_T = Auth_T - TK_{S1} - TK_{S2}$$
$$Auth_S = x_T R_1 + r_2 Z_T$$

$\{Auth_S\}$

$$Auth_S \stackrel{?}{=} x_T R_1 + x_T R_2$$

FIGURE 8.8 Zhao's scheme.

could get the private key put away in the tag. The list of symbols is shown in Table 8.3. Figure 8.8 shows the Zhao et al.'s [50] proposed scheme.

8.5.4 Shehzad Ashraf Chaudhry Scheme

Telecare clinical data frameworks have progressively been utilized. Recently, Zhang et al. [54] and Zhao both independently proposed two validation plans for telecare clinical data frameworks under RFID innovation. Figure 8.9 shows the Shehzad et al. [51] proposed scheme. The authors guaranteed

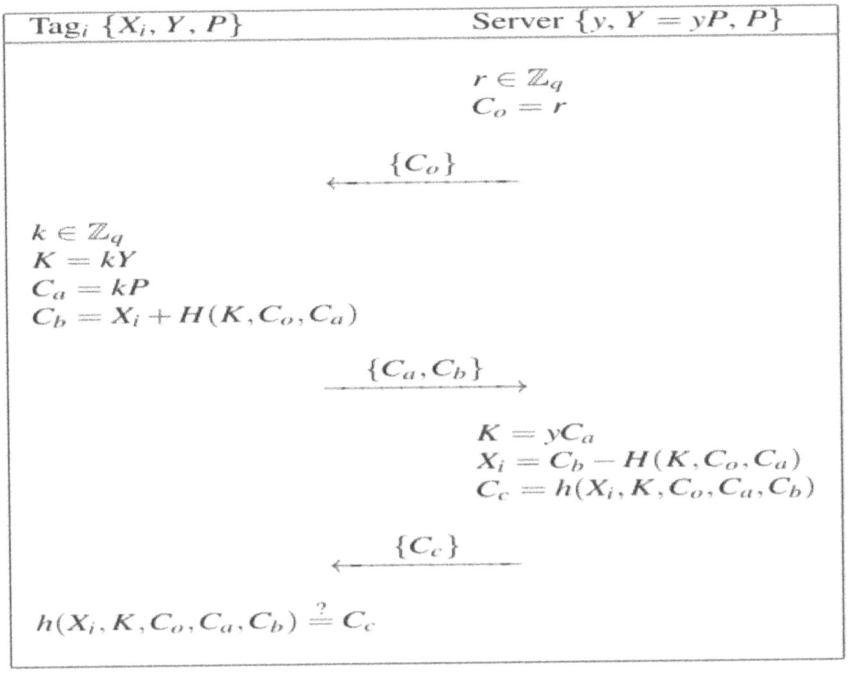

$$\text{Tag}_i \{X_i, Y, P\} \qquad\qquad \text{Server} \{y, Y = yP, P\}$$

$$r \in \mathbb{Z}_q$$
$$C_o = r$$

$$\xleftarrow{\quad \{C_o\} \quad}$$

$$k \in \mathbb{Z}_q$$
$$K = kY$$
$$C_a = kP$$
$$C_b = X_i + H(K, C_o, C_a)$$

$$\xrightarrow{\quad \{C_a, C_b\} \quad}$$

$$K = yC_a$$
$$X_i = C_b - H(K, C_o, C_a)$$
$$C_c = h(X_i, K, C_o, C_a, C_b)$$

$$\xleftarrow{\quad \{C_c\} \quad}$$

$$h(X_i, K, C_o, C_a, C_b) \stackrel{?}{=} C_c$$

FIGURE 8.9 Shehzad Ashraf Chaudhry scheme.

that their algorithms will accomplish all security necessities, including forwarding traceability. This paper [52] shows that both Zhang and Qi's plan and Zhao's method couldn't give a forward solution for traceability attacks.

This protocol is an improved version of Zhao's protocol, and it is based on the ECC method. The symbols are the same as in Table 8.3. Table 8.4 shows the symbols and Figure 8.10 shows the proposed scheme used in the K. Fan et al. [52] scheme.

8.5.5 K. Fan Scheme

TABLE 8.4 SYMBOLS USED IN THE K. FAN SCHEME

SYMBOL	INTERPRETATION
TID, RID	The private ID of the tag and reader
N_R, N_T, N_S	Random numbers generated by the reader, tag, and server
K_i	i^{th} session key
K1, k2	The subkeys in which $Ki = k2 \| k1$
Ev(Z), od(Z)	Even and odd number of the string
IDX_i, IDC_i	The ith session index value and index content
PRNG(.)	The pseudorandom number generation
Pi(.), pi(.)	Permutation functions
Cro(x.y)	Cross functions
Rot(x,y)	Circular shift operations on the strings x and y
MRotK(x,y)	Modular rotate function

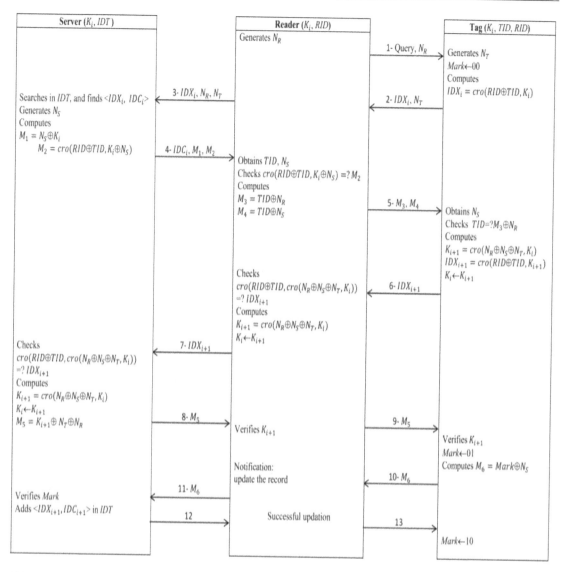

FIGURE 8.10 K. Fan scheme [52].

8.5.6 Shaohao Xie Scheme

RFID innovation is one of the center advancements of the Internet of things; it has been generally used in medical services and brings a great deal of comfort for individuals in day-to-day life. This notwithstanding, the security and protection difficulties of RFID verification algorithms are increasing every day.

The main problem in current RFID algorithms typically utilize a back-end server to accumulate the data of labeled articles, leading to data leakage if the user is hacked. To address this problem, the authors propose a security improved RFID verification convention for social insurance condition utilizing the method of indistinctness jumbling, which keeps the spillage of delicate information from the back-end worker. The list of symbols is shown in Table 8.5. Figure 8.11 shows the proposed scheme by Shaohao et al. [53].

Table 8.6 shows the advantages and security flaws in the RFID authentication protocols scheme that are explained in this chapter. In conclusion, the Shaohao Xie et al. scheme [53] shows less vulnerabilities to many attacks in medical IoT devices.

TABLE 8.5 Symbols Used in Shaohao Xie Scheme

SYMBOL	INTERPRETATION
ID_i, IDr	Identity of the ith tag and reader
K_i	The confidential key for ith tag
S_k	The confidential session key for all tags
K_{or}	The confidential key for the reader
R_1, R_2, R_3	Pseudorandom number generators
$F(K, m)$	Pseudo function of m computed using key K
Enc(K, m), Dec(K, c)	Encryption and decryption of m, c under key K

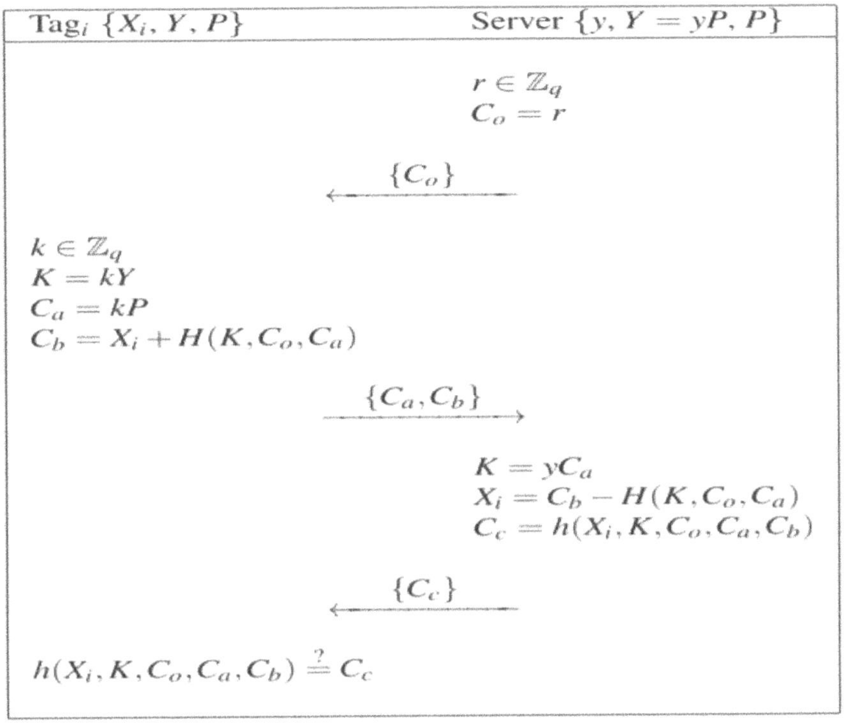

FIGURE 8.11 Shaohao Xie scheme.

TABLE 8.6 Comparative Analysis of RFID Authentication Protocols Scheme

ATTACKS	CHUNHUA ET AL, [48]	ZHANG ET AL, [49]	ZHAO ET AL, [50]	SHEHZAD ET AL, [51]	FAN ET AL, [52]	SHAOHAO ET AL, [53]
Tag's impersonation	Yes	–	–	No	Yes	–
Server spoofing	Yes	–	–	–	–	No
Replay	Yes	–	–	No	–	–
Denial of service	Yes	–	–	–	–	No
Modification	Yes	–	–	–	–	No
Cloning	Yes	–	–	–	–	No

(Continued)

TABLE 8.6 (Continued)

ATTACKS	CHUNHUA ET AL, [48]	ZHANG ET AL, [49]	ZHAO ET AL, [50]	SHEHZAD ET AL, [51]	FAN ET AL, [52]	SHAOHAO ET AL, [53]
Desynchronization	Yes	–	–	–	–	–
Man in the middle	Yes	–	–	No	–	No
Forward untraceability	–	Yes	No	yes	No	–
Secure disclosure	–	–	–	–	Yes	–
Location privacy	–	–	–	Yes	–	No
Mutual authentication	–	–	–	Yes	–	–
Eavesdropping and tracking	–	–	–	–	–	No

8.6 CONCLUSION

Medical data information is confidential, and it needs to be protected. This chapter explains the security of authenticated RFID protocols for medical IoT. First, this chapter presents a brief introduction about proposed medical IoT and RFID protocols and their features. Second, this chapter explains the details about attacks in RFID protocols and an analysis of security requirements.

In conclusion, medical IoT devices cannot be fully secured until a strong foundation is created with hardware components. The efforts and cost of manufacturing the devices could lead to waste if the concerned device has a hardware Trojan inserted. This hardware Trojan will be triggered anytime in the network and can quickly destroy or damage the device. The IoT will be part of daily lives in the upcoming generation for people. Extraordinary steps or procedures need to be taken by the device manufacturers or advisers or third party sources to ensure user information privacy. In this chapter, the security challenges of overall Medical IoT systems are explored, focusing on vulnerabilities of RFID authentication protocols.

REFERENCES

1. Finkenzeller, K., RFID handbook: Fundamentals and applications in contactless smart cards, radio frequency identification and near-field communication. Wiley, Hoboken, 2010.
2. Anandhi, S., Anitha, R., and Sureshkumar, V., IoT enabled RFID authentication and secure object tracking system for smart logistics. Wireless Personal Communications, 104(2):543–560, 2019.
3. Fan, K., Jiang, W., Li, H., and Yang, Y., Lightweight RFID protocol for medical privacy protection in IoT. IEEE Transactions on Industrial Informatics, 14(4):1656–1665, 2018.
4. Feldhofer, M., Dominikus, S., and Wolkerstorfer, J., Strong authentication for RFID systems using the AES algorithm. In International workshop on cryptographic hardware and embedded systems (pp. 357–370). Springer, Berlin, Heidelberg, 2004.
5. Tsudik, G., A family of dunces: Trivial RFID identification and authentication protocols. In International workshop on privacy-enhancing technologies (pp. 45–61). Springer, Berlin, Heidelberg, 2007.
6. Atzori, L., Iera, A., and Morabito, G., The Internet of Things: A survey. Computer Networks, 54(15):2787–2805, 2010.
7. The Internet of Things, ITU international reports. Technical Report, International Telecommunications Union, 2005.
8. Tewari, A., and Gupta, B.B., Cryptanalysis of a novel ultra-lightweight mutual authentication protocol for IoT devices using RFID tags. Journal of Supercomputing, 2016. doi:10.1007/s11227-016-1849-x

9. Juels, A., and Weis, S., Defining strong privacy for RFID. Cryptology ePrint Archive Report 2006/137, 2006.
10. Song, B., and Mitchell, C.J., Scalable RFID security protocols supporting tag ownership transfer. Computer Communications, 34(4):556–566, 2011.
11. Niu, B., X. Zhu, H. Chi, and H. Li, Privacy, and authentication protocol for mobile RFID systems. Wireless Personal Communications, 2014. doi:10.1007/s11277-014-1605-6.
12. Ning, H., Liu, H., Mao, J., and Zhang, Y., Scalable, and distributed key array authentication protocol in radio frequency identification based sensor systems. IET Communications, 5(12):1755–1768, 2011.
13. Batina, L., Lee, Y.K., Seys, S., Singele, D., and Verbauwhede, I., Extending ECC-based RFID authentication protocols to privacy, preserving multi-party grouping proofs. Personal and Ubiquitous Computing, 16(3):323–335, 2012.
14. Peeters, R., and Hermans, J., Attack on Liao and Hsiao's secure ECC-based RFID authentication scheme integrated with ID-verifier transfer protocol. http://eprint.iacr.org/2013/399, 2013.
15. Safkhani, M., Peris-Lopez, P., Hernandez-Castro, J.C., and Bagheri, N., Cryptanalysis of the Cho others. Protocol: A hash-based RFID tag mutual authentication protocol. Journal of Computational and Applied Mathematics, 259(1):571–577, 2014.
16. Kuo, W.C., Chen, B.L., and Wuu, L.C., Secure indefinite-index RFID authentication scheme with challenge-response strategy. Information Technology and Control, 42(2):124–130, 2013.
17. Farash M.S., Cryptanalysis, and improvement of an efficient mutual authentication RFID scheme based on elliptic curve cryptography. Journal of Supercomputing, 70(2):987–1001, 2014.
18. Lee, Y.K., Sakiyama, K., Batina, L., and Verbauwhede, I., Elliptic curve based security processor for RFID. IEEE Transactions on Computers, 57(11):1514–1527, 2008.
19. Ning, H., Liu, H., Mao, J., and Zhang, Y., Scalable and distributed key array authentication protocol in radio frequency identification based sensor systems. IET Communications, 5(12):1755–1768, 2011.
20. Alomair, B., Clark, A., Cuellar, J., and Poovendran, R., Scalable RFID systems: A privacy-preserving protocol with constant-time identification. IEEE Transactions on Parallel and Distributed Systems, 23(8):1536–1550, 2012.
21. Alomair, B., and Poovendran, R., Privacy versus scalability in radio frequency identification systems. Computer Communications, 33(18):2155–2163, 2010.
22. Batina, L., Seys, S., Singelée, D., and Verbauwhede, I., Hierarchical ECC-based RFID authentication protocol. In International workshop on radio frequency identification: Security and privacy issues (pp. 183–201). Springer, Berlin, Heidelberg, 2011.
23. Guo, P., Wang, J., Li, B., and Lee, S., A variable threshold value authentication architecture for wireless mesh networks. Journal of Internet Technology, 15(6):929–936, 2014.
24. Atamli, A.W., and Martin, A., Threat-based security analysis for the Internet of Things. In 2014 international workshop on secure Internet of Things, IEEE, 2014, pp. 35–43.
25. Miorandi, D. et al., Internet of Things: Vision, applications and research challenges. Ad Hoc Networks, 10(7):1497–1516, 2012.
26. Federal Trade Commission, Internet of Things: Privacy & security in a connected world. FTC Staff Report, 2016.
27. United States Department of Health and Human Services Office for Civil Rights, Summary of the HIPAA privacy rule. www.hhs.gov/hipaa/ for-professionals/privacy/index.html, n.d.
28. Feamster, Nick, 2016. Who will secure the Internet of Things? https://freedomto-tinker.com/2016/01/19/who-will-secure-the-internet-of-things/, 2016.
29. www.medicaldesignbriefs.com/component/content/article/mdb/features/articles/25545.
30. Mosley, Brian, NSF funded IoT security research excites at the 2017 CNSF exhibition. http://cra.org/govaffairs/blog/2017/05/2017-cnsf-exhibition/, 2017.
31. Cha, Seunghun, and Kim, Hyoungshick, Detecting encrypted traffic: A machine learning approach. Vol. 10144. Springer, Cham. Information Security Applications, 2017.
32. https://nordicapis.com/securing-medical-iot-devices/
33. Hopper, N.J., and Blum, M., Secure human identification protocols. In International conference on the theory and application of cryptology and information security (pp. 52–66). Springer, 2001.
34. Peris-Lopez, P., Hernandez-Castro, J.C., Estevez-Tapiador, J.M., and Ribagorda, A., EMAP: An efficient mutual-authentication protocol for low-cost RFID tags. In OTM confederated international conferences "on the move to meaningful internet systems" (pp. 352–361). Springer, 2006.
35. Kulseng, L., Yu, Z., Wei, Y., and Guan, Y., Lightweight mutual authentication and ownership transfer for RFID systems. In INFOCOM, 2010 proceedings IEEE, IEEE, 2010, pp. 1–5.

36. Kardas, S., Akgün, M., Kiraz, M.S., and Demirci, H., Cryptanalysis of lightweight mutual authentication and ownership transfer for RFID systems. In Lightweight security & privacy: Devices, protocols, and applications (LightSec), 2011 workshop on, IEEE, 2011, pp. 20–25.
37. Cheng, Z.-Y., Liu, Y., Chang, C.-C., and Chang, S.-C., Authenticated RFID security mechanism based on chaotic maps. Security and Communication Networks, 6(2):247–256, 2013.
38. Akgün, M., Uekae, T., and Caglayan, M.U., Vulnerabilities of RFID security protocol based on chaotic maps. In 2014 IEEE 22nd international conference on network protocols, IEEE, 2014, pp. 648–653.
39. Benssalah, M., Djeddou, M., and Drouiche, K., Security enhancement of the authenticated RFID security mechanism based on chaotic maps. Security and Communication Networks, 7(12):2356–2372, 2014.
40. Akgün, M., Bayrak, A.O., and Çalayan, M.U., Attacks, and improvements to chaotic map-based RFID authentication protocol. Security and Communication Networks, 8(18):4028–4040, 2015.
41. Zhu, W., Yu, J., and Wang, T., A security and privacy model for mobile RFID systems in the Internet of Things. In Communication technology (ICCT), 2012 IEEE 14th international conference on, IEEE, 2012, pp. 726–732.
42. Erguler, I., A potential weakness in RFID-based internet-of-things systems. Pervasive and Mobile Computing, 20:115–126, 2015.
43. Fan, K., Gong, Y., Liang, C., Li, H., and Yang, Y., Lightweight and ultralightweight RFID mutual authentication protocol with cache in the reader for IoT in 5G. Security and Communication Networks, 9(16):3095–3104, 2015.
44. Zhao, Z., A secure RFID authentication protocol for healthcare environments using elliptic curve cryptosystem. Journal of Medical Systems, 38(5):1–7, 2014.
45. Srivastava, K., Awasthi, A.K., Kaul, S.D., and Mittal, R.C., A hash-based mutual RFID tag authentication protocol in the telecare medicine information system. Journal of Medical Systems, 39(1):153, 2014.
46. He, D., Kumar, N., Chilamkurti, N., and Lee, J.-H., Lightweight ECC based RFID authentication integrated with an ID verifier transfer protocol. Journal of Medical Systems, 38(10):1–6, 2014.
47. Wu, F., Xu, L., Kumari, S., Li, X., Das, A.K., and Shen, J., A lightweight and anonymous RFID tag authentication protocol with cloud assistance for e healthcare applications. Journal of Ambient Intelligence and Humanized Computing 9(4):919–930, 2018.
48. Jin, C., Xu, C., Zhang, X., and Zhao, J., A secure RFID mutual authentication protocol for healthcare environments using elliptic curve cryptography. Journal of Medical Systems, 39(3):24, 2015.
49. Zhang, Z., and Qi, Q., An Efficient RFID Authentication protocol to enhance patient medication safety using elliptic curve cryptography. Journal of Medical Systems, 38(5):47, 2014. doi:10.1007/s10916-014-0047-8.
50. Zhao, Z., A secure RFID authentication protocol for healthcare environments using elliptic curve cryptosystem. Journal of Medical Systems, 38(5):46, 2014. doi:10.1007/s10916-014-0046-9.
51. Farash, Mohammad Sabzinejad, Nawaz, Omer, Mahmood, Khalid, Chaudhry, Shehzad Ashraf, and Khan, Muhammad Khurram, A provably secure RFID authentication protocol based on elliptic curve for healthcare environments. Journal of Medical Systems, 40:165, 2016. doi:10.1007/s10916-016-0521-6.
52. Fan, K., Jiang, W., Li, H., and Yang, Y. Lightweight RFID protocol for medical privacy protection in IoT. IEEE Transactions on Industrial Informatics, 14(4):1656–1665, 2018.
53. Xie, Shaohao, Fangguo Zhang, and Rong Cheng. "Security-enhanced RFID authentication protocols for healthcare environment." Wireless Personal Communications, 1–16, 2020.
54. Zhang, Z., and Qi, Q., An efficient RFID authentication protocol to enhance patient medication safety using elliptic curve cryptography. Journal of Medical Systems, 38(5):47, 2014.

SOA-MZI-Based Nanoscale Optical Communication with various Modulation Formats

9

M. Margarat, Dr. B. Elizabeth Caroline, and S. Soumiya

Contents

9.1 INTRODUCTION

Advancement in the nano optical device design to support higher data rate communication is tremendous and demands higher bandwidth to prop up ever-increasing requirements of the 5G era. Recently,

DOI: 10.1201/9781003126645-9

all-optical communications have attracted a large interest due to the feasibility of higher spectral efficiency and large bandwidth. Success of emerging nano devices defines the intelligence and required information of the ultra-high speed communication channels, power efficient and portable super computing devices, ultra-fast switches, interconnects, and autonomous smart systems, and its size is restricted to a maximum of 100 nm. If semiconductor materials are fabricated with 1 to 100 nm, at least in one dimension, the device shares the properties of nano devices. Ever-increasing demand of connectivity in the telecommunication industry requires ultra-fast data rates, which in turn lead to increased memory and smart computing capabilities of communication devices with less power consumption. The prototype of all-optical devices based on high bit rate optical logic gates are coming out from the laboratories to avoid inefficient opto-electronics conversation in the existing systems. A variety of all-optical logic gates is available to explore all-optical signal processing using nonlinear waveguides [1], semiconductor optical amplifier (SOA) [2–4], an erbium doped fiber amplifier [5], and photonic crystal fiber [6–8]. An SOA-based design is significant because of its compact size, low power consumption, good wavelength conversion, smaller footprint, polarization sensitivity, and integration capability. SOA-based reporting work such as a magnitude comparator, half adder, and latches [9–15] using SOA exhibits good performance. The nonlinearity of SOA [16] includes self-phase modulation (SPM), cross-gain modulation (XGM), cross-phase modulation (XPM), four wave mixing (FWM), and cross polarization modulation (XPolM). SOA-based gates can be designed along with interferometric configurations such as ultra nonlinear interferometer (UNI) gates, Sagnac interferometer (SI) gates, Michelson interferometer (MI) gates, delay interferometer (DI) gates and Mach-Zehnder interferometer (MZI) gates.

All-optical gates can be designed with UNI configuration [17], where the orthogonally polarized clock pulse will be delayed by passing across polarization maintained fiber. All-optical logic gates implemented with Sagnac interferometers have a bulky structure but are comparatively faster [18–22]. The amplitude and phase modulated probe and pump signal along the SOA leads to destructive and constructive interference between the pump and probe signal, which consequently generates a logic 0 and 1. The 2×2 coupler in the SI method joins input and output port. The total phase shift of $\pi/2$ is produced in the Sagnac gate by applying the clock signal along the polarization maintained fiber. TWSOA is used in to design all-optical AND logic gate successfully at 20 Gbps [23]. Then, based on the constructive and the destructive superposition of the pump and probe signal, logic 0 and 1 are produced. The SOA-MI structure is used along with symmetrical Fiber Bragg grating to produce logic output based on XGM of SOA. The MI arrangement is very simple, and with proper arrangements, the results will be accurate in producing the required logic. In MI, superposition principle between the probe and pump signal will result in logic 0 and 1. Comparatively, the MI arrangement [24–26] requires much less power than MZI. The DI configuration [27] requires less power, as it possesses only one SOA for a nonlinear operation to achieve a logic function.

In the all-optical gates designed using MZI methods, the output of all-optical gates is realized when two SOAs are placed on the arms of the interferometer. Phase modulation can be changed to amplitude modulation when the phase difference is introduced in the arms of the interferometer. Phase difference can be obtained either by 2×2 coupler or using phase shifters. There are two types of MZI configuration, which includes co-propagating MZI and counter-propagating MZI. A co-propagating MZI uses filters to separate the probe and the pump signal, while there is no need of a filter in counter-propagating MZI because the probe and the pump signals travel in opposite direction.

In this chapter, an SOA–MZI-based secured all-optical communication systems with all-optical XOR gates is proposed, and investigation of various modulation formats over the proposed system is carried out with an aim to support higher data rate requirements of nanoscale communication devices. Section 9.2 presents the design of the SOA–MZI-based secured transmission link. Section 9.3 describes the impact of various modulation schemes over the proposed design, and Section 9.4 explains the results and calculated performance metrics. Section 9.5 concludes the chapter.

9.2 DESIGN OF SOA–MZI-BASED COMMUNICATION LINK

The block diagram of the proposed system is given in Figure 9.1. It consists of a transmitter, all-optical XOR gate, and receiver. An XOR operation is achieved with the XPM technique [28]. Choice of operating wavelength and source defines the performance of the system [29]. The operating wavelength of the pump and the probe signal at 10 Gbps are 1,552.52 nm and 1,558 nm, respectively. Two data signals are generated with a pseudorandom generator, and 80 ps delay is used to generate other data input. A Mach–Zehnder modulator is used to modulate the data and carrier signals. Modulated output is transmitted through an XOR gate to generate encrypted output. Message security is enhanced with encryption, which uses right key in both encryption and decryption process. In encryption, XOR logic is widely used because of its advantages such as computational speed, particularly in hardware, analysis feasibility, and commutative and associative property. An all-optical XOR gate is the basic building block of the encryption process in the proposed system, and it is used for secure data transmission in support of data hiding when no particular security is required. To achieve all-optical XOR operations, SOA-MZI configuration is used, where, the active region of SOA is altered by varying the carrier intensity of the pump signal. Optimized phase shift of π radians occurs when the lower frequency probe signal and higher frequency pump signal travels along the upper and lower arms of the SOA-MZI configuration, which in turn leads to XOR operation.

The Boolean function of XOR logic gives 1 at the output when nonuniform input data combines at SOA (combination of $A = 0$, $B = 1$, and $A = 1$, $B = 0$). On the other hand, when both the inputs are the same, (combination of $A = 1$, $B = 1$, and $A = 0$, $B = 0$), logic output turns into 0; 1 logic is represented by the presence of optical pulses, and 0 logic is represented by an absence of optical pulse at the receiver output. SOA parameters values used to realize proposed XOR gate are given in Table 9.I.

9.2.1 SOA

The semiconductor optical amplifier is highly nonlinear in nature and handles optical signals at its input and output. Its implementation model is based on the wave equation and carrier density evolution.

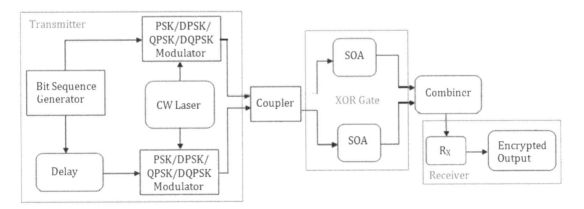

FIGURE 9.1 Block diagram of secured optical communication link.

TABLE 9.1 SOA Parameter Values

PARAMETERS	SOA 1 & 2
Bias current	180 mA
Confinement factor	0.3
Length	400 µm
Width	3 µm
Thickness	25 nm
Saturation power	28.48 mW
Line width enhancement factor	300/m
Spontaneous carrier lifetime	0.15 ns
Transparency carrier lifetime	1×10^{24}/m
Material gain constant	2×10^{-20}/m
Material loss	1050/m
I/P and O/P insertion loss	3 dB

Polarization of optical field is considered in the gain saturation process, and polarization dependence is considered negligible in the XGM and XPM process of SOA. The electric field propagation in the z-axis is given by the standard wave equation:

$$\frac{\partial^2 E(z,t)}{\partial z^2} - \frac{n^2 \partial^2 E(z,t)}{C^2 \partial t^2} = \frac{1}{\varepsilon_0 C^2} \frac{\partial^2 P}{\partial^2 t} \tag{9.1}$$

The relationship between polarization P and susceptibility X is given as $P = X E$ and

$$X(\mathbf{N}) = \frac{nc}{\omega}(\alpha - j)g(N) \tag{9.2}$$

Carrier dependent density $g(N)$ is calculated by

$$g(N) = \Gamma\left[\alpha(\mathbf{N} - \mathbf{N_0})\right] - \alpha_p \tag{9.3}$$

Wave equation is given by

$$\frac{\partial E}{\partial z} = \frac{1 + j\alpha}{2} g(z,t) E(z,t) \tag{9.4}$$

Integrating in z direction

$$E(L,t) = E(0,t) \exp\left[\frac{1 + j\alpha}{2} \int_0^L g(z,t)dz\right] \tag{9.5}$$

Carrier density rate equation is given as

$$\frac{dN}{dt} = \frac{I}{qV} - \frac{N}{\tau_s} - \frac{g(n)}{h\omega}|E(t,z)|^2 \tag{9.6}$$

$$\tau_s \frac{dg(z,t)}{dt} = -\left[g(z,t) - g_0\right] - \frac{1}{P_s}\frac{\partial}{\partial z}\left|E(t,z)\right|^2 \tag{9.7}$$

Where we have introduced the following parameters:

$$P_s = \frac{Ah\omega}{\Gamma a \tau_s}\left(saturation\ power\right) \tag{9.8}$$

$$\overline{N_0} = \frac{I\tau_s}{qV}\left(transparant\ carrier\ density\right) \tag{9.9}$$

$$g0 = \Gamma a\left[\overline{N_0} - N_0\right] \tag{9.10}$$

By integrating

$$\left[1 + \tau_s \frac{d}{dx}\right]\left[G(t) - G_0\right] = -\left[\frac{\left|E(L,t)\right|^2}{P_s} - \frac{\left|E(0,t)\right|^2}{P_s}\right] \tag{9.11}$$

Where $G_0 = g_0 L$ and $G(t) = \int_0^L g(z,t)dz$ $\qquad$ (9.12)

9.2.2 Bit Sequence Generator

A pseudorandom sequence generator is used to produce maximal length pseudorandom binary sequences. The general PRBS pattern is represented as $2^n - 1$, Where n indicates the length of the shift register used in creating the pattern. The pattern included all possible combinations of required n number of input binary bits, ones, and zeros, other than null value pattern.

9.2.3 Modulator

Classifications of various types of digital modulation techniques are given in Figure 9.2, and phase modulators are considered for the proposed design for its ability to support the higher data rate requirement of optical systems.

9.2.3.1 Phase Shift Keying (PSK) Modulator

PSK uses two phases separated by 180°, so it can also be called binary phase shift keying (BPSK). The constellation points for BPSK are positioned at 0° and 180° on the real axis, so it handles the highest distortion level and hence is the most robust out of all PSKs. But it could modulate only one bit per symbol and sets limitation over high data rate handling.

The BPSK general equation is given as

$$s_n(t) = \sqrt{\frac{2E_b}{T_b}}\cos\left(2\pi ft + \pi(1-n)\right); n = 0,1 \tag{9.13}$$

Which results in two phases given as

$$s_0(t) = \sqrt{\frac{2E_b}{T_b}}\cos\left(2\pi ft + \pi\right) = -\sqrt{\frac{2E_b}{T_b}}\cos\left(2\pi ft\right); for\ bit\ 0 \tag{9.14}$$

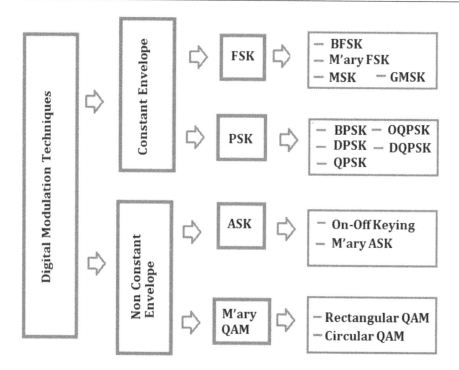

FIGURE 9.2   Classification of digital modulation techniques.

$$s_1(t) = \sqrt{\frac{2E_b}{T_b}} \cos(2\pi ft); for\, bit\, 1 \tag{9.15}$$

$$\text{And } \varphi(t) = \sqrt{\frac{2}{T_b}} \cos(2\pi ft) \tag{9.16}$$

9.2.3.2 Quadrature PSK (QPSK) Modulator

QPSK is a type of PSK technique with four states and uses four equispaced points on the constellation diagram. QPSK supports increased data rate by encoding two bits per symbol. Input data $b_k(t) = b_0, b_1, b_2, \ldots,$ is being divided into two data-streams: in-phase data, $b_i(t)$ and quadrature data, $b_q(t)$:

$$b_I(t) = b_0, b_2, b_{4,\ldots} \tag{9.17}$$

$$b_q(t) = b_1, b_3, b_{5,\ldots} \tag{9.18}$$

$b_I(t)$ and $b_q(t)$ each possess half of the bit rate with respect to $b_k(t)$.

Corresponding QPSK waveform, $S(t)$, is given as

$$S(t) = \frac{1}{\sqrt{2}} * b_I(t)\cos\left(2\pi f_0 t + \frac{\pi}{4}\right) + \frac{1}{\sqrt{2}} * b_q(t)\sin\left(2\pi f_0 t + \frac{\pi}{4}\right) \tag{9.19}$$

With trigonometric identities,

$$S(t) = \cos\left(2\pi f_0 t + \theta(t)\right) \tag{9.20}$$

TABLE 9.2 Truth Table of DQPSK Precoder Circuit

(I_{k-1}, Q_{k-1})		(0,0)		(0,1)		(1,0)		(1,1)	
A_K	B_K	I_K	Q_K	I_K	Q_K	I_K	Q_K	I_K	Q_K
0	0	0	0	0	1	1	0	1	1
0	1	1	0	0	0	1	1	0	1
1	0	0	1	1	1	0	0	1	0
1	1	1	1	1	0	0	1	0	0

The four combinations of in-phase data $b_I(t)$ and quadrature data $b_q(t)$ for the value of $\Theta(t)$ is calculated with

$$\theta(t) = 0°, \pm 90°, or\ 180° \tag{9.21}$$

9.2.3.3 DQPSK Modulator

Differential encoding of in-phase data and quadrature data is required before phase modulation in DPSK modulation, where the output depends on the current and previous stage inputs a_k, b_k and I_{k-1}, Q_{k-1}, respectively, and the corresponding truth table is given in Table 9.2.

$$I_k = \left(\overline{a_k \oplus b_k}\right).\left(a_k \oplus I_{k-1}\right) + \left(a_k \oplus b_k\right).\left(b_k \oplus Q_{k-1}\right) \tag{9.22}$$

$$Q_k = \left(\overline{a_k \oplus b_k}\right).\left(b_k \oplus Q_{k-1}\right) + \left(a_k \oplus b_k\right).\left(a_k \oplus I_{k-1}\right) \tag{9.23}$$

9.2.3.4 DPSK Modulator

DPSK follows PSK format definition, where the optical input stream is produced by the modulating phase of the input. Phase difference is applied to the carrier in accordance with consecutive data bits, which is the information part of DPSK; that is, phase shift is applied only when the previous bit value is 1; otherwise, no shift will be applied to produce DPSK encoded data.

9.2.4 Coupler

Each port of coupler gets optical inputs and couples the signals together; the coupler model is described in equation 9.24, where A is the amplitude of the complex optical field:

$$\begin{bmatrix} A_{01} \\ A_{02} \end{bmatrix} - \begin{bmatrix} \sqrt{1-\alpha} & j\sqrt{\alpha} \\ j\sqrt{\alpha} & \sqrt{1-\alpha} \end{bmatrix} \begin{bmatrix} A_{i1} \\ A_{i2} \end{bmatrix} \tag{9.24}$$

9.2.5 Combiner

The combiner is used to combine both the given inputs, and the coupler model is described in equations 9.24 and 9.25, where α, φ_1, φ_2, O, I_i and O_i are input power ratio coefficient, phase shift for input 1 and 2, output signal and i^{th} input, and output signal, respectively

$$o = I_1\sqrt{\alpha}\,e^{j\varphi 1} + I_2\sqrt{(1-\alpha)}\,e^{j\varphi 2} \tag{9.25}$$

$$\left|\frac{\sqrt{\alpha}e^{j\varphi 1}}{\sqrt{(1-\alpha)}e^{j\varphi 2}}\right| I = \left|\begin{matrix} O_1 \\ O_2 \end{matrix}\right| \tag{9.26}$$

9.2.6 Receiver

Optical receiver sensitivity is measured based on the test conditions, and it assumes direction detection with negligible ASE noise. Sensitivity of the receiver depends on the pulse shape of the transmitted signal, post detection filter, and noise power.

The Q-factor calculated at the output of receiver satisfies equation 9.27, when shot noise and receiver noise is available.

$$\bar{P}_R = \frac{Q^2}{2\eta}h\nu G^x\frac{w_0}{r_0^2} + \frac{Q}{G\eta}h\nu\frac{\sigma_{th}}{qr_0} \tag{9.27}$$

Where r_0 is calculated from $r(t)$, with the sampling time-instant equal to t_0. The approximation used in calculation of performance metrics includes negligible pattern noise, sensitivity not closer to the quantum limit, Gaussian electrical noise, and negligible intersymbol interference.

9.2.7 Simulation Circuit of Secured Optical Communication Link

The simulation circuit of the proposed design is given in Figure 9.3, and it consists of three main sections, including transmitter, fiber link, and receiver. Binary sequences are given as input at 10 Gbps for RZ driving signals. During this simulation sequence, the length is 700 μm. Then the laser source is input for modulating a symbol with 1558.2 nm of its wavelength and set as 1 mw of input power for the given system. Another input is given from the source, and it gets delayed by 80 ps connected to the modulator. These two signals are combined by the coupler, and each of them is split by a beam splitter. One output of each splitter is directly connected to the SOA. The two SOAs are combined with the Mach–Zehnder interferometer to control the amplitude of the optical wave, providing a much better performance. The interferometer could also be a tool that is used to make precise measurements by

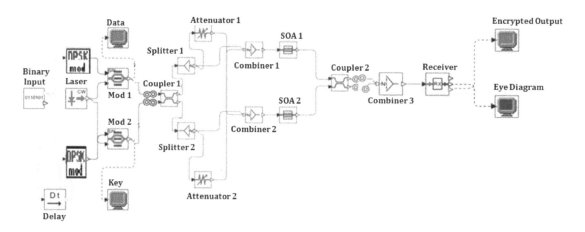

FIGURE 9.3 Simulation circuit of secured optical communication link.

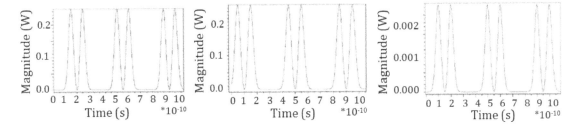

FIGURE 9.4 (a) First input to XOR gate. (b) Second input to XOR gate. (c) Coded XOR output.

interferencing the two beams of sunshine. It will split the signal from the output and is connected to the SOA. The phase of the signal gets changed because of change in intensity modulation. The SOA-MZI could also be a tool; it is used to determine the phase shift variation between two correlated beams that are exposed from the splitting of sunshine. The SOA receiver section obtaining a logical output from the encrypted signal and the resultant output for the inputs are given in Figure 9.4. By analyzing and evaluating the performances of these systems, we obtain an eye diagram, bit error rate, Q-value, and extinction ratio.

9.3 IMPACT OF VARIOUS MODULATION SCHEMES OVER PROPOSED SYSTEM

Contemporary and current all-optical devices have the ability to handle data rates up to 10 Gbps, but high-capacity extended-distance transmission links require an improved bandwidth to tackle the requirement. Optical researchers are interested in investigating various types of modulation to support a higher data rate [30–31]. Different modulation techniques have encountered likelihood studies to resolve nonlinear effect tolerance. Modulation formats are important in optical communication in reducing errors due to nonlinearities and cross-talks [32]. In any modulation method, the baseband signal is superimposed on the carrier wave, and input binary ones and zeroes are transmitted with corresponding changes based on the type of modulation technique. In amplitude shift keying (ASK), carrier wave amplitude is alternated between two discrete values with reference to the input data. In frequency shift keying (FSK), transmission of binary data is accomplished by changing its frequency. But these two methods suffer with power inefficiency. The next most power-efficient modulation types are phase shift keying (PSK); the phase of the carrier wave is changed between discrete values at the beginning of the pulse. Encoding information is based on the phase of the signal. The phase of the carrier signal is used as the reference to change the phase of the baseband signal. Bandwidth utilization is the highest limiting factor in the modulation techniques such as ASK, FSK, and PSK and becomes less efficient to handle higher data rates in optical communication systems. The introduction of multilevel modulation formats such as QPSK, DQPSK, and DPSK ensured higher spectral efficiency. In QPSK, bit rate is twice that of symbol rate, and it carries two continuous bits in one symbol in succession, which results in better bandwidth utilization equivalent to one-quarter of bit rate. This allows QPSK to have noticeably reduced dispersion. However, the complexity of the system increases proportionately with the increased SNR requirement. Alternatively, DPSK provides a greater than 1 bps-Hz boost in spectral efficiency and also ensures robust transmission. In DQPSK, to encode symbol information, symbol to symbol phase changes are used instead of absolute phase. In DQPSK, as in QPSK, one symbol encompasses two bits, and thus the bit rate is twice the symbol rate. The reduced system intricacy, PMD tolerance, increased chromatic dispersion, and spectral efficiency of DQPSK facilitate ultra-long distance transmission possibilities. Despite all these advantages with respect to QPSK, DQPSKs incorporate optical transceivers, which increases the size of the system and its power requirements. So to overcome this challenge, DPSK can be considered alternatively because of the absence of local carrier frequency generation at the receiver. In differential phase shift keying (DPSK), the phase of the previous bit is used as a reference of relative phase for avoiding the necessity for a reference coherent phase at the receiver. Apart from this, other modulation formats are susceptible to XPM in saturated SOA that turns on crosstalk penalty, but the pulse pattern effect is zero in a DPSK-modulated optical pulse and supports XPM [33]. Recent research articles report various advanced modulation formats and explore various applications in the design of all-optical devices [34–40]. Among them, PSK, QPSK, DQPSK, and DPSK are considered in this research work to investigate the proposed SOA-MZI-based all-optical secured communication system.

9.4 RESULTS AND DISCUSSION

A BER pattern along with eye diagram and spectrum output analysis is essential to ensure the overall signal integrity of the transmission path. The eye pattern of the received signal gives a quantitative analysis of the measured performance in terms of attenuation, channel bandwidth, jitter, and fall/rise time variations. The eye pattern begins to close when the signal frequency increases. The sampling time interval of the received signal is represented by the width of eye opening. Noise margin can be calculated with the height of the eye pattern, used to calculate the immunity to noise or noise margin. The obtained eye pattern along with calculated values is given in Figure 9.5. The spectrum of the encrypted signal is given in Figure 9.6 and describes qualitative analysis of power spectral density, signal intensity, group, phase, and polarization dispersion. Differential modulation formats exhibit significantly broader spectrum and results in a clear diagram. Obtained clear eye pattern and wider spectrum of DPSK-modulated output describes its suitability for SOA-based optical systems comparatively among considered multilevel modulation formats.

The optimum value of length, width, and thickness of the SOA in design are obtained by calculating various performance metrics such as BER, Q-factor and extinction ratio using equations 9.28–9.30, and the resultant graph is given in Figures 9.7 through 9.9). The phase shift modulated signal finds its suitability for optical communication systems because of its advantages like spectral efficiency, transmission reach, and BER. The obtained values describe that differential modulation schemes are the best choice in the design of SOA-based optical systems due their reduced BER and higher ER and QF values. System performances at a data rate of 10 Gbps, 50 Gbps, and 100 Gbps under various modulation formats are compared and given in Table 9.2.

$$ER = 10 \, \text{Log} \frac{P_1}{P_2} \, (\text{dB}) \tag{9.28}$$

$$QF = \omega_0 \frac{W}{P_{loss}} \tag{9.29}$$

$$BER = 0.5 \, \text{erfc} \frac{QF}{\sqrt{2}} \tag{9.30}$$

Where P_1 is OFF power and P_2 is ON power. ω_0, w, P_{loss} are frequency, stored energy, and overall power loss, respectively.

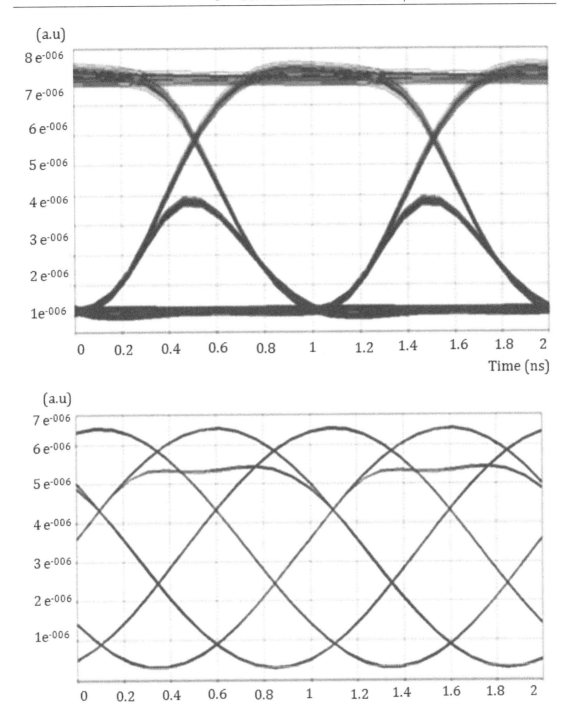

FIGURE 9.5   Eye diagram of (a) PSK, (b) QPSK, (c) DQPSK, and (d) DPSK.

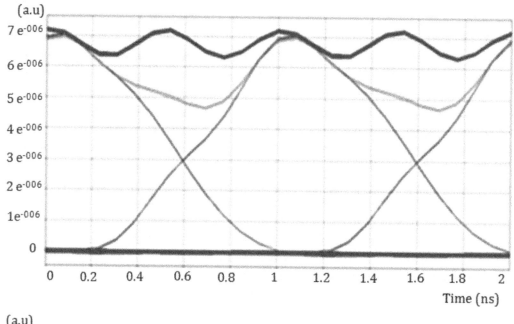

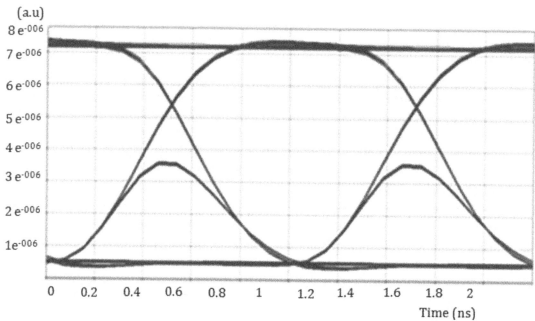

FIGURE 9.5   (Continued)

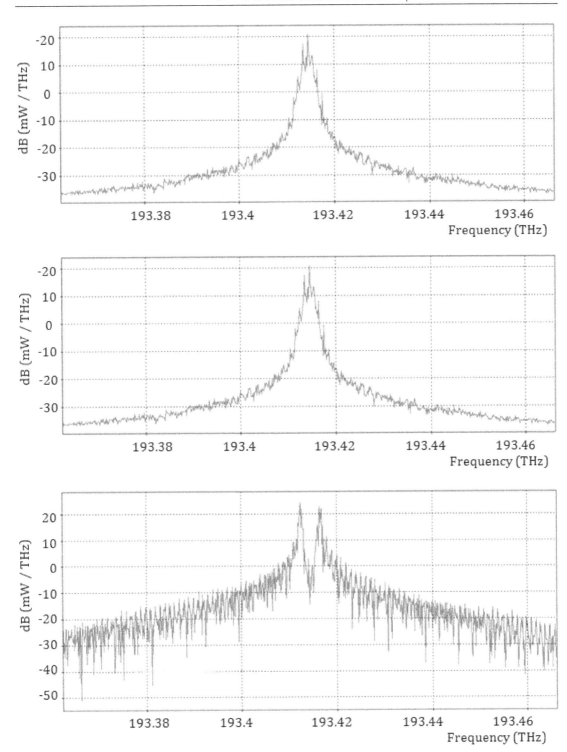

FIGURE 9.6   Spectrum of (a) PSK, (b) QPSK, (c) DQPSK, and (d) DPSK.

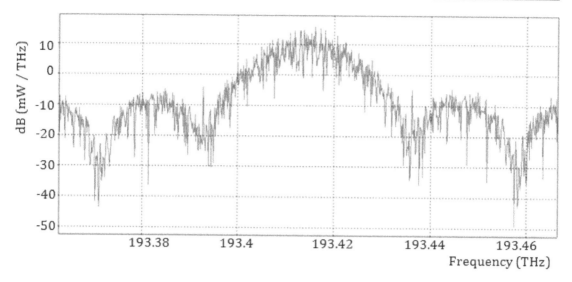

FIGURE 9.6   (Continued)

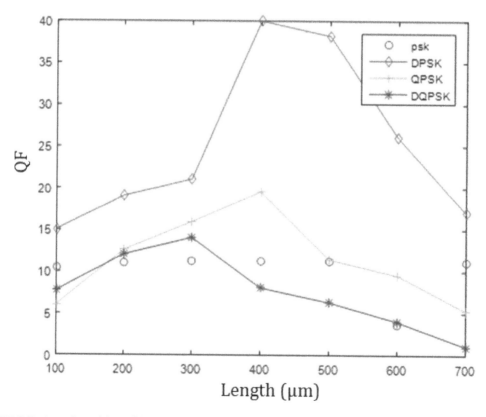

FIGURE 9.7   Length vs. (a) *QF*, (b) *BER*, and (c) *ER*.

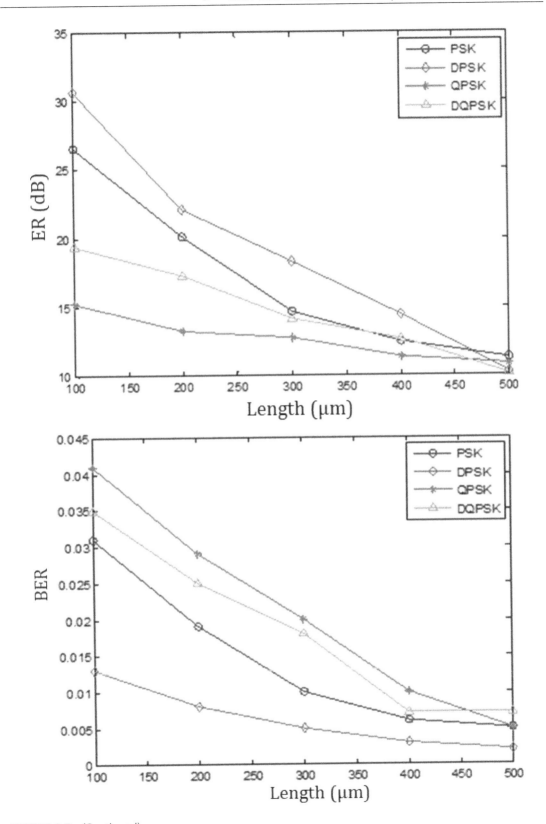

FIGURE 9.7   (Continued)

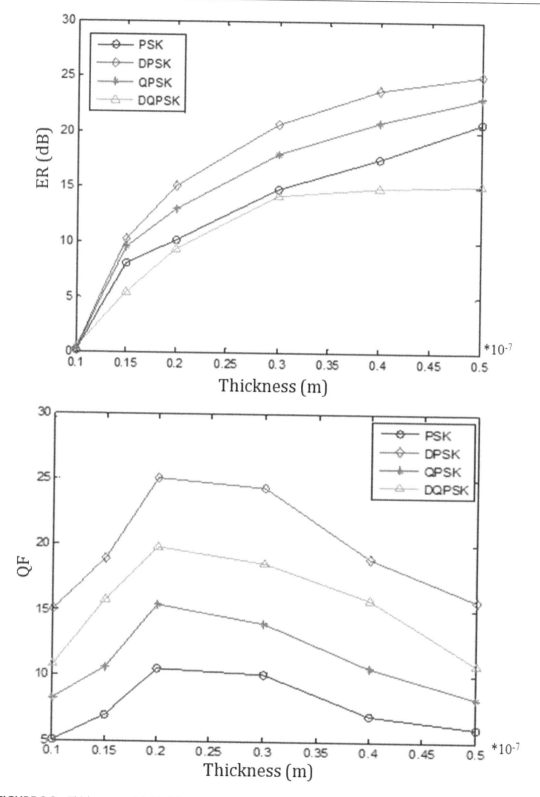

FIGURE 9.8 Thickness vs. (a) *QF*, (b) *BER*, and (c) *ER*.

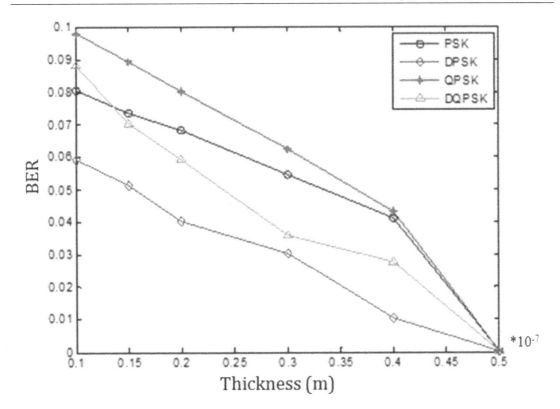

FIGURE 9.8   (Continued)

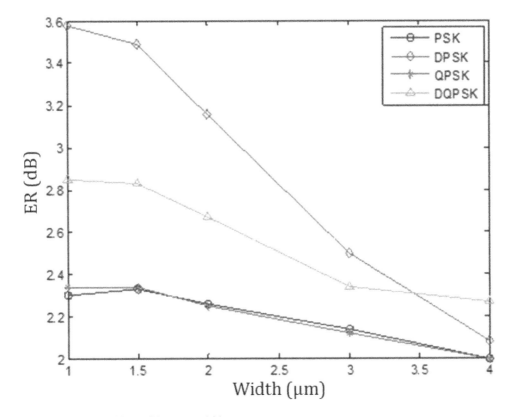

FIGURE 9.9   Width vs. (a) *QF*, (b) *BER*, and (c) *ER*.

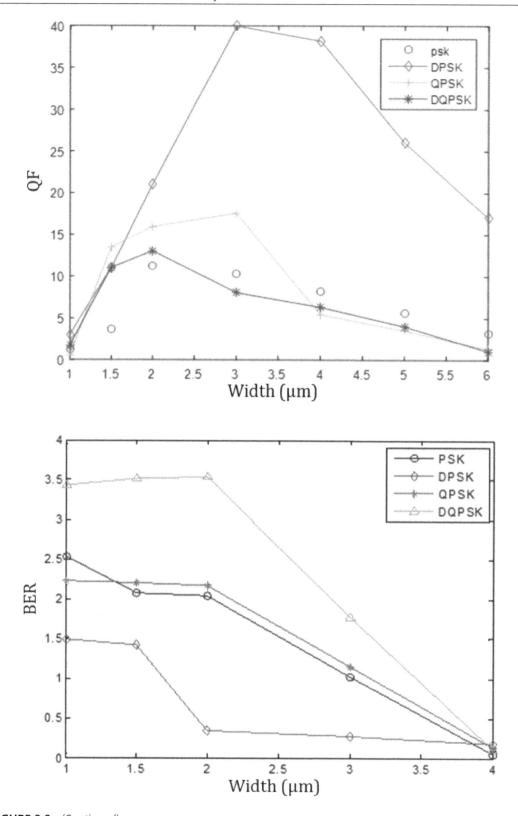

FIGURE 9.9   (Continued)

TABLE 9.3   Comparison of PSK, QPSK, DQPSK, DPSK Modulation Formats over Proposed System

DATA RATE	MODULATION FORMAT	BER	QF	ER (DB)
10 Gbps	PSK	$0.1493\ e^{-3}$	11	16
	QPSK	$0.999\ e^{-39}$	35.66	20
	DPSK	$1.0\ e^{-40}$	40	21
	DQPSK	$1.0\ e^{-40}$	33.93	26
50 Gbps	PSK	$0.001\ e^{-1}$	7.12	11
	QPSK	$0.9990\ e^{-40}$	16	18
	DPSK	$1.0\ e^{-40}$	40	21
	DQPSK	$0.455\ e^{-3}$	10	15
100 Gbps	PSK	$0.002\ e^{-1}$	6	9
	QPSK	$0.453\ e^{-39}$	7.45	10
	DPSK	$1.0\ e^{-40}$	40	21
	DQPSK	$0.2275\ e^{-1}$	6	8

9.5 CONCLUSION

In this chapter, analysis of various phase shifted modulation formats over a SOA-MZI-based secured optical communication system is done in which the security is ensured with all-optical XOR gate. The *QF*, *BER*, and *ER* values of the proposed design with respect to various modulation schemes over proposed all-optical devices are investigated. The obtained values ensure that differential modulation schemes are best suited for the design of SOA-based optical communication system. Focus on future research would be in advanced multilevel modulation formats and for increased transmission rates.

REFERENCES

[1] Yabu, Tetsuro, Masahiro Geshiro, Toshiaki Kitamura, Kazuhiro Nishida, and Shinnosuke Sawa. "All-optical logic gates containing a two-mode nonlinear waveguide." IEEE Journal of Quantum Electronics 38, no. 1 (2002): 37–46.

[2] Kim, Jae Hun, Young Min Jhon, Young Tae Byun, Seok Lee, Deok Ha Woo, and Sun Ho Kim. "All-optical XOR gate using semiconductor optical amplifiers without additional input beam." IEEE Photonics Technology Letters 14, no. 10 (2002): 1436–1438.

[3] Shanmugapriya, P., B. Elizabeth Caroline, and M. Margarat. "Design of various all-optical code converters using SOA-MZI." In 2020 7th International Conference on Smart Structures and Systems (ICSSS), pp. 1–6. IEEE, 2020.

[4] Margarat, M., B. Elizebeth Caroline, and M. Vinothini. "Design of all-optical universal gates and verification of Boolean expression using SOA-MZI." In 2018 IEEE International Conference on System, Computation, Automation and Networking (ICSCA), pp. 1–6. IEEE, 2018.

[5] Singh, Simranjit, and Rajinder S. Kaler. "Performance investigation of Raman erbium-doped fiber amplifier hybrid optical amplifier in the scenario of high-speed orthogonal-modulated signals." Optical Engineering 53, no. 3 (2014): 036102.

[6] Caroline, B. Elizabeth, M. Margarat, J. Vidhya, D. Purushothaman, and R. Jayasri. "Design of superficial half adder with 2D photonic crystals multi resonance effect." In 2020 7th International Conference on Smart Structures and Systems (ICSSS), pp. 1–5. IEEE, 2020.

[7] Araújo, Antônio, Antônio Oliveira, Francisco Martins, Amarílio Coelho, Wilton Fraga, and José Nascimento. "Two all-optical logic gates in a single photonic interferometer." Optics Communications 355 (2015): 485–491.

[8] Christina, X. Susan, A. P. Kabilan, and P. Elizabeth Caroline. "Ultra compact all optical logic gates using photonic band-gap materials." Journal of Nanoelectronics and Optoelectronics 5, no. 3 (2010): 397–401.

[9] Gayen, Dilip Kumar, Arunava Bhattachryya, Tanay Chattopadhyay, and Jitendra Nath Roy. "Ultrafast all-optical half adder using quantum-dot semiconductor optical amplifier-based Mach-Zehnder interferometer." Journal of Lightwave Technology 30, no. 21 (2012): 3387–3393.

[10] Chakraborty, Bikash, and Sourangshu Mukhopadhyay. "All-optical method of developing half and full subtractors by the use of phase encoding principle." Optik 122, no. 24 (2011): 2207–2210.

[11] Chattopadhyay, Tanay, and Dilip Kumar Gayen. "Reconfigurable all-optical delay flip flop using QD-SOA assisted Mach – Zehnder interferometer." Journal of Lightwave Technology 32, no. 23 (2014): 4571–4577.

[12] Kumar, Santosh, Ashish Bisht, Gurdeep Singh, Kuldeep Choudhary, K. K. Raina, and Angela Amphawan. "Design of 1-bit and 2-bit magnitude comparators using electro-optic effect in Mach – Zehnder interferometers." Optics Communications 357 (2015): 127–147.

[13] Li, Wenbo, Shaozhen Ma, Hongyu Hu, and Niloy K. Dutta. "All-optical latches based on two-photon absorption in semiconductor optical amplifiers." JOSA B 29, no. 9 (2012): 2603–2609.

[14] Wang, Jing, Gianluca Meloni, Gianluca Berrettini, Luca Potì, and Antonella Bogoni. "All-optical clocked flip-flops exploiting SOA-based SR latches and logic gates." In International Workshop on Optical Supercomputing, pp. 5–18. Springer, Berlin, Heidelberg, 2009.

[15] Li, Wenbo, Hongyu Hu, and Niloy K. Dutta. "Optical latches using optical amplifiers." In Photonic Applications for Aerospace, Commercial, and Harsh Environments IV, vol. 8720, p. 87200O. International Society for Optics and Photonics, SPIE Digital Library, 2013.

[16] Srivastava, Vikrant K., Devendra Chack, Vishnu Priye, and Chakresh Kumar. "All-optical XOR gate based on XGM properties of SOA." In 2010 International Conference on Computational Intelligence and Communication Networks, pp. 544–547. IEEE, 2010.

[17] Webb, R. P., X. Yang, R. J. Manning, and R. Giller. "All-optical 40 Gbit/s XOR gate with dual ultrafast nonlinear interferometer." Electronics Letters 41, no. 25 (2005): 1396–1397.

[18] Ferreira, A. C., A. G. Coêlho Jr, J. R. R. Sousa, C. S. Sobrinho, F. T. C. B. Magalhães, G. F. Guimarães, J. C. Sales, J. W. M. Menezes, and A. S. B. Sombra. "PAM – ASK optical logic gates in an optical fiber Sagnac interferometer." Optics & Laser Technology 77 (2016): 116–125.

[19] Zoiros, K. E., G. Papadopoulos, T. Houbavlis, and G. T. Kanellos. "Theoretical analysis and performance investigation of ultrafast all-optical Boolean XOR gate with semiconductor optical amplifier-assisted Sagnac interferometer." Optics Communications 258, no. 2 (2006): 114–134.

[20] Chattopadhyay, Tanay, and Jitendra Nath Roy. "Semiconductor optical amplifier (SOA)-assisted Sagnac switch for designing of all-optical tri-state logic gates." Optik 122, no. 12 (2011): 1073–1078.

[21] Li, Qiliang, Mengyun Zhu, Dongqiang Li, Zhen Zhang, Yizhen Wei, Miao Hu, Xuefang Zhou, and Xianghong Tang. "Optical logic gates based on electro-optic modulation with Sagnac interferometer." Applied Optics 53, no. 21 (2014): 4708–4715.

[22] Singh, Karamdeep, and Gurmeet Kaur. "Interferometric architectures based all-optical logic design methods and their implementations." Optics & Laser Technology 69 (2015): 122–132.

[23] Oliveira, Jackson Moreira, Hudson Afonso Batista Silva, Fabio Barros Sousa, Jorge Everaldo Oliveira, and Marcos Benedito Caldas Costa. "Design of all-optical AND logic gate at 20 Gb/s based on SOA-MI with optimum injection current and length of TWA-SOA." Scientia Plena 15, no. 7 (2019).

[24] Mikkelsen, B., T. Durhuus, C. Joergensen, R. S Pedersen, S. L. Danielsen, K. E. Stuhkjaer, M. Gustavsson, W. van Berlo, and M. Janson, "10 Gbit/s wavelength converter realised by monolithic integration of semiconductor optical amplifiers and Michelson interferometer." In Proc. ECOC'94, Firenze, Italy, September 1994, vol. 4, pp. 67–70, 1994.

[25] Kim, Jae Hun, Young Tae Byun, Young Min Jhon, Seok Lee, Deok Ha Woo, Sun Ho Kim, and Kwang Nam Kang. "All-optical XOR gate by using semiconductor optical amplifiers." U.S. Patent 6,930,826, issued August 16, 2005.

[26] Stubkjaer, Kristian E. "Semiconductor optical amplifier-based all-optical gates for high-speed optical processing." IEEE Journal of Selected Topics in Quantum Electronics 6, no. 6 (2000): 1428–1435.

[27] Wang, Q., H. Dong, G. Zhu, H. Sun, J. Jaques, A. B. Piccirilli, and N. K. Dutta. "All-optical logic OR gate using SOA and delayed interferometer." Optics Communications 260, no. 1 (2006): 81–86.

[28] Kotb, Amer, and Fahad Alhashmi Alamer. "Dispersion on all-optical logic XOR gate using semiconductor optical amplifier." Optical and Quantum Electronics 48, no. 6 (2016): 327.

[29] Caroline, B. Elizabeth, M. Margarat, V. Gunaseeli, and M. Keerthana. "Performance analysis and comparison of various optical sources with optical soliton." In 2019 IEEE International Conference on System, Computation, Automation and Networking (ICSCAN), pp. 1–7. IEEE, 2019.

[30] Khan, MD Shahrukh Adnan, Mirza Mursalin Iqbal, Khandaker Sultan Mahmood, Anjuman Ara Anee, Md Sazzadur Rahman, and Shoaib Mahmud. "differential quadrature phase shift keying modulation in optical fiber communication-modelling, design, case implementation and limitation." In 2018 Fourth International Conference on Advances in Computing, Communication & Automation (ICACCA), pp. 1–6. IEEE, 2018.

[31] Khan, Md Shahrukh Adnan, Md Masum Howlader, Muhammad Ahad Rahman Miah, Sakhawat Hossen Rakib, Abdullah Al Amin, and Sharsad KaraKuni. "Performance analysis of receiver power sensitivity of advanced modulation formats in WDM based standard mode fibre for next generation data rate." In 2017 4th International Conference on Advances in Electrical Engineering (ICAEE), pp. 395–399. IEEE, 2017.

[32] Agalliu, Rajdi, and Michal Lucki. "Benefits and limits of modulation format for optical communications." Advances in Electrical and Electronic Engineering 12, no. 2 (2014).

[33] Cho, Pak S., and Jacob B. Khurgin. "Suppression of cross-gain modulation in SOA using RZ-DPSK modulation format." IEEE Photonics Technology Letters 15, no. 1 (2003): 162–164.

[34] Cavalcante, D. N. S., J. S. Negreiros, L. R. Marcelino, J. I. S. Miranda, R. R. Barboza, L. S. P. Maia, G. M. Medeiros, and G. F. Guimarães. "Experimental AND OR logic gates with MZI and SOA using PAM modulation." IEEE Photonics Technology Letters 31, no. 1 (2018): 11–14.

[35] Michael, Margarat, B. Elizabeth Caroline, and Susan Christina Xavier. "M-ary DPSK coded binary to gray, BCD to gray, and octal to binary all-optical code converters based on SOA-MZI configuration at 500 Gb/s." Applied Optics 59, no. 27 (2020): 8126–8135.

[36] Mishina, Ken, Daisuke Hisano, and Akihiro Maruta. "All-optical modulation format conversion and applications in future photonic networks." IEICE Transactions on Electronics 102, no. 4 (2019): 304–315.

[37] Kong, Deming, Yan Li, Hui Wang, Xinping Zhang, Junyi Zhang, Jian Wu, and Jintong Lin. "All-optical XOR gates for QPSK signals based on four-wave mixing in a semiconductor optical amplifier." IEEE Photonics Technology Letters 24, no. 12 (2012): 988–990.

[38] Kotb, Amer. "Analysis of all-optical logic XOR gate for 100 Gb/s phase-shift keying modulated data signals in semiconductor optical amplifier-based Mach-Zehnder interferometer." Optical and Quantum Electronics 47, no. 5 (2015): 1063–1070.

[39] Matsuura, Motoharu, Keita Mizusaka, and Naoya Oka. "All-optical OOK to multiple-level PSK format conversion using a self-generating optical clock." IEEE Photonics Technology Letters 28, no. 14 (2016): 1577–1580.

[40] Mao, Yaya, Bo Liu, Rahat Ullah, Tingting Sun, and Lilong Zhao. "All-optical XOR function accompanied with OOK/PSK format conversion with multicast functionality based on cascaded SOA configuration." Optics Communications 466 (2020): 125421.

Effect of Dielectric Material on Electrical Parameters Present near Source Region in Hetero Gate Dielectric TFET

10

Rajesh Saha, Suman Kumar Mitra, and Deepak Kumar Panda

Contents

10.1 INTRODUCTION

To follow Moore's law, the consistent scaling down of the nanoscale MOSFETs leads to a degradation in power consumption and various short channel performance effects [1–2]. It is very difficult to maintain the switching ratio (I_{ON}/I_{OFF}) with the scale of supply voltage, and also, the subthreshold swing (SS)

DOI: 10.1201/9781003126645-10

of MOSFET is not scalable below 60 mV/dec. at 300^0K temperature [3]. To overcome these barriers, researchers have proposed a device that operates on the band-to-band tunneling (BTBT) principle, which is popularly identified as a TFET (TFET) [4–5]. It is promising from a low power perspective due to thermal SS (< 60 mV/dec.) as well as low value of leakage current [6]. However, one major limitation of it is the lower value of the ON current. The performance of TFET is further enhanced by various techniques like gate engineering, structural engineering, dielectric engineering, and many more. The various device architectures proposed by researchers are double gate (DG), triple gate (TG), dual material gate (DMG), circular gate (CG), heterojunction, gate-all-around (GAA), cylindrical gate, gate underlap, and gate modulated TFETs [7–14], etc. In terms of channel engineering, the materials beyond Si are used as channel material to improve the performance of TFET [15–16]. Researchers have proposed a hetero-gate dielectric TFET (HG-TFET), and this device has improved SS and ON current compared to TFETs with SiO_2 as a gate dielectric [17]. Mitra et al. have included a back gate in HG-TFET, and the influence of back gate bias on electrical characteristics is highlighted [18].

In this chapter, we have considered three different gate dielectric materials (Al_2O_3, HfO_2, and La_2O_3) near the source region in HG-TFET. The effects of these dielectric materials on drain current characteristics, switching performance (I_{ON}/I_{OFF}), and V_T are highlighted in HG-TFET. We have reported the horizontal field and impact ionization rate taking Al_2O_3, HfO_2, and La_2O_3 as gate dielectric material near the source region in HG-TFET. Further, we have investigated DC and the hot carrier effect, taking drain bias as parameter in HG-TFET. Finally, we have shown such analysis considering L_{h-k} as a parameter in HG-TFET.

10.2 LITERATURE SURVEY

The first p–i–n structure, which is popularly known as TFET, was proposed in 1978 at Brown University, and the effectiveness of TFET for spectroscopy was advised [19]. Banerjee et al. have reported a TFET structure considering p-type channel in place of an i-region under the gate [20], and they have considered gate-source overlap in this structure, which imposed a line tunneling. The presence of line tunneling improves the SS and switching behavior of TFET. Reddick from Cambridge reported experimentally that TFET resolves the issue of the down-scaling dimension of MOSFET as well as punch-through [21].

Zhang et al. [6] theoretically investigated that SS of TFETs can be scaled down to less than 60 mV/dec., and thus TFET is a feasible device for applications of low-power. S. H. Kim fabricated a heterojunction TFET with Ge source having a switching ratio in the order of 10^6, and such device was fabricated using the established CMOS process flow [8]. Sandow et al. investigated the behavior of an ultra-thin body SOI TFET experimentally for the variation in channel length, thickness of gate oxide, and specification of doping in source/drain [22]. They have concluded that electrical parameters are strong function oxide thickness and doping concentration, with less dependence on gate length. The BTBT rate in TFET is enhanced by increasing the number of gate-like DG, TG TFETs [7], which leads to enhanced ON current. Saurabh and Kumar have reported the electrical parameters of DMG TFET, where lower and higher work function material is placed near the source and drain regions, respectively [9]. The DMG TFET poses better drain current, short channel, RF, and analog performance compared to conventional TFET. However, ON current of TFET is improved by introducing line tunneling along with horizontal tunneling [23–24]. The vertical tunneling produces a vertical field perpendicular to the channel, which enhanced the BTBT rate of TFET.

A compact analytical model for SOI TFET is reported in [25]. The parameters such as threshold voltage, channel charge, gate capacitance, drain current, SS, transconductance, and output conductance have been modeled analytically. The model for surface potential and transfer characteristic in a circular gate (CG)-TFET is reported, and this model is validated with the TCAD simulator [14]. Likewise, the various electrical parameters for triple material GAA TFET is modeled analytically, and it shows agreement with the simulator [13]. An analytical model for drain current in TFET considering a SiGe layer between source and channel region is presented in the literature [26]. This model is verified with a simulator for different mole fraction, gate bias, and dielectric constant.

The ON current (I_{ON}) and *SS* of a TFET are enhanced using a high-k as a gate dielectric material, with the expense of degradation in leakage and ambipolar current [27]. K. Boucart [28] reported that the tunneling mechanism in a double gate TFET is improved by using high-k gate dielectrics at the tunnel junction even with the scaled device dimensions. The electrical parameters of a hetero-gate dielectric TFET (HG-TFET) is highlighted in literature, and this device has improved *SS* and ON current compared to conventional TFETs [17]. A HG-TFET with back gate is reported, and the subthreshold characteristic of this device is not affected with the scaling of the dimensions [18]. The research reported that ON current of the TFET is not a function of back gate bias. The DC, RF/analog and linearity performance of a junctionless HG-TFET (JL HG-TFET) due to interface trap is compared with JL TFET [29]. Results reveal that HG engineering in TFET increases the transconductance and provides better linearity and lower distortion compared to JL TFET. The RF and linearity performance of hetero dielectric BOX TFET (HDB TFET) was investigated through a TCAD simulator, and results conclude that it has superior characteristic in terms of transconductance, cut-off frequency, and distortion parameters [30].

S. Mookerjea et al. have reported the HCE in TFET, and results show a large homogeneous horizontal field at the tunnel junction leads to a significant amount of carrier heating [31]. The HCE in high-k based TFET was investigated experimentally in the presence of trap charge, and the transfer characteristic degraded compared to TFET with SiO_2 as the gate dielectric [32]. A dual functional n-TFET and p-type impact ionization FETs (p-IFETs) was designed using the ATLAS simulator [33]. The hot carrier injection (HCI) stress in TFET is studied at the drain region having p/n$^+$ junction [34], and analysis reveals that the device performance degrades due to the presence of an interfacial layer at the drain side.

10.3 DEVICE STRUCTURE AND COMPUTATIONAL DETAILS

The 2D representation of the hetero-gate dielectric TFET (HG-TFET) designed in this research is portrayed in Figure 10.1. To improve the ON current and subthreshold characteristic of HG-TFET, the high-k gate dielectric is considered near the tunnel junction, while SiO_2 ($k = 3.9$) is taken at the drain side. The various high-k materials considered are Al_2O_3 ($k = 10$), HfO_2 ($k = 22$), and La_2O_3 ($k = 27$). The length of high-k and SiO_2 are represented by L_{h-k} and L_{SiO2}, respectively. The intrinsic channel is made of silicon, and the gate length is represented by L_G. The channel thickness (t_{Si}) = 10 nm is taken for HG-TFET. The

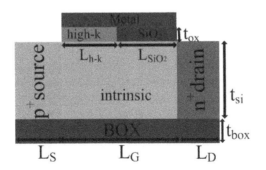

FIGURE 10.1   A 2D cross-section of HG-TFET.

p$^+$ source and n$^+$ drain regions are of equal length ($L_S = L_D$) = 20 nm. The thickness of gate dielectric (t_{ox}) = 2 nm, while BOX thickness (t_{box}) = 30 nm. The metal with work function (ϕ_M) = 4.1 eV is considered as gate material. The uniform doping concentrations are: p$^+$ source (= 10^{20} cm^{-3}), n$^+$ drain (= 5 × 10^{18} cm^{-3}), and intrinsic channel (= 10^{15} cm^{-3}).

Simulation of the HD-TFET is implemented through a 2D TCAD device simulator [35]. The transport of the carriers is analyzed through employing nonlocal BTBT Kane's model, having experimentally calibrated coefficients of A_{path} = 1.63 × 10^{14} cm^{-3} s^{-1}, B_{path} = 1.47 × 107 V cm^{-1}, and P_{path1} = 0.0567 eV [36]. Fermi–Dirac statistics and a bandgap narrowing model are assumed due to degenerate source and drain regions. An SRH model is assumed to account recombination of the carriers and as device dimensions are in nanoscale, the concentration-dependent Masetti model is activated. The high-field saturation model is activated in the simulator. The impact ionization rate is studied by enabling the avalanche model in the simulator.

10.4 RESULTS AND DISCUSSION

10.4.1 Effect of Dielectric Materials

This section presents the effect of dielectric materials near the tunnel junction on DC, and HCE performance is presented at V_{DS} = 1 V of HG-TFET. Here, we have considered L_G = 30 nm ($L_{h-k} = L_{SiO2}$ = 15 nm).

The effects of dielectric material near the source region in linear and log scale on transfer characteristic are shown in Figures 10.2a and b, respectively. When SiO$_2$ is taken as dielectric material, it will act as a conventional TFET. It is seen that ON current improves with insignificant change of ambipolar current as the dielectric material near the source region changed from SiO$_2$ to La$_2$O$_3$. This is because ON-state is extracted by tunnel junction, which is covered by high-k dielectric material, and the ambipolar behavior due to drain-channel junction is covered by SiO$_2$. On the other hand, the presence of high-k gate material degrades the OFF current of HG-TFET due to an increase in tunneling rate as dielectric material is changed from SiO$_2$ to La$_2$O$_3$.

In order to elaborately explain the behavior of transfer characteristics, we have plotted the BTBT rate and energy band of HD-TFET, taking dielectric material as the parameter, as shown in Figures 10.3 and 10.4, respectively. The higher value of the BTBT rate indicates enhancement in the ON current. It is perceived from Figure 10.3 that the tunneling rate at source/channel junction improves as the dielectric constant of gate material increases, and thus ON current of the device is improved. Likewise, it is observed from Figure 10.4 that the bandgap between the conduction (E_C) and valence (E_V) band reduces as the gate material dielectric constant increases. This reduction in energy bandgap leads to enhancement in tunneling rate, and thus, ON current is improved.

The switching ratio ($I_r = I_{ON}/I_{OFF}$) and threshold voltage (V_T) as a function of gate dielectric material is summarized in Figure 10.5. It is visualized that I_r of HG-TFET reduces as the dielectric material is changed, and this is due to significant degradation in OFF current compared to improvement in ON current. As the gate dielectric constant rises, the ON current increases, and thus less gate bias is required to create channel in HG-TFET, which indicates lower V_T as the gate dielectric constant increases.

The I_{ON}/I_{AMB} ratio at fixed V_{DS} = 1 V as a function of dielectric material is summarized in Figure 10.6. It is understood that the I_{ON}/I_{AMB} ratio improves as the dielectric constant of gate dielectric near the source junction rises, and this is due to increase in ON current.

The effect of dielectric materials near the source region on the horizontal field and impact ionization rate are presented in Figures 10.7a and b, respectively. It is observed in Figure 10.7a that the horizontal field increases as the dielectric material changed from SiO$_2$ to La$_2$O$_3$, which indicates channel is more

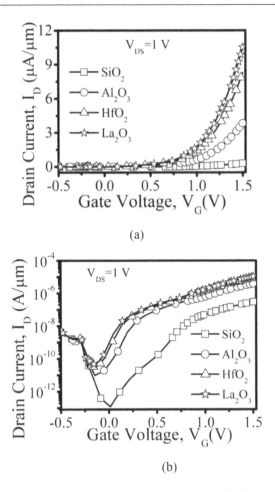

(a)

(b)

FIGURE 10.2 Effect of dielectric materials on transfer characteristic in (a) linear and (b) log scale.

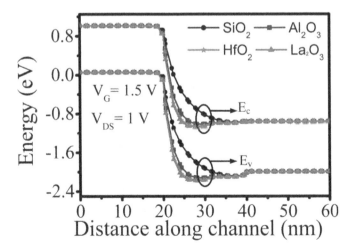

FIGURE 10.3 Effect of dielectric materials on eBTBT rate.

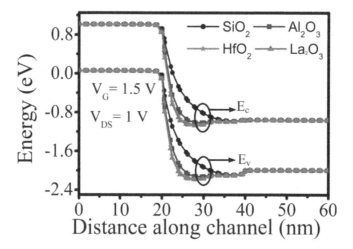

FIGURE 10.4   Effect of dielectric materials on energy bandgap.

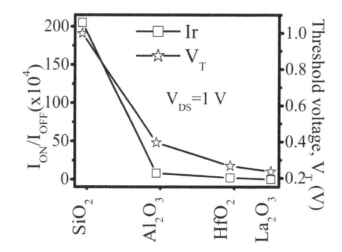

FIGURE 10.5   Effect of dielectric materials on V_T and I_{ON}/I_{OFF}.

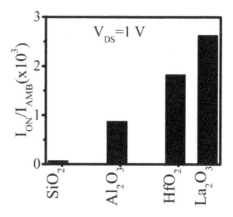

FIGURE 10.6   Effect of dielectric materials on I_{ON}/I_{AMB}.

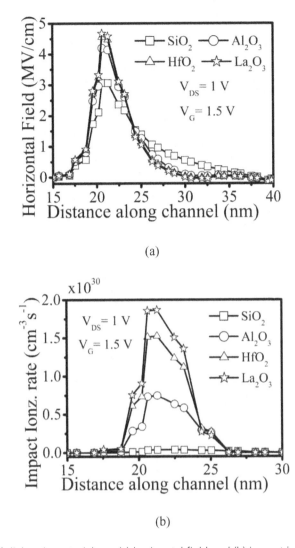

FIGURE 10.7   Effect of dielectric materials on (a) horizontal field and (b) impact ionization rate.

controlled by drain bias instead of gate bias. This increased in horizontal field increases the impact ionization rate for La_2O_3 as dielectric material compared to conventional TFETs, as summarized in Figure 10.7b. This increase in horizontal field and ionization rate indicates the degradation in HCEs of HG-TFET.

10.4.2 Effect of Drain Bias

This section presents the DC and HCE performance in HG-TFET, taking V_{DS} as a parameter for HfO_2 as dielectric material near the source region. We have taken $L_G = 30$ nm ($L_{h-k} = L_{SiO2} = 15$ nm).

The influence of drain bias (V_{DS}) in linear and log scale on transfer characteristic for HG-TFET are shown in Figures 10.8a and b, respectively. It is perceived that ON current of HG-TFET is increased with insignificant changes in OFF current as V_{DS} changes from 0.5 to 1.2 V. The log scale in Figure 10.8b summarized that I_{AMB} current is degraded significantly as V_{DS} is increased from 0.5 to 1.2 V. This behavior of drain current can be better explained from Figure 10.9, where we have shown BTBT rate for the variation

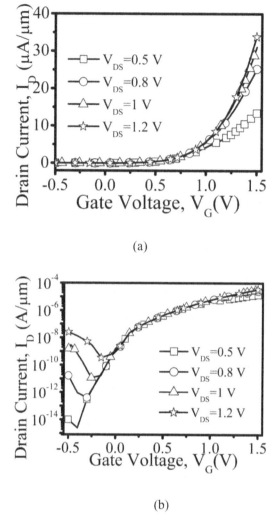

(a)

(b)

FIGURE 10.8   Effect of V_{DS} on transfer characteristic in (a) linear and (b) log scale.

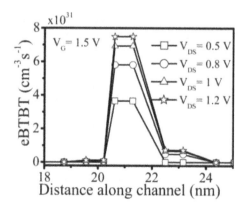

FIGURE 10.9   Effect of V_{DS} on eBTBT rate.

in V_{DS}. It is realized that the tunneling rate at the tunnel junction increases by a significant amount as V_{DS} is increased, and thus ON current of HG-TFET is improved. Further, the energy bandgap, taking V_{DS} as a parameter, is shown in Figure 10.10, and it is visualized that the increase in V_{DS} reduces the energy gap at tunnel junction. This leads to enhanced ON current at $V_{DS} = 1.2$ V compared to $V_{DS} = 0.5$ V.

The effect of V_{DS} on V_T and I_{ON}/I_{OFF} for HG-TFET considering HfO_2 as a gate dielectric is shown in Figure 10.11. As discussed in Figure 10.8, the ON current increases, while OFF current changes insignificantly, which in turn increases I_{ON}/I_{OFF} ratio as V_{DS} increases. As ON current of the device is increased with a rise in V_{DS}, less threshold voltage is needed to activate the channel in HG-TFET. The effect of V_{DS} on I_{ON}/I_{AMB} considering HfO_2 as gate dielectric is portrayed in Figure 10.12. It is perceived that as

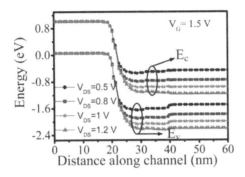

FIGURE 10.10   Effect of V_{DS} on energy bandgap.

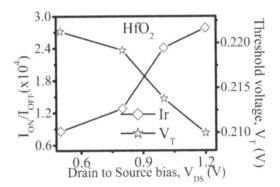

FIGURE 10.11   Effect of V_{DS} on V_T and I_{ON}/I_{OFF}.

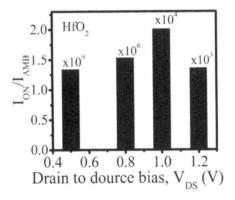

FIGURE 10.12   Effect of V_{DS} on I_{ON}/I_{AMB}.

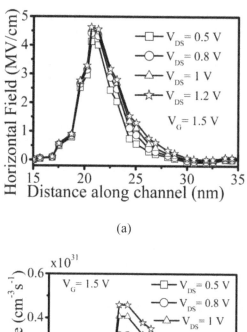

(a)

(b)

FIGURE 10.13   Effect of V_{DS} on (a) horizontal field and (b) impact ionization rate.

V_{DS} rises, the value of I_{ON}/I_{AMB} ratio decreases, which indicates degradation in ambipolar behavior of HG-TFET. Thus, a higher value of V_{DS} improves the I_{ON}/I_{OFF} ratio, while it degrades the I_{ON}/I_{AMB} ratio.

The effect of V_{DS} on horizontal field and impact ionization rate are demonstrated in Figures 10.13a and b, respectively. It is understood that both these components are increased with the increases in V_{DS}, pointing out that charge carriers become hot, which indicates degradation in HCE as V_{DS} changes from 0.5 to 1.2 V.

10.4.3 Effect of Length of High-k Gate Dielectric

Here, the DC and HCE performance for the variation in L_{h-k} is highlighted in HG-TFET. We have considered $V_{DS} = 1$ V, $L_{SiO2} = 15$ nm, and HfO_2 is used as the gate dielectric near the source region.

The effect of L_{h-k} on the transfer characteristic in linear and log scale is portrayed in Figures 10.14a and b, respectively. In this regard, we have considered $L_{h-k} = 5$, 10, 15, and 20 nm. It is seen that the ON current of HG-TFET is highest at $L_{h-k} = 15$ nm and lowest at $L_{h-k} = 10$ nm. However, the lower value of OFF current and ambipolar current are obtained at $L_{h-k} = 10$ nm in HG-TFET. In order to elaborately explain the behavior of the transfer characteristic, we have plotted the BTBT rate and energy bandgap, taking L_{h-k} as a parameter; this is summarized in Figures 10.15 and 10.16, respectively. It is seen that at L_{h-k} = 15 nm, the BTBT rate is increased and the energy bandgap is reduced compared to other values of L_{h-k}, and this leads to enhanced ON current.

The effect of L_{h-k} on I_{ON}/I_{OFF} ratio and V_T in HG-TFET is presented in Figure 10.17. It is realized that the switching ratio (I_r) decreases as L_{h-k} is increased, and this is pointed out in the transfer characteristic

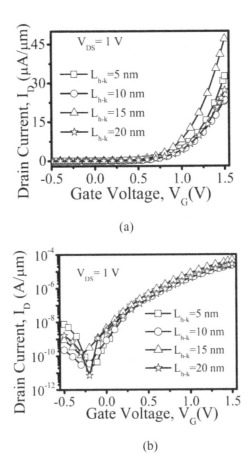

(a)

(b)

FIGURE 10.14   Effect of L_{h-k} on transfer characteristic in (a) linear and (b) log scale.

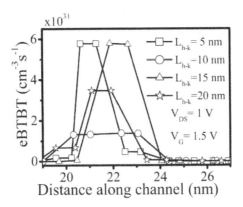

FIGURE 10.15   Effect of L_{h-k} on eBTBT rate.

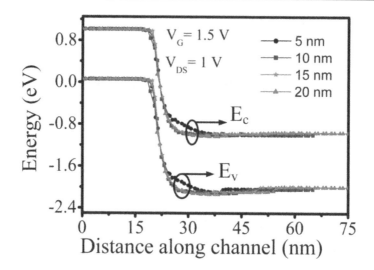

FIGURE 10.16   Effect of L_{h-k} on energy bandgap.

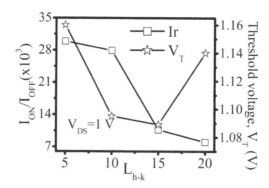

FIGURE 10.17   Effect of L_{h-k} on V_T and I_{ON}/I_{OFF}.

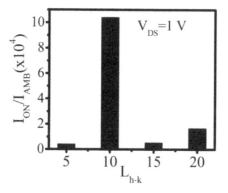

FIGURE 10.18   Effect of L_{h-k} on I_{ON}/I_{AMB}.

(Figure 10.14). The value of V_T is lowest at $L_{h-k} = 15$ nm, and this is due to enhanced ON current, which signifies low value of V_G is needed to turn on the HG-TFET. Figure 10.18 represents the I_{ON}/I_{AMB} as a function of L_{h-k} in HG-TFET. It is perceived that highest value of I_{ON}/I_{AMB} is obtained at $L_{h-k} = 10$ nm, and this is because of the lower value of ambipolar current.

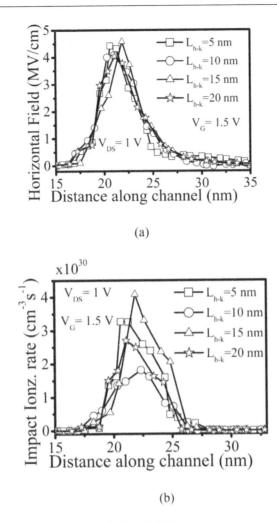

FIGURE 10.19   Effect of L_{h-k} on (a) horizontal field and (b) impact ionization rate.

To explain the HCEs as a function of L_{h-k} in HG-TFET, we have plotted the horizontal field and impact ionization rate, as summarized in Figures 10.19a and b, respectively. It is visualized that maximum peak of horizontal field is obtained at $L_{h-k} = 15$ nm compared to other values of L_{h-k}. Likewise, the peak value of ionization rate is obtained at $L_{h-k} = 15$ nm, as shown in Figure 10.19b. Thus, at $L_{h-k} = 15$ nm, the carriers become hot, which leads to significant amount of HCEs.

10.5 CONCLUSIONS

We have reported the influence of the gate dielectric material near the source region in HG-TFET on transfer characteristic, threshold voltage (V_T), I_{ON}/I_{OFF} ratio, I_{ON}/I_{AMB} ratio, and HCE. Furthermore, these parameters for the variation in V_{DS} considering HfO$_2$ as a dielectric are also highlighted. We have also reported the DC and HCEs for the variation in L_{h-k} in HG-TFET. It is perceived that for HG-TFET, the ON current is increased, while OFF current changes insignificantly as the dielectric material is changed from SiO$_2$ to La$_2$O$_3$. The HCEs also degraded while using a high-k dielectric near the source region. Analysis also reports that increasing V_{DS} improves the ON current but degrades the ambipolar current by a significant amount. Moreover, HCE degrades as V_{DS} increases from 0.5 to 1.2 V. The highest value of I_{ON} and improved I_{AMB} are obtained at L_{h-k} = 15 and 10 nm, respectively, for HG-TFET. The degradation in HCE is realized at L_{h-k} = 15 nm in HG-TFET.

ACKNOWLEDGMENT

The authors acknowledge the funding by Science & Engineering Research Board, Govt. of India (Sanction Reference. No. SRG/2019/000628).

REFERENCE

[1] A. C. Seabaugh and Q. Zhang, "Low-voltage tunnel transistors for beyond CMOS logic," *Proc. IEEE*, vol. 98, no. 12, pp. 2095–2110, Dec. 2010. doi:10.1109/JPROC.2010.2070470

[2] A. M. Ionescu and H. Riel, "Tunnel field-effect transistors as energy-efficient electronic switches," *Nature*, vol. 479, no. 7373, pp. 329–337, Nov. 2011. doi:10.1038/nature10679

[3] W. Y. Choi, B.-G. Park, J.-D. Lee, and T.-J. K. Liu, "Tunneling field-effect transistors (TFETs) with sub-threshold swing (SS) less than 60 mV/dec," *IEEE Electron Device Lett.*, vol. 28, no. 8, pp. 743–745, Aug. 2007. doi:10.1109/LED.2007.901273

[4] S. O. Koswatta, M. S. Lundstrom, and D. E. Nikonov, "Performance comparison between p–i–n tunneling transistors and conventional MOSFETs," *IEEE Trans. Electron Devices*, vol. 56, no. 3, pp. 456–465, Mar. 2009. doi:10.1109/TED.2008.2011934

[5] A. Pal, A. B. Sachid, H. Gossner, and V. R. Rao, "Insights into the design and optimization of tunnel-FET devices and circuits," *IEEE Trans. Electron Devices*, vol. 58, no. 4, pp. 1045–1053, Apr. 2011. doi:10.1109/TED.2011.2109002

[6] Q. Zhang, W. Shao, and A. Seabaugh, "Low-subthreshold-swing tunnel transistors," *IEEE Electron Device Lett.*, vol. 27, no. 4, pp. 297–300, Apr. 2006. doi:10.1109/LED.2006.871855

[7] L. Liu, D. Mohata, and S. Datta, "Scaling length theory of double-gate interband tunnel field-effect transistors," *IEEE Trans. Electron Devices*, vol. 59, no. 4, pp. 902–908, Apr. 2012. doi:10.1109/TED.2012.2183875

[8] S. H. Kim, H. Kam, C. Hu, and T.-J. K. Liu, "Ge-source tunnel field effect transistors with record high ION/IOFF," *VLSI Symp. Tech. Dig.*, pp. 178–179, 2009.

[9] S. Saurabh and M. J. Kumar, "Novel attributes of a dual material gate nanoscale tunnel field-effect transistor," *IEEE Trans. Electron Devices*, vol. 58, no. 2, pp. 404–410, Feb. 2011. doi:10.1109/TED.2010.2093142

[10] S. Dash and G. P. Mishra, "A 2D analytical cylindrical gate tunnel FET (CG-TFET) model: Impact of shortest tunneling distance," *Adv. Nat. Sci. Nanosci. Nanotechnol.*, vol. 6, no. 3, May 2015. doi:10.1088/2043-6262/6/3/035005

[11] L. De Michielis, L. Lattanzio, P. Palestri, L. Selmi, and A. M. Ionescu, "Tunnel-FET architecture with improved performance due to enhanced gate modulation of the tunneling barrier," *69th Device Research Conference*, Santa Barbara, CA, 2011, pp. 111–112. doi:10.1109/DRC.2011.5994440

[12] D. B. Abdi and M. Jagadesh Kumar, "Controlling ambipolar current in tunneling FETs using overlapping gate-on-drain," *IEEE J. Electron Devices Soc.*, vol. 2, no. 6, pp. 187–190, Nov. 2014. doi:10.1109/JEDS.2014.2327626

[13] N. Bagga and S. Dasgupta, "Surface potential and drain current analytical model of gate all around triple metal TFET," *IEEE Trans. Electron Devices*, vol. 64, no. 2, pp. 606–613, Feb. 2017. doi:10.1109/TED.2016.2642165

[14] R. Goswami and B. Bhowmick, "An analytical model of drain current in a nanoscale circular gate TFET," *IEEE Trans. Electron Devices*, vol. 64, no. 1, pp. 45–51, Jan. 2017. doi:10.1109/TED.2016.2631532

[15] S. Mookerjea and S. Datta, "Comparative study of Si, Ge and InAs based steep subthreshold slope tunnel transistors for 0.25 V supply voltage logic applications," *2008 Device Research Conference*, Santa Barbara, CA, 2008, pp. 47–48. doi:10.1109/DRC.2008.4800730

[16] R. Saha, B. Bhowmick, and S. Baishya, "Impact of WFV on electrical parameters due to High-k/Metal gate in SiGe channel tunnel FET," *Microelectron. Eng.*, vol. 214, 2019. doi:10.1016/j.mee.2019.04.024

[17] W. Y. Choi and W. Lee, "Hetero-gate-dielectric tunneling field-effect transistors," *IEEE Trans. Electron Devices*, vol. 57, no. 9, pp. 2317–2319, Sept. 2010. doi:10.1109/TED.2010.2052167

[18] S. K. Mitra, R. Goswami, and B. Bhowmick, "A hetero-dielectric stack gate SOI-TFET with back gate and its application as a digital inverter," *Superlattice. Microst.*, vol. 92, pp. 37–51, 2016. doi:10.1016/j.spmi.2016.01.040

[19] J. Quinn, G. Kawamoto, and B. McCombe, "Sub band spectroscopy by surface channel tunneling," *Surf. Sci.*, vol. 73, 1978. doi:10.1016/0039-6028(78)90489-2

[20] S. Banerjee, W. Richardson, J. Coleman, and A. Chatterjee, "A new three-terminal tunnel device," *IEEE Electron Device Lett.*, vol. 8, pp. 347–349, 1987. doi:10.1109/EDL.1987.26655

[21] W. Reddick and G. Amaratunga, "Silicon surface tunnel transistor," *Appl. Phys. Lett.*, vol. 67, pp. 494–496, 1995. doi:10.1063/1.114547

[22] C. Sandow and J. Knoch, "Impact of electrostatic and doping concentration on the performance of silicon tunnel field-effect transistors," *Solid-State Electron.*, vol. 53, pp. 1120–1129, 2009. doi:10.1016/j.sse.2009.05.009

[23] P. G. Der Agopian, J. A. Martino, A. Vandooren, R. Rooyackers, E. Simoen, A. Thean, and C. Claeys, "Study of line- TFET analog performance comparing with other TFET and MOSFET architectures," *Solid-State Electron.*, vol. 128, pp. 43–47, Feb. 2017. doi:10.1016/j.sse.2016.10.021

[24] S. W. Kim, W. Y. Choi, M.-C. Sun, H. W. Kim, and B.-G. Park, "Design guideline of Si-based L-shaped tunneling field-effect transistors," *Jpn. J. Appl. Phys.*, vol. 51 pp. 06FE09-1–06FE09-4, 2012. doi:10.1016/j.sse.2016.10.021

[25] B. Bhushan, K. Nayak, and V. R. Rao, "DC compact model for SOI tunnel field-effect transistors," *IEEE Trans. Electron Devices*, vol. 59, no. 10, pp. 2635–2642, 2012. doi:10.1109/TED.2012.2209180

[26] R. Goswami, B. Bhowmick, and S. Baishya, "Physics-based surface potential, electric field and drain current model of a δp^+ Si$_{(1-x)}$Ge$_x$ gate – drain underlap nanoscale n-TFET," *Int. J. Electron.*, vol. 103, no. 9, 2016. doi:10.1080/00207217.2016.1138514

[27] A. Vandooren, R. Rooyackers, D. Leonelli, F. Iacopi, E. Kunnen, D. Nguyen, M. Demand, P. Ong, L. Willie, J. Moonens, O. Richard, A. S. Verhulst, W. G. Vandenberghe, G. Groeseneken, S. D. Gendt, and M. Heyns, "A 35 nm diameter vertical silicon nanowire short-gate tunnel FET with high-k/metal gate," *Proceed. of IEEE Silicon Nano Electron.* Workshop, pp. 21–22, 2009.

[28] K. Boucart and A. M. Ionescu, "Length scaling of double gate tunnel FET with a high-k gate dielectric," *Solid-State Electronic*, vol. 51, pp. 1500–1507, 2007. doi:10.1016/j.sse.2007.09.014

[29] S. Gupta, K. Nigam, S. Pandey, D. Sharma, and P. N. Kondekar, "Effect of interface trap charges on performance variation of heterogeneous gate dielectric junctionless-TFET," *IEEE Trans. Electron Devices*, vol. 64, no. 11, pp. 4731–4737, Nov. 2017, doi:10.1109/TED.2017.2754297

[30] S. Narwal and S. S. Chauhan, "Investigation of RF and linearity performance of electrode work-function engineered HDB vertical TFET," *Micro & Nano Lett.*, vol. 14, no. 1, pp. 17–21, 2019. doi:10.1049/mnl.2018.5307

[31] S. Mookerjea and S. Datta, "Band-gap engineered hot carrier tunnel transistors," *2009 Device Research Conference*, University Park, PA, 2009, pp. 121–122. doi:10.1109/DRC.2009.5354869

[32] L. Ding, et al., "Investigation of hot carrier stress and constant voltage stress in High- κ Si-Based TFETs," *IEEE Trans. Device Mater. Reliab.*, vol. 15, no. 2, pp. 236–241, June 2015. doi:10.1109/TDMR.2015.2423095

[33] M. Kim, Y. Jeon, Y. Kim, and S. Kim, "Impact-ionization and tunneling FET characteristics of dual-functional devices with partially covered intrinsic regions," *IEEE Trans. Nanotechnol.*, vol. 14, no. 4, pp. 633–637, July 2015. doi:10.1109/TNANO.2015.2427453

[34] S. C. Kang, et al., "Hot-carrier degradation estimation of a silicon-on-insulator tunneling FET using ambipolar characteristics," *IEEE Electron Device Lett.*, vol. 40, no. 11, pp. 1716–1719, Nov. 2019. doi:10.1109/LED.2019.2942837

[35] Sentaurus Device User Guide, Synopsys, Inc., 2013.

[36] A. Biswas, S. S. Dan, C. L. Royer, W. Grabinski, and A. M. Ionescu, "TCAD simulation of SOI TFETs and calibration of non-local band-to-band tunneling model," *Microelectron. Eng.*, vol. 98, pp. 334–337, 2012. doi:10.1016/j.mee.2012.07.077

Quantum-Dot Cellular Automata-Based Encoder Circuit Using Layered Universal Gates

11

Birinderjit Singh Kalyan and Balwinder Singh

Contents

11.1 INTRODUCTION

Current technology for silicon transistors faces significant drawbacks, such as high power consumption. Another acute problem is the large size of the function, which makes it virtually impossible to fit in small devices. Nanotechnology is proposing a sustainable solution to this problem, which leads

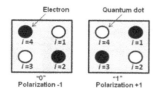

FIGURE 11.1   (a) Schematics of binary QCA cells. (b) Schematics of a QCA wire.

to quantum-dot cellular automata (QCA) growth. QCA is used to incorporate small-size designs, high packing density, and low power dissipation [1]. The designs in QCA are implemented as an array of cells where the cells undergo coulomb forces because of their neighboring cells [2]. QCA introduces a new mode of computation where the digital information is transmitted by polarization rather than current. Cells thus play the role of traditional wires in the circuit and allow for the transfer of information. The three-input majority gate, cable, and inverter are the rudimentary QCA logic circuits [3]. Further circuits that involve a combination of these basic gates can be built. The QCA stores rationale is reliant on singular electrons [2] positioned as appears in Figure 11.1a and not as voltage levels. The QCA cell comprises four quantum-specks situated toward the edges of a square [4,5]. There are two extra portable electrons for a cell that are allowed to burrow between close-by destinations. The coulombic aversion happening between two electrons makes them move their situation in inverse corners, the rising idea of polarization. In this way, there are just two stable states for the parallel information 1 and 0. On the off chance that polarization is −1, the rationale is 0, and when polarization is +1, the rationale is supposed to be 1. The customary QCA wire can be built essentially by falling a QCA cell chain, as appears in Figure 11.1b.

11.2 MAJORITY GATE AND CLOCK MECHANISM

QCA's fundamental building block is a gate to the majority. It is a five-cell 3-input gate, where the output will be high if two or more inputs are kept high. Consequently, if the inputs are called A, B, and C and output is Y, then the logical expression of Y can be written as

$$Y = AB + BC + CA \tag{11.1}$$

Thus, by fixing the value of one input to the majority gate, either to logic 1 or logic 0, we can design an AND or OR gate, as shown in Figure 11.2.

M (A, B, 0) = AB which is the AND operation
M (A, B, 1) = A + B which is the OR operation.

In this manner, any meaning of computerized rationale can be acknowledged by utilizing the greater part entryway and not the key. The check flags in different computerized circuits are key components that synchronize the advanced circuits and control the QCA circuit information stream. All through the QCA timing grouping, four distinctive progressive stages are partitioned into four clock zones, and a 90-degree stage delay happens between the particular check zones, as appears in Figure 11.3. The clock's four stages are: turn, hold, discharge, and unwind. As referenced in [6], commotion is significantly more of a worry in the format of the three information larger part entryway, in this way putting a separate clock zone at the information cells, center QCA cells, and dominant part yield.

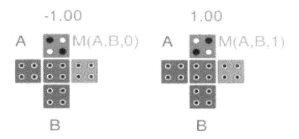

FIGURE 11.2   (a) The fundamental structure of QCA majority gate as a 2-input AND gate. (b) A 2-input OR gate.

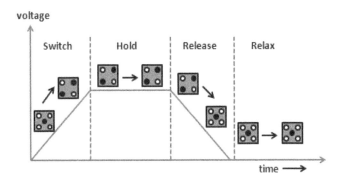

FIGURE 11.3   Clocking mechanism in QCA.

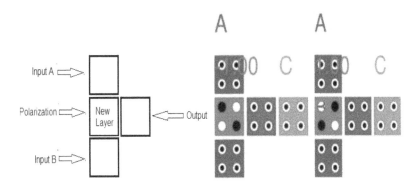

FIGURE 11.4   Basic design of NAND and NOR gate.

11.3 LAYERED NAND AND NOR GATE IN QCA

The encoder circuits are constructed using universal logic gates such as NAND and NOR gates, which were previously proposed in [7]. The design of the universal gate (UG) requires new-layer information, which makes it distinct from majority voting. Unlike majority voting, this gate design has a new layer block in the middle that is set to either +1 or −1 polarization. Polarization +1 is given to the new layer to create a NAND gate, and −1 to create a NOR gate, as shown in Figure 11.4.

Therefore, for

$$NAND\,Logic = UG(A, B, +1);$$

$$NOR\,Logic = UG(A, B, -1);$$

+1 or −1 are polarization in the new layer cell.

11.4 IMPLEMENTATION OF DIGITAL CIRCUITS USING UNIVERSAL GATES

11.4.1 Encoder Circuit

As shown in Figure 11.5, the encoder is a digital electronic circuit whose inputs are decimal digits, and the outputs are coded input representation. Table 11.1 displays the circuit's facts table. Just one of the input lines is kept high at one time, and the corresponding binary code is stored in the output pins. Hence an N bit encoder is made up of $2N$ input lines and N output lines. A 4 to 2 encoder, for example, has four inputs (i.e., A_0, A_1, A_2, and A_3), and two output bits (i.e., F_1, F_0).

11.4.2 Priority Encoder Circuit

The encoders mentioned to date function properly if only one of the input lines is held high. But sometimes it can happen that more than one input can be held high in an encoder. In that case, performance prediction gets difficult. Figure 11.6 shows the priority encoder circuit, which is based on the priority

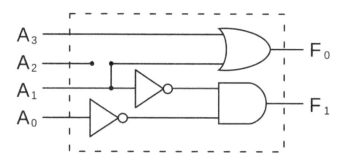

FIGURE 11.5   4 to 2 encoder circuit diagram.

TABLE 11.1   Truth Table of 4 to 2 Encoder Circuit

A_3	A_2	A_1	A_0	F_1	F_0
0	0	0	1	0	0
0	0	1	0	0	1
0	1	0	0	1	0
1	0	0	0	1	1

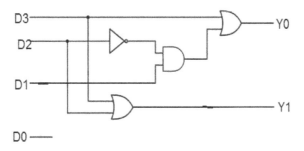

FIGURE 11.6 4 to 2 priority encoder circuit diagram.

TABLE 11.2 Truth Table of 4 to 2 Priority Encoder Circuit

D_3	D_2	D_1	D_0	Y_1	Y_0
0	0	0	0	X	X
0	0	0	1	0	0
0	0	1	X	0	1
0	1	X	X	1	0
1	X	X	X	1	1

values assigned to the inputs. The most popular set of priorities is based on the highest decimal input value that is held high. For example, if both D_0 and D_2 are high in a 4 to 2 priority encoder, the output will be 10, because it assigns the D_2 line of higher priority than the D_0 line and shows the output and the higher priority line, completely neglecting the lower priority line(s). Input D_3 has the highest priority, so if D_3 is high, output is 11 regardless of other input line status. So we can say from the table of the circuit and of truth,

$$Y_0 = D_3 + D_2'D_1$$
$$Y_1 = D_3 + D_2$$

11.4.3 Octal to Binary (8 to 3) Encoder

An octal to binary encoder accepts eight input lines and will produce 3-bit output based on the triggered input line. Its input lines range from D_0 to D_7, and the output is a 3-bit binary code ($Y_2 Y_1 Y_0$). The circuit diagram and the truth table for the octal to binary encoder are shown in Figure 11.7 and in Table 11.3.

From the truth table, it can be deduced that the most significant output bit (Y_2) will be 1 when any of D_4, D_5, D_6, and D_7 is high.

Therefore,

$$Y_2 = D_4 + D_5 + D_6 + D_7 \tag{11.2}$$

In the same way,

$$Y_1 = D_2 + D_3 + D_6 + D_7$$
$$\text{And } Y_0 = D_1 + D_3 + D_5 + D_7$$

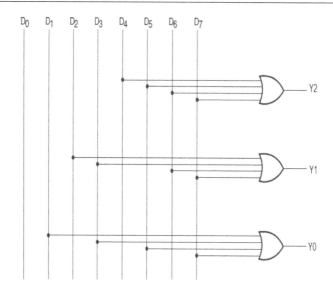

FIGURE 11.7   8 to 3 encoder circuit diagram.

TABLE 11.3   Truth Table of 8 to 3 Encoder Circuit

D_7	D_6	D_5	D_4	D_3	D_2	D_1	D_0	Y_2	Y_1	Y_0
0	0	0	0	0	0	0	1	0	0	0
0	0	0	0	0	0	1	0	0	0	1
0	0	0	0	0	1	0	0	0	1	0
0	0	0	0	1	0	0	0	0	1	1
0	0	0	1	0	0	0	0	1	0	0
0	0	1	0	0	0	0	0	1	0	1
0	1	0	0	0	0	0	0	1	1	0
1	0	0	0	0	0	0	0	1	1	1

11.5 POTENTIAL ENERGY IN QCA CELLS

In quantum-dot cellular automata, the cells are connected with one another. In each cell, there exist two electrons separated diagonally by a fixed distance. This configuration is made to avoid coulomb repulsion force. For practical calculation, it is assumed that dimensions of all cells are the same.

From QCA software, the width and height of each QCA cell is 18 nm, the dot diameter is 5 nm, and inter-cell spacing is 2 nm. For simplicity of calculation, it is assumed that only neighboring cells have an influence in determining the potential energy [8]. Therefore, in this way, the total potential energy of the design can be found for different input combinations [9, 10]. It is well known that for stability, the electrons should be aligned in such a way so as to reduce the potential energy as much as possible [11, 12]. The input state with least potential energy is the most stable and suitable for QCA operation [6].

11.6 QCA IMPLEMENTATION

11.6.1 QCA Layout and Simulation Results of Different Encoder Circuits

Figure 11.8 shows the QCA layout of a 4 to 2 encoder circuit and Figure 11.9 shows the simulated output of the circuit. The QCA layout of the 4 to 2 priority encoder circuit is shown in Figure 11.10, and Figure 11.11 shows the simulation results. Figures 11.12 and 11.13 show the octal to binary encoder circuit using QCA and the corresponding simulated output.

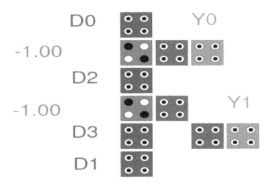

FIGURE 11.8   QCA layout of 4 to 2 encoder circuit.

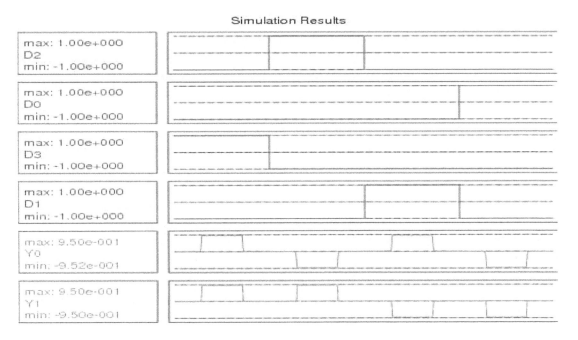

FIGURE 11.9   Simulated output of 4 to 2 encoder circuit.

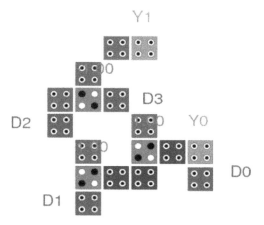

FIGURE 11.10   QCA layout of 4 to 2 priority encoder circuit.

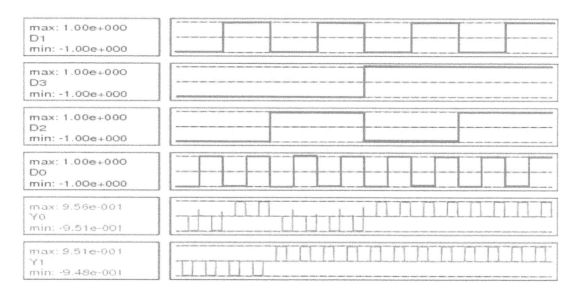

FIGURE 11.11   Simulated output of 4 to 2 priority encoder circuit.

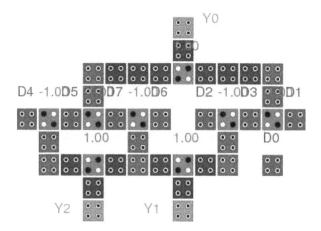

FIGURE 11.12   QCA layout of octal to binary encoder circuit.

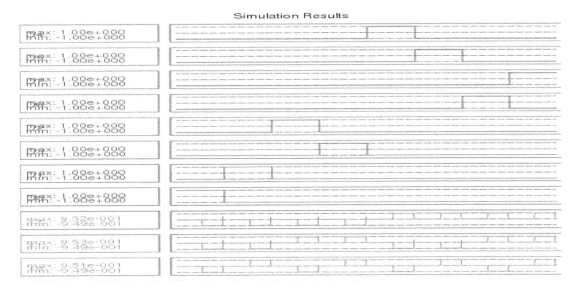

FIGURE 11.13   Simulated output of octal to binary encoder circuit.

TABLE 11.4   Cell Count, Area, Length, Breadth, and Latency of Simulated Encoder Circuits

DEVICE NAME	LENGTH (IN NM)	BREADTH (IN NM)	CELL COUNT	AREA (IN NM²)	OPERATION COST	LATENCY
4 to 2 Encoder	136	78	11	21,170	0.3928	0.25
4 to 2 Priority Encoder	138	122	17	24,764	0.4047	0.5
Octal to Binary Encoder	178	255	42	50,416	0.3589	0.5

11.7 OBSERVATION

From the following simulations, different values have been measured/ calculated and they are provided in the Table 11.4.

From this observation, it has be concluded that a 4 to 2 encoder is consuming 11 cell structures as compare to a 4 to 2 priority encoder and octal to binary encoder, which are 17 and 42 cell count, respectively. Hence, the 4 to 2 encoder is having lower complexities by 54% and 35% than the 4 to 2 priority encoder and octal to binary encoder.

11.8 COMPARATIVE STUDY

The design of the encoder using layered NAND and NOR gates helps the circuit use fewer cells, and the total covered area is significantly less than the previous design. For example, to realize 4 to 2 encoders, only 11 cells are used, and the area covered by the design is 21,170 nm². Different designs are available for priority encoders. Table 11.5 provides a comparative analysis of the numerous previously constructed priority encoder circuits.

TABLE 11.5   Comparison of Proposed Priority Encoder Design with Existing Designs

DESIGN REFERENCE	NUMBER OF CELLS	AREA REQUIRED (NM²)
Design as in [13]	183	292,042.14
Design as in [14]	178	210,000
Design as in [15]	100	130,000
Proposed Design	17	24,764

TABLE 11.6   Comparison of Proposed Octal to Binary Encoder Design with Existing Designs

DESIGN REFERENCE	NUMBER OF CELLS	AREA REQUIRED (NM²)
Design as in [15]	281	66,000
Design as in [16]	79	130,000
Proposed Design	42	50,416

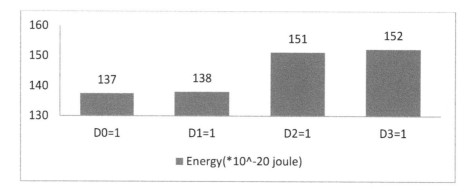

FIGURE 11.14   Potential energy of 4 to 2 encoder circuit for different input states.

Therefore, if the existing designs are compared, then it can be found that the design in [13] takes 183 cells to produce the circuit of the same efficiency as has been produced by the proposed design, which takes only 17 cells compared to designs [14] and [15], which took 178 and 100 cells, respectively.

Table 11.5 clearly describes the reduction in size and cell count in spite of giving the same output.

The comparative analysis with the previous designs of the octal to binary encoder is shown in Table 11.6.

Studying the comparative designs from Tables 11.5 and 11.6, the proposed designs are proved to be better than their predecessor in terms of the number of cells required and the area to realize the circuit.

11.9 POTENTIAL ENERGY OF ENCODER CIRCUITS FOR DIFFERENT INPUT STATES

In QCA, a circuit's potential energy also varies in relation to its input states and polarizations. The state with the least potential energy is the most stable one. The potential energies of various encoder circuits in different inputs are listed in Figures 11.14 through 11.16.

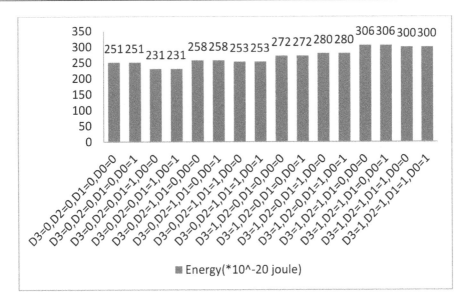

FIGURE 11.15 Potential energy of priority encoder circuit for different states.

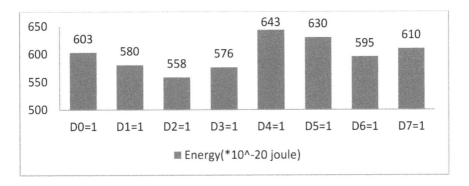

FIGURE 11.16 Potential energy of octal to binary encoder.

11.10 ENERGY DISSIPATION

The Hamiltonian matrix is used to calculate total energy and power in an array of QCA cells [4,5,17,18]. Hence, energy dissipation of a QCA cell in one clock cycle, $T_{CC} = [-T, T]$ is derived in respect to Hamiltonian and coherence vectors as described in equation 11.3.

$$E_{diss} = \frac{\hbar}{2} \int_{-T}^{T} \vec{\Gamma} \cdot \frac{d\vec{\lambda}}{dt} dt = \frac{\hbar}{2} \left(\left[\vec{\Gamma} \cdot \vec{\lambda} \right]_{-T}^{T} - \int_{-T}^{T} \vec{\lambda} \cdot \frac{d\vec{\Gamma}}{dt} dt \right) \tag{11.3}$$

This tool estimates the dissipated energy of whole circuit for each input combinations in various tunneling energy levels under non-adiabatic switching [19]. The upper bound power dissipation model [19] is shown in equation 11.4.

$$P_{diss} = \frac{E_{diss}}{T_{cc}} \left\langle \frac{\hbar}{2T_{cc}} \overrightarrow{\Gamma}_+ \times \left[-\frac{\overrightarrow{\Gamma}_+}{\left|\overrightarrow{\Gamma}_+\right|} \tanh\left(\frac{\hbar\left|\overrightarrow{\Gamma}_+\right|}{k_b T}\right) + \frac{\overrightarrow{\Gamma}_-}{\left|\overrightarrow{\Gamma}_-\right|} \tanh\left(\frac{\hbar\left|\overrightarrow{\Gamma}_-\right|}{k_b T}\right) \right] \right\rangle \qquad (11.4)$$

In Table 11.7 and Table 11.8, compare the various energy dissipation levels, such as average leakage energy dissipation, average switching energy dissipation, and average energy dissipation of previous priority encoder structure and the proposed priority encoder; the same for the previous 4 to 2 priority encoder structure and the proposed 4 to 2 priority encoder. It shows various energy dissipation levels, and from the results it is concluded that the proposed structure of both the priority encoder and the 4 to 2 priority encoder dissipate less energy as compare with the previous structures

TABLE 11.7   Energy Dissipation Results Priority Encoder Structure

DESIGN	AVERAGE LEAKAGE ENERGY DISSIPATION (MeV)			AVERAGE SWITCHING ENERGY DISSIPATION (MeV)			AVERAGE ENERGY DISSIPATION (MeV)		
	0.5 EK	1.0 EK	1.5 EK	0.5 EK	1.0 EK	1.5 EK	0.5 EK	1.0 EK	1.5 EK
Previous Priority Encoder Structure	7.51	51.91	67.69	51.78	40.41	38.66	37.28	52.3	60.34
Proposed Priority Encoder	6.37	40.7	52.78	43.08	33.73	28.05	30.81	45.49	54.83

TABLE 11.8   Energy Dissipation Results 4 to 2 Priority Encoder Structure

DESIGN	AVERAGE LEAKAGE ENERGY DISSIPATION (MeV)			AVERAGE SWITCHING ENERGY DISSIPATION (MeV)			AVERAGE ENERGY DISSIPATION (MeV)		
	0.5 EK	1.0 EK	1.5 EK	0.5 EK	1.0 EK	1.5 EK	0.5 EK	1.0 EK	1.5 EK
Previous 4 to 2 Priority Encoder Structure	5.1	31.4	51.4	38.4	51.1	61.66	41.3	59.3	63.1
Proposed 4 to 2 Priority Encoder	4.7	20.7	42.5	31.8	45.7	49.05	31.2	44.9	54.1

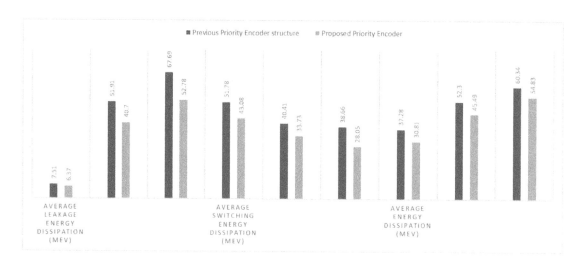

FIGURE 11.17   Comparative analysis of energy dissipation of priority encoder structure.

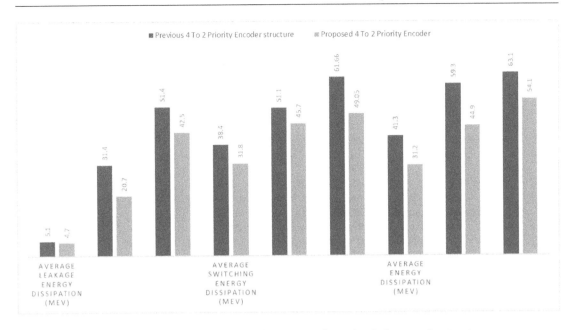

FIGURE 11.18 Comparative analysis of energy dissipation of 4 to 2 priority encoder structure.

11.11 CONCLUSION

Two inferences can be taken from those simulations. First, using the aforementioned universal gate logic, the number of cells and area needed to construct different encoder circuits is drastically reduced without compromising the effectiveness. The circuits therefore get better compared with other current designs, as seen in Tables 11.7 and 11.8. Second, the potential energy of the encoder circuits was measured using standard methods for various possible input combinations and is rounded off to their nearest possible integer values. These values are described in section 11.9 using bar-graphs. As the input state that has the least potential energy has already been identified as better suited for QCA operation, the most stable input state can be found from the graphs. From Figure 11.14 for 4 to 2 circuit encoder, the stable state occurs when $D_0 = 1$. Likewise, for the priority encoder circuit from Figure 11.13, the states with D_1 as 1 and all higher bits are considered to be 0 as the most stable state. For the octal to binary encoder circuit is shown in Figure 11.16, $D_2 = 1$ is the most stable state. This is concluded from Figure 11.17 and Figure 11.18, the comparison of energy dissipation between the two structure of the priority encoder structure and 4 to 2 priority encoder structure with the available previous structure.

REFERENCES

[1] C. Pérez-Delgado, and D. Cheung, "*Local Unitary Quantum Cellular Automata*", Physical Review A, Vol. 76, 032320, 2007.

[2] Hema Sandhya Jagarlamudi, Mousumi Saha, and Pavan Kumar Jagarlamudi, "*Quantum Dot Cellular Automata Based Effective Design of Combinational and Sequential Logical Structures*", World Academy of Science, Engineering and Technology, Vol. 5. No. 12, 2011.

[3] Rumi Zhang, Konrad Walus, Wei Wang, and Graham A. Jullien, *"Performance Comparison of Quantum-dot Cellular Automata Adders"*, in 2005 IEEE International Symposium on Circuits and Systems, pp. 2522–2526. IEEE, 2005. https://doi.org/10.1109/ISCAS.2005.1465139

[4] Birinderjit Singh Kalyan, and Balwinder Singh, *"Design and Simulation Equivalent Model of Floating Gate Transistor"*, in 2015 Annual IEEE India Conference (INDICON). IEEE, 2015.

[5] Birinderjit Singh Kalyan, and Balwinder Singh, *"Quantum Dot Cellular Automata (QCA) Based 4-bit Shift Register Using Efficient JK Flip Flop"*, International Journal of Pure and Applied Mathematics, Vol. 118, No. 19, 2018, pp. 143–157.

[6] Keivan Navil, Amir Mokhtar Chabil, and Samira Sayedsalehi, *"A Novel Seven Input Majority Gate in Quantum-dot Cellular Automata"*, IJCSI International Journal of Computer Science Issues, Vol. 9, No. 1, 2012.

[7] Chiradeep Mukherjee, A.S. Sukla, S.S. Basu, R. Chakrabarty, A. Khan, and D. De, *"Layered T Full Adder Using Quantum-dot Cellular Automata"*, IEEE International Conference on Electronics, Computing and Communication Technologies (CONECCT). 2015.

[8] Yaser Adelnia, and Abdalhossein Rezai, *"A Novel Adder Circuit Design in Quantum-Dot Cellular Automata Technology"*, International Journal of Theoretical Physics, Vol. 58, No. 1, 2019, pp. 184–200. https://doi.org/10.1007/s10773-018-3922-0

[9] Craig S. Lent, P. Douglas Tougaw, and Wolfgang Porod, *"Quantum Cellular Automata: The Physics of Computing with Arrays of Quantum Dot Molecules"*, in Proceedings of the Workshop on Physics and Computing, pp. 5–13, 1994. https://doi.org/10.1109/PHYCMP.1994.363705

[10] Sara Hashemi, and Keivan Navi, *"New Robust QCA D Flip Flop and Memory Structures"*, Microelectronics Journal, 2012, pp. 929–940. https://doi.org/10.1016/j.mejo.2012.10.007

[11] M. Mohammadi, M. Mohammadi, and S. Gorgin, *"An Efficient Design of Full Adder in Quantum-dot Cellular Automata (QCA) Technology"*, Microelectronics Journal, Vol. 50, 2016, pp. 35–43.

[12] Mohammad Torabi, *"A New Architecture for T Flip Flop Using Quantum-Dot Cellular Automata"*, in IEEE 3rd Asia Symposium on Quality Electronic Design, 2011. https://doi.org/10.1109/ASQED.2011.6111764

[13] P. Ilanchezhian, and Dr. R.M.S. Parvathi, *"Analysis and Design of Priority Encoder Circuit Using Quantum Dot Cellular Automata"*, International Journal of Engineering & Research Technology, Vol. 2, No. 3, 2013.

[14] Jun-Cheol Jeon, *"Quantum-Dot Cellular Automata Based Priority Encoder Using Multi-Layer Structure"*, Proceedings of The International Workshop on Future Technology FUTECH, Vol. 1, No. 1, 2017.

[15] Bahniman Ghosh, Shoubhik Gupta, Smriti Kumari, and Akshaykumar Salimath, *"Novel Design of Combinational and Sequential Logical Structures in Quantum Dot Cellular Automata"*, Journal of Nanostructure in Chemistry, Vol. 3, No. 15, 2013.

[16] Md. Sofeoul-Al-Mamun, Mohammad Badrul Alam Miah, and Fuyad Al Masud, *"A Novel Design and Implementation of 8–3 Encoder Using Quantum-dot Cellular Automata (QCA) Technology"*, European Scientific Journal Edition, Vol. 13, No. 15, 2017.

[17] Birinderjit Singh Kalyan, and Balwinder Singh, *"Performance Analysis of Quantum Dot Cellular Automata (QCA) Based Linear Feedback Shift Register (LFSR)"*, International Journal of Computing and Digital Systems, Vol. 9, No. 3, 2020.

[18] A.N. Bahar, M.S. Uddin, M. Abdullah-Al-Shafi, M.M.R. Bhuiyan, and K. Ahmed, *"Designing Efficient QCA Even Parity Generator Circuits with Power Dissipation Analysis"*, Alexandria Engineering Journal, Vol. 57, No. 4, 2018, pp. 2475–2484. https://doi.org/10.1016/j.aej.2017.02.002

[19] S. Srivastava, S. Sarkar, and S. Bhanja, *"Power Dissipation Bounds and Models for Quantum-dot Cellular Automata Circuits"*, IEEE Conference on Nanotechnology, Vol. 1, June 2006, pp. 375–378. https://doi.org/10.1109/NANO.2006.247655

Investigation of Nano-Structured Honeycomb Patch Antenna with Photonic Crystal Substrate

12

Sathish Kumar Danasegaran, Elizabeth Caroline Britto, A. Sridevi, and S. Poonguzhali

Contents

DOI: 10.1201/9781003126645-12

12.1 INTRODUCTION

The patch antenna with a photonic crystal (PhC) substrate is designed for various wireless technology applications requiring more precise control of photons for lossless transmission. Different shapes with various pattern patch antennas [1–12] have been designed due to their low cost and compatibility. The main problem is, its constraint in the bandwidth, low gain, and minimum efficiency due to propagation loss at the surface area. The photonic bandgap (PBG) crystal is mainly preferred because of its capability to confine the emission of light to the precise direction; subsequently, antennas with photonic crystal substrate are extensively designed. When the operational bandwidth of the patch antenna is matched with the PBG, it will enhance the efficiency of the antenna radiation pattern, minimize the side lobe level, and in turn avoid the coupling problems. The PBG [13] is a structure that varies sporadically in its dielectric property. Because of the PBG structure, the electromagnetic waves from the antenna surface will be uniform without any interference, noise, or losses. PhCs can be primed using multiple techniques such as mechanical drilling, layer by layer photolithography, and reactive ion beam etching. The remaining techniques are complex and time-consuming processes; for a greater number of layers of order 10, the layer required depends on the required bandgap. The process, which involves a colloidal self-assembly nanoparticle, is sprayed, coated, or electrodeposited, which makes it simple and inexpensive. The selection of assembly method and state of affairs for different dimensions and geometric shapes can be realized for the different bandwidth signal. The colloidal PhCs have an optimistic color structure due to bandgaps in the visible region, are easily optical tunable, have a flexible structure, and are tuned by electrical and magnetic fields. Every layer of a PhC's atomic structure is in a face-centered cubic lattice (FCC) arrangement. The FCC lattice has a very large PBG of every spherical crystalline structure due to Brillouin zones [14–16]. Figure 12.1 shows the structure of photonic crystal.

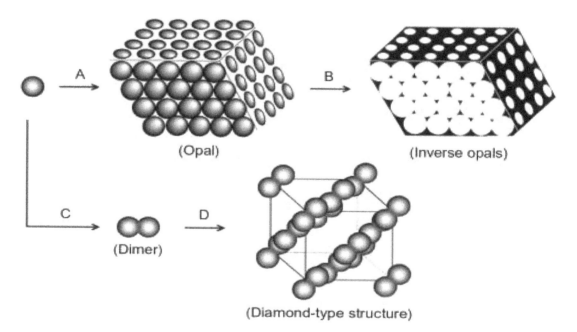

FIGURE 12.1   Structure of photonic crystal

12.2 PROPOSED MODEL

In the proposed system, to improve the efficiency of gain and radiation pattern, the loss of surface waves is reduced. This system uses a PhC substrate at terahertz frequency in the range from 10^{12}. Both a conventional antenna and a photonic crystal antenna are designed at THz frequency, and their various measuring parameters are compared. The performance parameters, such as resonant frequency, return loss (RL), directivity (D), voltage standing wave ratio (VSWR), and gain of the antenna will influence the antenna bandwidth and the quality of signal transmission. The directivity and gain is to be considered for the antenna at THz frequency. Both return loss and VSWR are important, since they gives us the amount of reflected power and impedance loss due to the mismatch of the feeder.

12.2.1 Design of Nano-Structured Honeycomb Pentagon Antenna

The configuration of the patch antenna is provided with an aperture couple feeding mechanism and designed using computer simulation technology (CST) software. There are various types of patches, such as square, rectangle, triangle, honeycomb, etc. The honeycomb patch antenna diminished the return loss and improved the gain as compared with the other patches. The coplanar feeding arrangements have physical separation between the feeding point and radiation point; due to this high isolation symmetry, a radiation pattern occurs. There are various substrate materials such as Rogers 4350, Duroid, FR4, epoxy, Bakelite, etc. The material used in this design is the FR-4 (flame retardant) substrate because it reduces the power loss and increases the gain and directivity more than the other substrates. Generally, the FR-4 substrate is utilized as electrical insulators that possess substantial mechanical strength and will resist moisture effectively. The material used in ground, feed and patch are copper, as it a good conductor of heat. In Table 12.1, the ground width is denoted as 'GW' and the ground length is denoted as 'GL'. The width of the substrate is denoted as 'SW', and the substrate length is denoted as 'SL'. The thickness of the design is denoted as 'T', and the height of the design is denoted as 'H'. Patch length is denoted as 'PL', and the patch width is denoted as 'PW'. The length of the feed is '7 μm', and the width of the feed is 0.6 μm. The dimension of the design is in micrometers. In this chapter, two types of antenna structures are designed. Figure 12.1 represent the conventional antenna and Figure 12.2 represents the honeycomb patch antenna on a PhC. Table 12.1 provides the dimension of the propounded honeycomb patch antenna. The patch antenna with the internal dimension ($72 \times 34\ \mu m^2$) resonates with the THz frequency

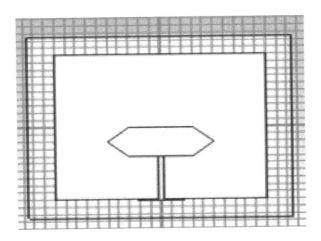

FIGURE 12.2 Conventional antenna.

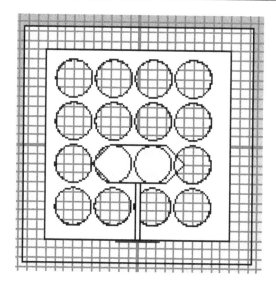

FIGURE 12.3 Patch antenna on a PhC substrate.

TABLE 12.1 Dimension of the Antenna

PARAMETERS	VALUES (IN μM)
GW	25
GL	30
T	3
SW	22
SL	22
H	3
PL	4
PW	10

The honeycomb patch antenna length and width is estimated using the following equations:

The patch width $W = \dfrac{C_0}{2fo}\sqrt{\dfrac{2}{\varepsilon_r + 1}}$ (12.1)

The patch length $L = \dfrac{C_0}{2fo\sqrt{\varepsilon_r}}$ (12.2)

The extension of the length can be done using the following equation:

$$\Delta L = 0.412h\frac{\left(\varepsilon_{reff} + 0.3\right)\left(\dfrac{W}{h} + 0.264\right)}{\left(\varepsilon_{reff} - 0.258\right)\left(\dfrac{W}{h} + 0.8\right)}$$ (12.3)

where c denotes the light velocity, h represent substrate height and f_0 denotes the operating frequency of the antenna.

12.3 SIMULATION RESULTS AND COMPARATIVE ANALYSIS OF CONVENTIONAL ANTENNA WITH A PHC ANTENNA

12.3.1 Analysis of Various Parameters of Conventional Patch Antenna

The return loss and directivity of the conventional patch antenna are plotted in the following figures. The existing system of this antenna is conventional, and the signals are generated in THz. Figure 12.4 provides the return loss of S_{11} as −35.449 dB for the conventional patch antenna at 11.3 THz frequency. Figure 12.5 displays the VSWR of the patch antenna, which is 1.03435 at 11.3 THz frequency.

Figure 12.6 shows the directivity plot of the conventional patch antenna; the value is 5.11 dBi at the resonant frequency of 11.3 THz. Figure 12.7 shows the gain of 3.5 dB of the conventional patch antenna at the resonant frequency of 11.3 THz.

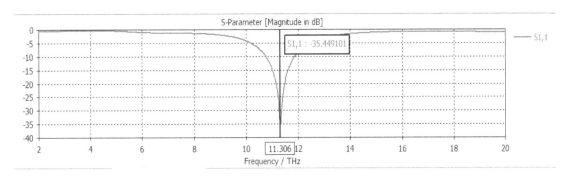

FIGURE 12.4   Return loss of conventional antenna.

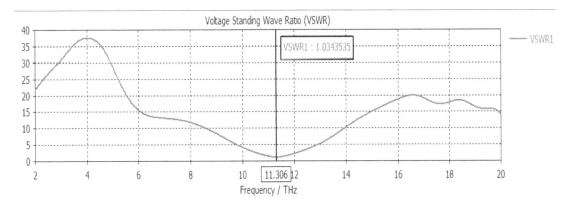

FIGURE 12.5   VSWR of conventional antenna.

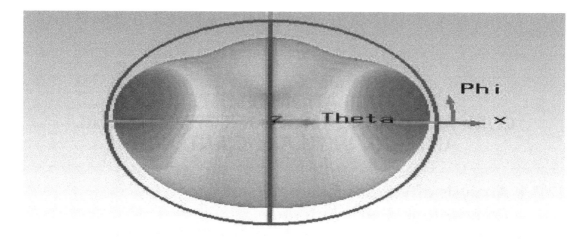

FIGURE 12.6   Directivity of conventional antenna.

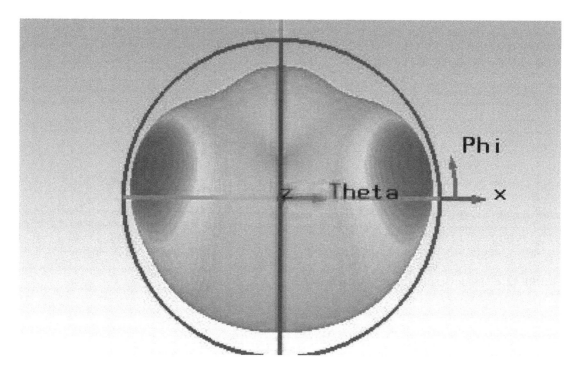

FIGURE 12.7   Gain of conventional antenna.

12.3.2 Analysis of Various Parameters of Nano-Structured Patch Antenna with Photonic Substrate

The design of the nano-structured honeycomb patch antenna is fully compactable; if the radius of the hollows is reduced, then it reduces the gain and bandwidth. To achieve the required gain and bandwidth, the maximum distance between the centers of the slots is evaluated for certain dimensional specifications. As this process was done, the hole radius $r = 2$ µm was determined as optimum value.

Figure 12.8 provides the return loss for the nano-structured honeycomb PhC substrate patch antenna. The return loss of the antenna on a PhC is −44.14 dB at 14.72 THz for the S_{11} parameter. For the S_{11} parameter, the conventional patch antenna's resonant frequency obtained is 11.30 THz, with the return loss of −35.44 dB. Therefore, the return loss is lower for higher resonant frequency for a nano-structured honeycomb patch antenna with a PhC substrate.

Figure 12.9 shows the VSWR of the nano-structured honeycomb PhC substrate patch antenna. The VSWR of the antenna on a PhC is 1.0124 at 14.726 THz. The VSWR of conventional patch antenna is 1.0345 at resonant frequency 11.30 THz. Therefore, the VSWR of the honeycomb patch antenna with a PhC substrate is less when compared to the conventional patch antenna.

Figure 12.10 shows the directivity of the honeycomb patch antenna; the directivity is 5.90 dBi at resonant frequency 14.726 THz, which is greater than the directivity of the conventional patch antenna.

Figure 12.11 displays the gain of the honeycomb patch antenna; the gain is 3.85 dB at resonant frequency 14.726 THz, which is higher than the gain of the conventional antenna.

The designed nano-structured patch antenna along with a PhC substrate has obtained high gain and directivity when compared to the conventional patch antenna. The Table 12.2 describes the comparison of resonant frequency, return loss, VSWR, directivity, and gain. From the table, the designed PhC antenna parameter is better than the conventional patch antenna.

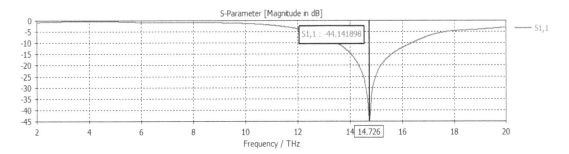

FIGURE 12.8   Return loss of a PhC substrate antenna.

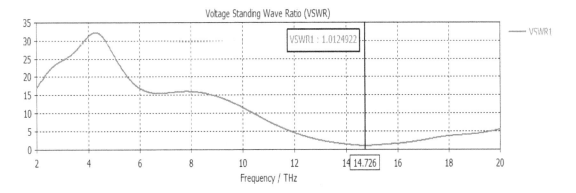

FIGURE 12.9   VSWR of a PhC substrate antenna.

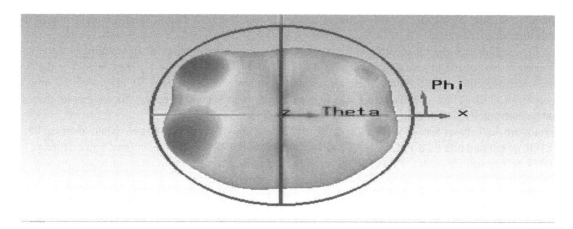

FIGURE 12.10   Directivity of a PhC substrate antenna.

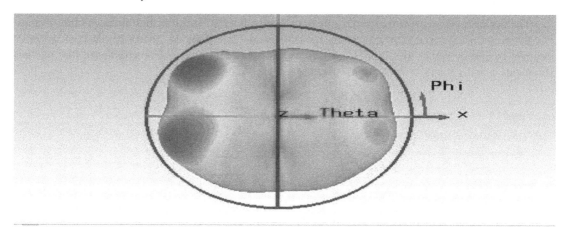

FIGURE 12.11   Gain of a PhC substrate antenna.

TABLE 12.2   Comparison of Simulated Output for Conventional and PhC Substrate Antenna

PARAMETERS	CONVENTIONAL ANTENNA	PhC ANTENNA
Resonant frequency	11.30 THz	14.72 THz
Return loss	−35.44 dB	−44.14 dB
VSWR	1.03	1.01
Directivity	5.11 dBi	5.90 dBi
Gain	3.51 dB	3.85 dB

The conventional honeycomb patch antenna resonates at the frequency of 11.30 THz, with the antenna characteristics of −35.44 dB return loss, 5.11 dBi directivity, and 3.5 dB of gain. When the conventional antenna is replaced by a PhC antenna with a square lattice structure, the antenna characteristics are enhanced. PhC helps to diminish the sidelobes and improve the antenna efficiency by forbidding the electromagnetic waves that propagate into the substrate. The dielectric PhC acts as a frequency filter and restricts the waves from propagating into the substrate. The PhC antenna with THz frequency reduces data loss, due to which higher data rate and bandwidth are attained. The output of the antenna on a PhC is better than the conventional antenna.

12.4 PHC PATCH ANTENNA FOR VARIOUS RADII OF HONEYCOMB STRUCTURE

In this section, the patch antenna on the PhC is designed for different radii such as 1.4 µm, 1.6 µm, 0.8 µm, 0.6 µm, 1.8 µm, and 1.5 µm, and the simulated outputs are discussed briefly as follows.

12.4.1 Design Parameters of a PhC Antenna with a Radius of 1.8 µm

Figure 12.12 illustrates the structure of a PhC honeycomb patch antenna with 1.8 µm radius.

Figure 12.13 displays the return loss measurement for a PhC honeycomb patch antenna with 1.8 µm radius. The return loss of the nano-structured honeycomb PhC patch antenna is −40.45 dB at 14.384 THz.

Figure 12.14 shows the VSWR loss measurement for a PhC honeycomb patch antenna for a radius of 1.8 µm. The VSWR of the nano-structured honeycomb PhC substrate patch antenna is 1.0191 at 14.384 THz.

Figure 12.15 shows the directivity for a PhC honeycomb patch antenna for a radius of 1.8 µm. The directivity is 5.439 dBi for a honeycomb PhC substrate patch antenna.

Figure 12.16 shows the gain of PhC honeycomb patch antenna for a radius of 1.8 µm. The gain is 3.258 dB for honeycomb photonic substrate patch antenna.

12.4.2 Design Parameters of a PhC Antenna with a Radius of 1.6 µm

Figure 12.17 shows the structure of a PhC honeycomb patch antenna for a 1.6 µm radius.

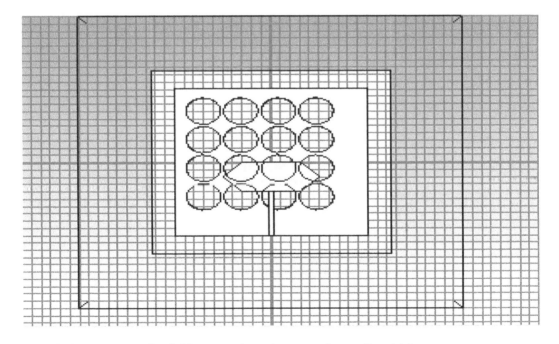

FIGURE 12.12 Structure of a PhC honeycomb patch antenna for a radius of 1.8 µm.

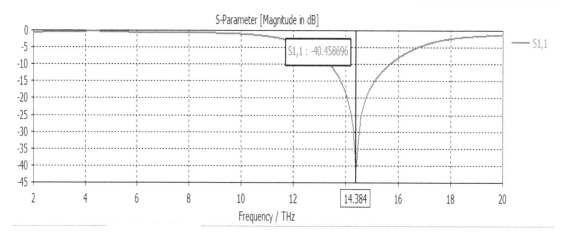

FIGURE 12.13   Return loss of the PhC patch antenna for radius 1.8 μm.

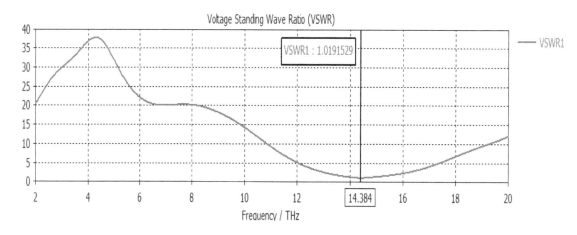

FIGURE 12.14   VSWR of PhC patch antenna for radius 1.8 μm

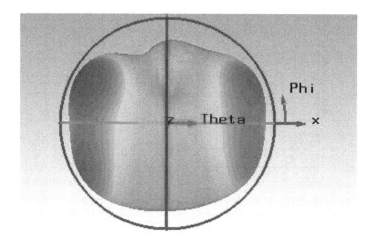

FIGURE 12.15   Directivity of a PhC patch antenna for radius 1.8 μm.

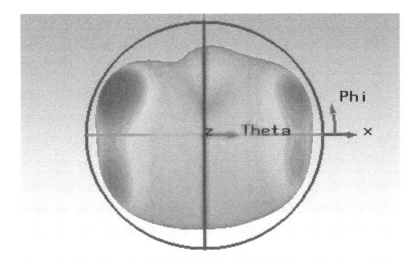

FIGURE 12.16   Gain of a PhC patch antenna for radius 1.8 μm.

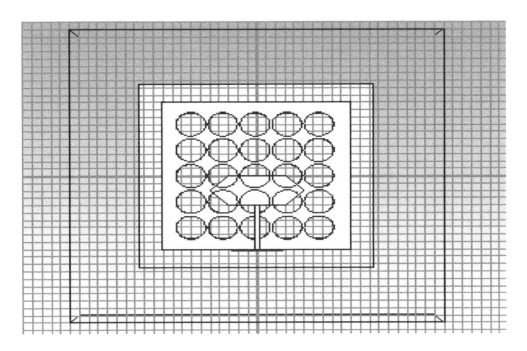

FIGURE 12.17   Structure of a PhC honeycomb patch antenna for a radius of 1.6 μm.

Figure 12.18 provides the return loss measurement for a PhC honeycomb patch antenna for a 1.6 μm radius. The return loss of the nano-structured honeycomb PhC substrate patch antenna is −38.70 dB at 19.694 THz.

Figure 12.19 shows the VSWR loss measurement for a PhC honeycomb patch antenna for a radius of 1.6 μm. The VSWR of the nano-structured honeycomb photonic substrate patch antenna is 1.02349 at 19.694 THz.

Figure 12.20 shows the directivity for a PhC honeycomb patch antenna for a radius of 1.6 μm. The directivity is 4.372 for a honeycomb PhC substrate patch antenna.

Figure 12.21 shows the gain for a PhC honeycomb patch antenna for a radius of 1.6 μm. The gain is 3.343 for a honeycomb PhC substrate patch antenna.

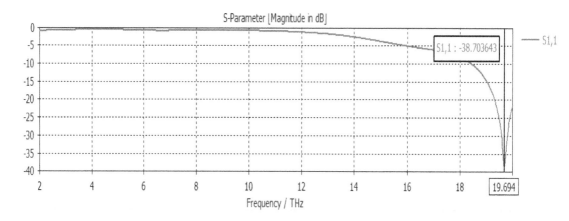

FIGURE 12.18   Return loss of a PhC patch antenna for a radius of 1.6 μm.

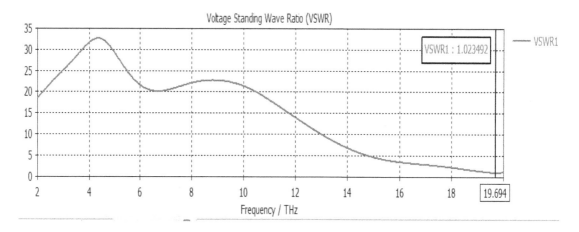

FIGURE 12.19   VSWR of a PhC patch antenna for a radius of 1.6 μm.

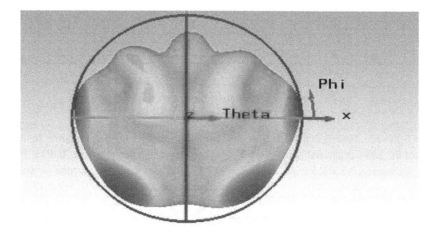

FIGURE 12.20   Directivity of a PhC patch antenna for a radius of 1.6 μm.

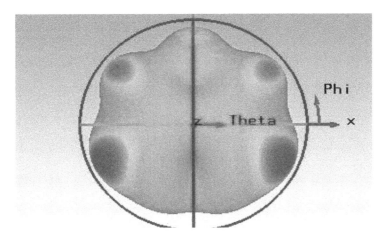

FIGURE 12.21   Gain of a PhC patch antenna for a radius of 1.6 μm.

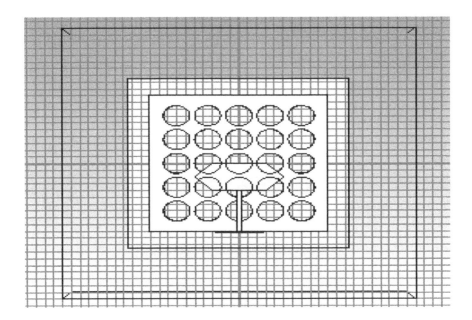

FIGURE 12.22   Structure of a PhC honeycomb patch antenna for a radius of 1.5 μm.

12.4.3 Design Parameters of a PhC Antenna with a Radius of 1.5 μm

Figure 12.22 shows the structure of a PhC honeycomb patch antenna for a radius of 1.5 μm.

Figure 12.23 shows the return loss measurement for a PhC honeycomb patch antenna for 1.5 μm radius. The return loss of the nano-structured honeycomb photonic substrate patch antenna is −34.417 dB at 18.74 THz.

Figure 12.24 shows the VSWR loss measurement for a PhC honeycomb patch antenna for a radius of 1.4 μm. The VSWR of the nano-structured honeycomb photonic substrate patch antenna is 1.03877 at 18.74 THz.

Figure 12.25 shows the directivity of a PhC honeycomb patch antenna for a radius of 1.4 μm. The directivity is 4.582 for a honeycomb PhC substrate patch antenna.

Figure 12.26 shows the gain for a PhC honeycomb patch antenna for a radius of 1.5 μm. The gain is 3.508 for a honeycomb PhC substrate patch antenna.

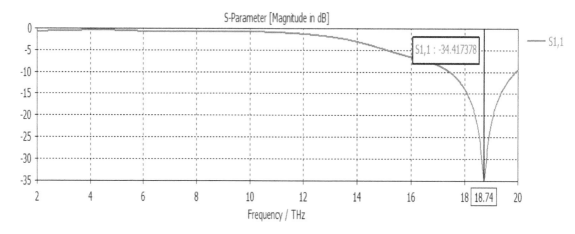

FIGURE 12.23   Return loss of a PhC patch antenna for a radius of 1.5 µm.

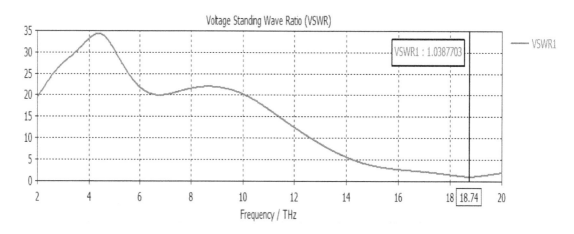

FIGURE 12.24   VSWR of a PhC patch antenna for a radius of 1.5 µm.

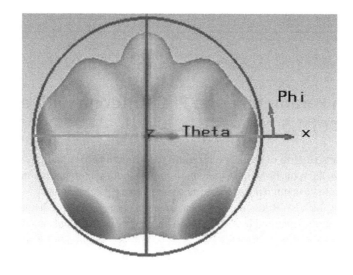

FIGURE 12.25   Directivity of a PhC patch antenna for a radius of 1.5 µm.

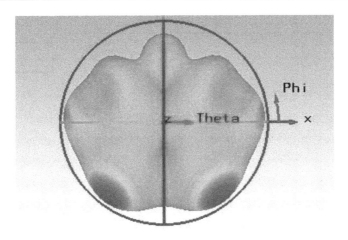

FIGURE 12.26   Gain of a PhC patch antenna for a radius of 1.5 μm.

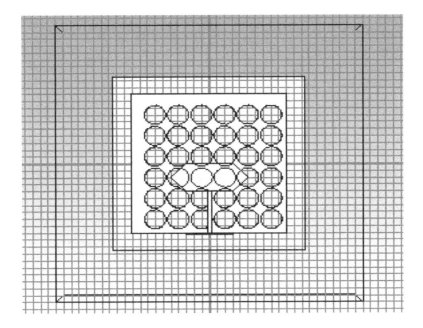

FIGURE 12.27   Structure of a PhC honeycomb patch antenna for a radius of 1.4 μm.

12.4.4 Design Parameters of a PhC Antenna with a Radius of 1.4 μm

Figure 12.27 shows the structure of a PhC honeycomb patch antenna for a radius of 1.4 μm.

Figure 12.28 shows the return loss measurement for a PhC honeycomb patch antenna for 1.4 μm radius. The return loss of the nano-structured honeycomb PhC substrate patch antenna is −33.234 dB at 18.668 THz.

Figure 12.29 shows the VSWR loss measurement for a PhC honeycomb patch antenna for a radius of 1.4 μm. The VSWR of the nano-structured honeycomb PhC substrate patch antenna is 1.0445 at the frequency of 18.66 THz.

Figure 12.30 shows the gain of a PhC honeycomb patch antenna for a radius of 1.4 μm. The gain and directivity is 3.52 dB and 5.01 dBi for a honeycomb PhC substrate patch antenna.

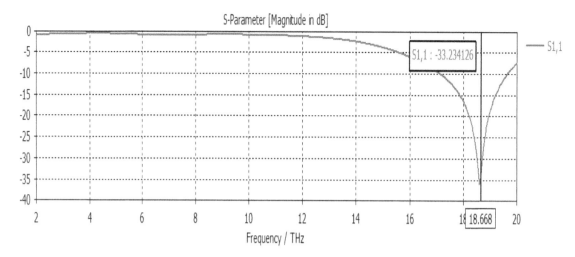

FIGURE 12.28   Return loss of a PhC patch antenna for a radius of 1.4 μm.

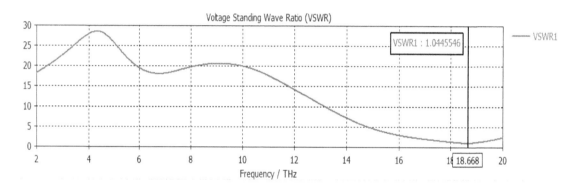

FIGURE 12.29   VSWR of PhC patch antenna for a radius of 1.4 μm.

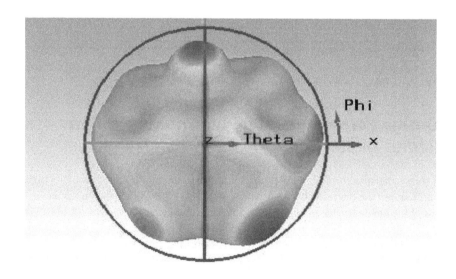

FIGURE 12.30   Gain of a PhC patch antenna for a radius of 1.4 μm.

12.4.5 Design Parameters of a PhC Antenna with a Radius of 0.8 μm

Figure 12.31 shows the structure of a PhC honeycomb patch antenna for a radius of 0.8 μm.

Figure 12.32 shows the return loss measurement for a PhC honeycomb patch antenna for 0.8 μm radius. The return loss of the nano-structured honeycomb PhC substrate patch antenna is −30.980 dB at 15.53 THz.

Figure 12.33 shows the VSWR loss measurement for a PhC honeycomb patch antenna for a radius of 0.8 μm. The VSWR of the nano-structured honeycomb PhC substrate patch antenna is 1.0581 at 15.53 THz.

Figure 12.34 shows the directivity of a PhC honeycomb patch antenna for a radius of 0.8 μm. The directivity is 4.922 for a honeycomb PhC substrate patch antenna.

Figure 12.35 shows the gain for a PhC honeycomb patch antenna for a radius of 0.8 μm. The gain is 2.5 for a honeycomb photonic substrate patch antenna.

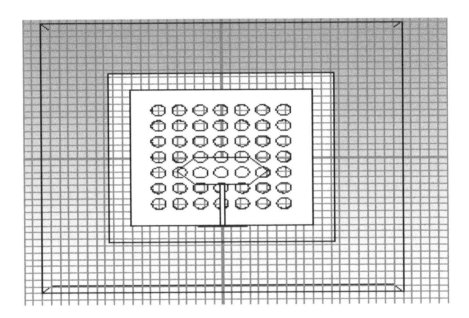

FIGURE 12.31   Structure of a PhC honeycomb patch antenna for a radius of 0.8 μm.

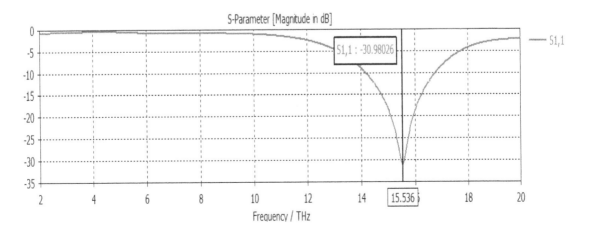

FIGURE 12.32   Return loss of a PhC patch antenna for a radius of 0.8 μm.

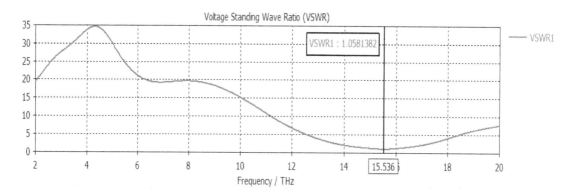

FIGURE 12.33 VSWR of a PhC patch antenna for a radius of 0.8 μm.

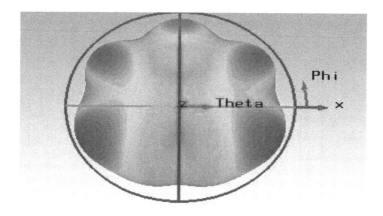

FIGURE 12.34 Directivity of a PhC patch antenna for a radius of 0.8 μm.

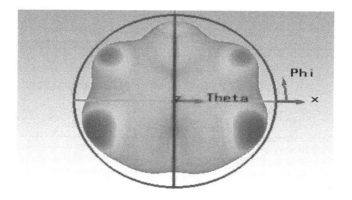

FIGURE 12.35 Gain of a PhC patch antenna for a radius of 0.8 μm.

12.4.6 Design Parameters of a PhC Antenna with a Radius of 0.6 μm

Figure 12.36 shows the structure of PhC honeycomb patch antenna for a radius of 0.6 μm.

Figure 12.37 shows the return loss measurement for a PhC honeycomb patch antenna for a radius of 0.6 μm. The return loss of the nano-structured honeycomb PhC substrate patch antenna is −33.775 dB at 13.97 THz.

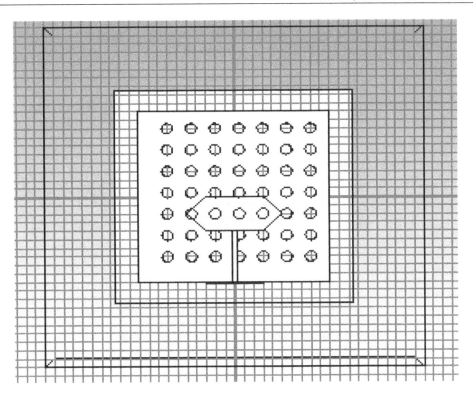

FIGURE 12.36   Structure of a PhC honeycomb patch antenna for a radius of 0.6 μm.

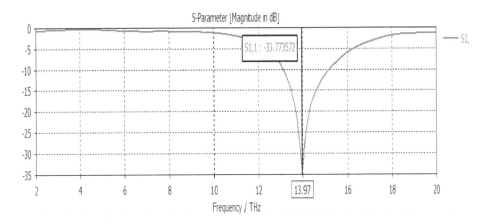

FIGURE 12.37   Return loss of a PhC patch antenna for a radius of 0.6 μm.

Figure 12.38 shows the VSWR loss measurement for a PhC honeycomb patch antenna for 0.6 μm radius. The VSWR of the nano-structured honeycomb PhC substrate patch antenna is 1.0418 at 13.97 THz.

Figure 12.39 shows the directivity of a PhC honeycomb patch antenna for a radius of 0.6 μm. The directivity is 5.013 for a honeycomb PhC substrate patch antenna.

Figure 12.40 shows the Gain for a PhC honeycomb patch antenna for a radius of 0.6 μm. The gain is 2.739 for a honeycomb PhC substrate patch antenna.

Table 12.3 represents the directivity and gain values of the respective design of antennas. These are the output of other design structures with different radii of the antenna on photonic crystals. Here,

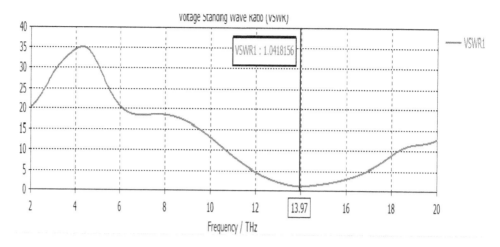

FIGURE 12.38   VSWR of a PhC patch antenna for a radius of 0.6 μm.

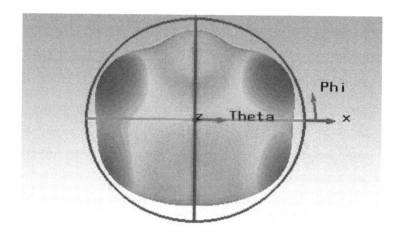

FIGURE 12.39   Directivity of a PhC patch antenna for a radius of 0.6 μm.

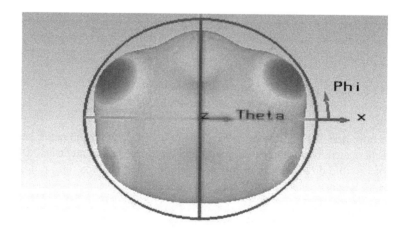

FIGURE 12.40   Gain of PhC patch antenna for radius 0.6 μm.

TABLE 12.3 PhC Characteristics for Different Hole Radius Values

HOLE RADIUS (µM)	DIRECTIVITY (dBi)	GAIN (dB)
1.8	5.439	3.258
1.6	4.372	3.343
1.5	4.582	3.508
1.4	5.011	3.520
0.8	4.922	2.500
0.6	5.013	2.739

the PhC hole with 0.6 µm radius produces the directivity 5.013 dBi and gain 2.739 dB; the PhC hole with 0.8 µm radius produces the directivity 4.922 dBi and gain 2.500 dB; the PhC hole with 1.4 µm radius produces the directivity 5.011 dBi and gain 3.520 dB; the PhC hole with 1.5 µm radius produces the directivity 4.582 dBi and gain 3.508 dB; the PhC hole with 1.6 µm radius produces the directivity 4.372 dBi and gain 3.343 dB; and the PhC hole with 0.6 µm radius produces the directivity 5.439 dBi and gain 3.258 dB. On the analysis, we observed that the PhC hole with a radius of 1.8 µm produces high gain and reduces the power loss. When compared with the conventional antenna, the antenna designed with a PhC crystal produces a high gain and also diminishes the power loss by using the THz frequency.

12.5 APPLICATIONS OF MICROSTRIP PATCH ANTENNA

Due to the high-performance design features of microstrip patch antennas, they are used in many applications such as biomedical instruments, wireless and satellite communication [17] and aircraft, and rocket and missile tracking. Because of various advantages such less weight and ease of design and fabrication, the microstrip patch antenna is used in many commercial applications. Certain applications are listed here.

12.5.1 Wireless Cellular and Satellite Communication Applications

This cellular application requires less weight, less cost, a small shape, and low dimension antennas. The microstrip patch antenna can be easily designed with various polarizations, which is the basic need of satellite communication [18]. In satellite communication, both square and circular patch antennas are employed with too many feeder systems.

12.5.2 GPS Applications

The fabricated microstrip patch antennas [19] have a substrate with high permittivity, which is the major requirement of Global Position System applications. The permittivity depends on the material types used, and it will be less expensive. In an automated vehicle with route map applications, GPS receivers should be able to track and capable enough to adapt circular polarization.

12.5.3 Radar Application

Radar is used to estimate the moving target, which is used in real-time application. Primarily it is employed in a moving vehicle, and with its speedy fabrication process, bulk production is possible.

12.5.4 Worldwide Interoperability for Microwave Access (WiMAX)

The nano-structured patch antenna is made; it is used for WiMax [20] applications with IEEE 802.16 standard. WiMax uses 70 Mbps as a data rate, transmits data at the rate of more than 70 Mbps, and operates in three resonant modes, from the 2.7 to 5.3 GHz range up to a 30-mile radius.

12.6 CONCLUSION

The honeycomb PhC patch antenna was built and simulated by utilizing CST software. The PhC antenna characteristics are found to be loftier than the conventional antenna. Table 12.2 represents the variance between the conventional antenna and the PhC antenna performances. The determined return loss of the patch antenna on PhC is −44.14 dB on 14.72 THz. This work explains the impact of PhC with the square lattice structure, and the analysis is done by changing the various hole radii in the PhC structure. The simulated output proved that the antenna performances of PhC are more enhanced than that of the conventional antenna structure. This PhC antenna is applicable for high-speed satellite and wireless communication.

12.7 FUTURE SCOPE

With the help of advanced techniques and high performance matching systems, we can enhance the bandwidth of the antenna; moreover, the directivity can be improved. The finite element method can be employed to resolve and contrive all the designs. It is intended to try to reduce the size of the antenna by scaling it to the optimum configuration.

REFERENCES

[1] M. Rostamzadeh, S. Mohamadi, J. Nourinia, C. Ghobadi, and M. Ojaroudi, "Square Monopole Antenna for UWB Applications with Novel Rod-Shaped Parasitic Structures and Novel V-Shaped Slots in the Ground Plane," IEEE Antennas and Wireless Propagation Letters, vol. 11, pp. 446–449, 2012.
[2] Babu Lal Shahu, Srikanta Pal, and Neela Chattoraj, "Design of Super Wideband Hexagonal-shaped Fractal Antenna with Triangular Slot," Microwave and Optical Technology Letters, vol. 57, no. 7, pp. 1659–1662, 2015.
[3] S. Tripathi, A. Mohan, and S. Yadav, "Hexagonal Fractal Ultra-wideband Antenna Using Koch Geometry with Bandwidth Enhancement," IET Microwaves, Antennas & Propagation, vol. 8, no. 15, pp. 1445–1450, 2014. doi:10.1049/iet-map.2014.0326.

[4] Elizabeth Caroline Britto, Sathish Kumar Danasegaran, and William Johnson, "Design of Slotted Patch Antenna Based on Photonic Crystal for Wireless Communication," International Journal of Communication Systems, e4662, Wiley, 2020.

[5] K. RamaDevi, A. Mallikarjuna Prasad, and A. Jhansi Rani, "Design of a Pentagon Microstrip Antenna for Radar Altimeter Application," International Journal of Web & Semantic Technology (IJWesT), vol. 3, no. 4, October 2012.

[6] D. Sathish Kumar, B. Elizabeth Caroline, K. Sagadevan, G. Sakthiganesh, and R. Saravanan, "Investigation of High Directional Gain Pentagonal Shaped Patch Antenna," 7th International Conference on Smart Structures and Systems (ICSSS), IEEE, 2020.

[7] Sanyog Rawat, Ushaben Keshwala, and Kanad Ray, "Compact Design of Modified Pentagon-shaped Monopole Antenna for UWB Applications," International Journal of Electrical and Electronic Engineering & Telecommunications, vol. 7, no. 2, April 2018.

[8] D. Sathish Kumar, K. Shobana, P. Venmathi, S. Kalpana, S. Shanmuga Priya, and G. Swetha, "Gain Parameter for Various Microstrip Patch Antenna – Survey," 7th International Conference on Smart Structures and Systems (ICSSS), IEEE, 2020.

[9] S. Rawat, U. Keshwala, and K. Ray, "Compact Design of Modified Pentagon Shaped Monopole Antenna for UWB Applications," International Journal of Electrical and Electronic Engineering & Telecommunications, vol. 7, no. 2, pp. 66–69, 2018.

[10] D. Sathish Kumar, P. Prithika, and B. Elizabeth Caroline, "Investigating the Performance of Microstrip Patch Antenna With Photonic Crystal on Different Substrate," Proceeding of International Conference on Systems Computation Automation and Networking 2019. https://doi.org/10.1109/ICSCAN.2019.8878861.

[11] S. S. Bhatia, A. Sahni, and S. B. Rana, "A Novel Design of Compact Monopole Antenna with Defected Ground Plane for Wideband Applications," Progress in Electromagnetic Research M, vol. 70, pp. 21–31, 2018.

[12] J. Liu, S. Zhong, and K. P. Esselle, "A Printed Elliptical Monopole Antenna with Modified Feeding Structure for Bandwidth Enhancement," IEEE Transactions on Antennas and Propagation, vol. 59, no. 2, pp. 667–670, 2011.

[13] Eli Yablonovitch, "Photonic Band Gap Structures," Journal of the Optical Society of America B, vol. 10, no. 2, pp. 283–295, 1994.

[14] John D. Joannopoulos, Steven G. Johnson, Joshua N. Winn, and Robert D. Meade, Photonic Crystals – Molding the Flow of Light, Second Edition, Princeton University Press, Princeton, NJ, Chapter 5, pp 66–75, 2008.

[15] Sathish Kumar Danasegaran, Elizabeth Caroline Britto, and William Johnson, "Investigation of the Influence of Fluctuation in Air Hole Radii and Lattice Constant on Photonic Crystal Substrate for Terahertz Applications," Optical Engineering, vol. 59, no. 8, p. 087102, 2020. https://doi.org/10.1117/1.OE.59.8.087102.

[16] Ritesh Kumar Kushwaha, P. Karuppanan, and L. D. Malviya, "Design and Analysis of Novel MPA on Photonic Crystal in THz," Elsevier – Physics B: Condensed Matter, pp. 107–112, 2018.

[17] R. Bhatoa Roopan, S. S. Saini, S. Sharma, and E. Sidhu, "Novel High Gain Honeycomb Shaped Slotted Ground Microstrip Patch Antenna Design for Broadcasting Fixed Satellite, Mobile Satellite and Downlink Frequency Applications," 2016 International Conference on Global Trends in Signal Processing, Information Computing and Communication (ICGTSPICC), Jalgaon, 2016, pp. 348–352, doi:10.1109/ICGTSPICC.2016.7955326.

[18] Raad H. Thaher, and Saif Nadhim Alsaidy, "New Compact Pentagonal Microstrip Patch Antenna for Wireless Communications Applications," American Journal of Electromagnetics and Applications, vol. 3, no. 6, 2015, pp. 53–64.

[19] Sumanpreet Kaur Sidhu, and Jagtar Singh Sivia, "Comparison of Different Types of Microstrip Patch Antennas," International Conference on Advancements in Engineering and Technology (ICAET 2015), International Journal of Computer Applications (0095-8887).

[20] Syamala Killi, K. T. P. S. Kumar, D. Naresh Kumar, Sahithi Parasurampuram, and Saikrishna Reguluvalasa, "Design and Analysis of Pentagonal Patch Antenna with H- Tree Fractal Slots for S-Band and Wi-Max Applications," International Journal of Engineering Science and Computing, vol. 6, no. 4, April 2016, pp. 4537–4540.

Silicon Micromachining for Submillimeter

13

Wave and THz-Technology

A. Karmakar, B. Biswas, and M.K. Hooda

Contents

13.1 INTRODUCTION

Submillimeter wave (submmW) or terahertz (THz) research has gained unprecedented interest and momentum over the last few years due to huge demands for a higher data-rate in pursuit of a high-speed THz link. This has been further propelled by diversified applications in other fronts such as (i) THz imaging in surveillance applications, thanks to its high resolution capability, and (ii) THz imaging for medical applications [1, 2]. THz rays (T-rays) are non-ionizing radiation, which means it does not trigger

DOI: 10.1201/9781003126645-13

any harmful chemical reaction in human tissue. With this attractive feature, THz technology is becoming attractive in biomedical applications, too. This chapter would present a generic overview on this emerging THz technology, inherent challenges associated with it, and potential applications. It will also deal with various passive building blocks that are the backbone to establishing a communication link at the THz frequency band. The fundamental limitation of the THz band is extremely high atmospheric attenuation, which has surged the development of high gain antennas over the last few years. The present chapter also deals with the design and development of high gain antennas for THz applications. A comprehensive look on the national status of present THz research and associated activities along with future directions is also presented here. As THz frequency range falls on the submillimeter regimes, fabrication of the extremely small devices and system components calls for two main sophisticated technologies: (i) CNC (computer numerical control) milling and (ii) silicon micromachining. Currently, THz systems are predominantly manufactured by CNC milling. In recent years, CNC milling has achieved a high level of precision and reliability. However, it lacks the ability to manufacture at volume, which makes the overall process time consuming as well as costly. In contrast, silicon micromachining technology is useful for volume production, and at the same it offers fabrication precision on the order of micrometers or even below that (depending upon the technology node and type of lithography process used). Moreover, the micromachining process yields a good surface profile with the surface roughness of finished product in the range of nanometers (therefore reducing losses associated with surface roughness at THz). Furthermore, this process is also capable of fabricating high-aspect ratio geometries that, using any other fabrication technology, are hardly possible to fabricate. Due to these inherent advantages of Si-micromachining, many physical, inertial, biological sensors, microphones, and RF/microwave devices are being fabricated with this technology. The present chapter, as will be elaborated in the next few sections, deals with the design and development of various passive components and high gain antennas using silicon micro-machining technology for THz applications. The passive components and antennas are designed and optimized using Ansys HFSS suite. The finalized designs are under fabrication using the well-established DRIE (deep reactive ion etching) micromachining technique at Semi-Conductor Laboratory (SCL), Chandigarh, India.

13.2 POTENTIALS AND CHALLENGES ASSOCIATED WITH THZ TECHNOLOGY

THz is the range of frequency from roughly 300 GHz to 3 THz. In terms of wavelength, this band refers to 1 mm to 100 μm, because of which the THz band is also known as the submillimeter wave band. Figure 13.1 shows a very popular electromagnetic spectrum ranging from microwave to X-rays with embedded

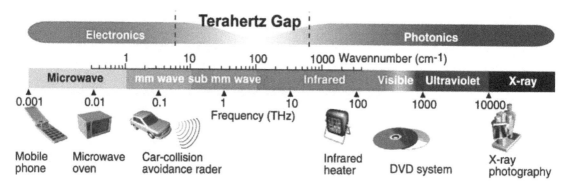

FIGURE 13.1   EM spectrum with application highlighting the THz gap [3].

applications at various bands. As evident from the Figure 13.1, due to a lack of practical technologies for generating and detecting the radiation over mm-waves and below optical bands, there exists a gap in the spectrum above 100 GHz and up to few THz. This spectrum hole is popularly known as 'the THz-gap'.

In spite of overall challenges, the THz band is gradually unraveling various attractive features in diversified frontiers of applied electromagnetic. One of these features is its non ionizing nature, meaning its photons are not energetic enough to knock electrons off from atoms/molecules in human tissue, thus eliminating the triggering of harmful chemical reactions in the body from the THz-beam, which makes it suitable for bio-medical application [4]. Another characteristic of THz waves is having the potential to detect hazardous gases and materials remotely. Various natural as well as manmade molecules in their gaseous state, like NH_4 (ammonia), CO (carbon monoxide), H_2S (hydrogen sulfide), and CH_3OH (methanol), etc., absorb photons when stimulated at terahertz frequencies. Those absorption bands serve as chemical fingerprints for these harmful gases. Even in our universe, which is more or less a black box to us, is radiating most at submillimeter wavelengths and far-infrared (100 GHz–10 THz) radiation along with cosmic microwave background (CMB) radiation [5]. Thus, this spectral range contains information about the origin of the planets, stars, galaxies, and clusters. With this ample presence in the universe, the THz-band offers a wide application range in sensors development for medical imaging, astronomy and astrophysics, gas detection, monitoring of the earth's atmosphere, and homeland security and surveillance.

As we can see from the atmospheric absorption curve in Figure 13.2, atmospheric attenuation is the main barrier to using the terahertz band for long-range communication and radar. Typically, if we generate a 1 W signal with a frequency of 1 THz, it will be diminished gradually to almost nothing (it will retains about 10 to 30% of its original strength) after traveling just 1 km [4].

As we have previously discussed, THz waves are nonionizing radiation, a feature which would be extremely useful in bio-medical imaging such as tumor detection, brain imaging, and full body scans for surveillance purpose. The resolution of THz imaging is much higher than existing technology, and it is

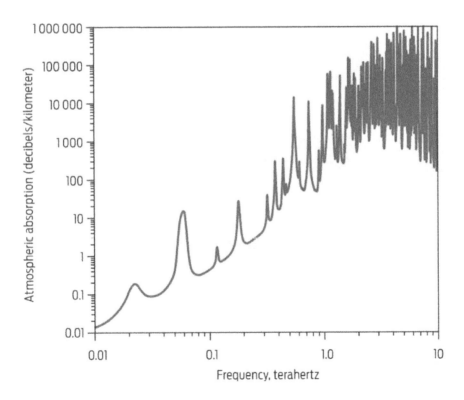

FIGURE 13.2 Atmospheric attenuation versus frequency curve [4].

also completely safe. But it has been observed that the terahertz signal's power decreases to 0.0000002% of its original strength after traveling just 1 mm in saline solution. (Saline solution is a good approximation for body tissue.) So for now at least, terahertz medical devices will be useful only for surface imaging applications like skin cancer and tooth decay and laboratory tests on thin tissue samples.

13.3 THZ TECHNOLOGY: NATIONAL SCENARIO

The THz frequency band possess a great potential to accommodate larger bandwidth and higher data rates to meet the rapidly growing data demand for the communication industries. One of the interesting facts is that the 275 GHz to 3,000 GHz band is not yet allocated for any kind of service. This huge bandwidth cannot be used because of higher atmospheric losses associated in between it, as we can see from Figure 13.2. So wide-range THz band communications simply cannot be established. We can observe 'low-loss' transmission windows at the THz range in the atmospheric attenuation curve. These windows are listed in Table 13.1 with potential applications. Instead of conventional high bandwidth communication links, one can think of short bandwidth communication links. So there are a few visions listed here that will serve as a road map for THz communication.

To achieve these objectives, various Indian institutes have started working in the THz communication domain, as listed in Table 13.2 [6].

TABLE 13.1   Different Windows in THz with POTENTIAL APPLICATIONS

WINDOW	FREQUENCY RANGE	LOSS (dB/m)	POTENTIAL APPLICATION
Window-I	625–725 GHz	0.02	Indoor application
Window-II	780–910 GHz	0.02	Indoor application
Window-III	1210–1410 GHz	0.125	Device-to-device communication
Window-IV	1420–1590 GHz	0.125	Device-to-device communication

TABLE 13.2   Different Indian Institute with THz Research Area

APPLICATION	INSTITUTE	AREA OF RESEARCH
Sources and detectors	IIT, Delhi	THz waveguide, time domain, and CW THz set up
	IISER Trivandrum	Semiconductor hetero-structure alloy as THz source
	TIFR Mumbai	SI-GaAs THz source
Channels	CEERI Pilani	THz waveguide in silicon
	CEERI Chennai	Wireless transmission, mm-wave remote and atmospheric sensing
Signal processing	IIT, Guwahati	THz meta-material structures for active and passive devices
	IIT–Mumbai	Mm wave signal processing
Mixer and antenna	SCL, Chandigarh	Silicon micromachined passive building blocks including antenna

13.4 EFFECT OF SURFACE ROUGHNESS AT THZ FREQUENCY

As outlined in the introduction, due to intrinsic and unavoidable tolerances, the fabricated structure is not perfectly finished and often possesses irregularities such as surface roughness. As skin depth of the conductor becomes comparable with thickness of the conductor at THz frequency, it becomes important to study the surface roughness. This study has been done through two analytical models, namely (i) the Hammerstad–Bekkadal model [7–8] and (ii) the Huray model [9–11]. The simulation results are also systematically presented in this chapter.

Although the metallization process involves some microscopic level of surface roughness in any fabrication method, it doesn't affect device performance significantly in DC or low frequency application. However, in the high frequency domain it invites some detrimental effects. Conductivity and surface impedance of the metal layer gets changed. This work reports the effect of surface roughness in a metalized wall of a rectangular waveguide (WR-2.2) etched using silicon micromachining. Two well-known mathematical models have been implemented to quantify the effect of roughness in device performance.

13.4.1 Hammerstad–Bekkadal Model

Here, surface roughness is treated as a series of peaks and valleys. as shown in Figure 13.3a. This model can predict well, while the RMS (root mean square) value of roughness (h_{RMS}) is comparable to the skin depth (δ_{skin}) of the conducting layer. The roughness factor (f_r) is calculated from h_{RMS} and δ_{skin} using the closed-form equation 13.1,

$$f_r = 1 + \frac{2}{\pi} \tan^{-1} \left[1.4 \left(\frac{h_{RMS}}{\delta_{skin}} \right)^2 \right] \tag{13.1}$$

Where

$$\delta_{skin} = \left(\frac{2}{\sigma_{smooth} \omega \mu} \right)^{1/2}$$

Consequently, the conductivity of the thin metal layer is altered as a function of f_r and operating frequency as follows

$$\sigma_{rough} = \frac{\sigma_{smooth}}{f_r} \tag{13.2}$$

With the modified conductivity (Figure 13.4a) of the metallic wall of the waveguide structure, its propagation or loss characteristics are also changed, as depicted in Figure 13.4b.

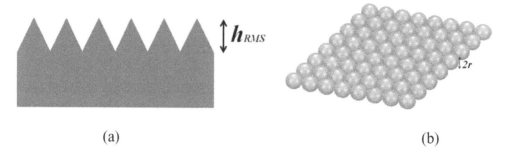

FIGURE 13.3   Roughness profiles of surface assumes for the (a) Hammerstad–Bekkadal model and the (b) Huray model.

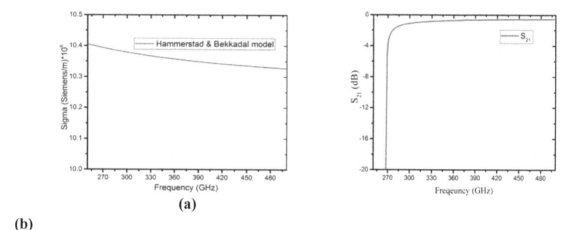

FIGURE 13.4   (a) Variation of conductivity with respect to frequency and (b) S_{21} plot for Hammerstad–Bekkadal model.

13.4.2 Huray Model

This model treats the surface roughness profile as a cluster of a 3D 'snowball' structure (Figure 13.3b), where the roughness factor (f_r) is expressed mathematically by equation 13.3 [11].

$$f_r = 1 + \frac{3}{2} \frac{\dfrac{4\pi r^2 N}{A}}{1 + \dfrac{\delta_{skin}}{r} + \dfrac{\delta_{skin}^2}{2r}} \tag{13.3}$$

Where, r is the radius of snowball; A is surface area of conductor (for the present case, it will be internal surface area of all the four side walls of the waveguide) and N is the ratio between the volume of the surface under consideration to the volume of each snowball.

Using equations 13.2 and 13.3, the modified conductivity of the waveguide wall can be obtained. Considering a practical wall thickness of 3 μm for WR-2.2 waveguide, the radius of the snowball can typically vary between 0.1 to 1.5 μm. Figure 13.5a depicts the modified conductivity plot for various r values. Change in conductivity of the waveguide wall reflects in the propagation characteristics, which is noticed in Figure 13.5b.

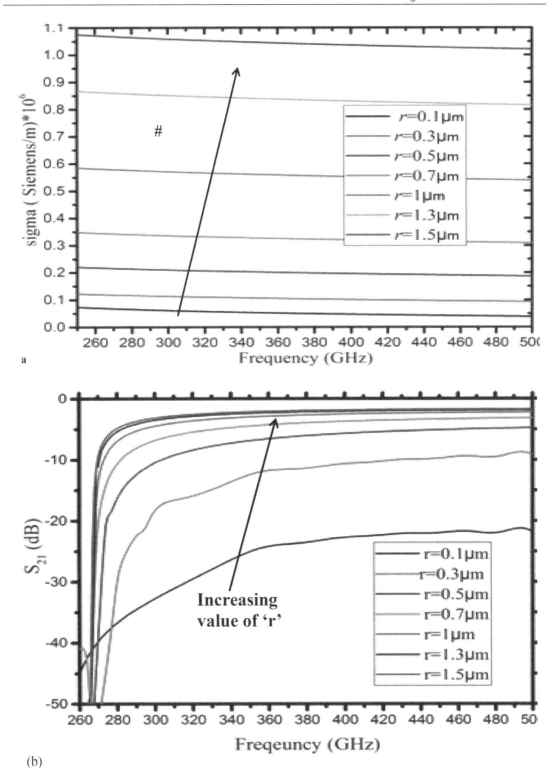

FIGURE 13.5 (a) Variation of conductivity with respect to frequency and (b) S_{21} plot for different radii of snowballs using the Huray model.

13.5 SILICON MICROMACHINING-BASED PASSIVE COMPONENTS FOR TRANSCEIVER MODULES AT THZ FREQUENCY

The term 'micromachining' was coined in the 1980s. It describes the process of the selective removal of substrate material, either by means of wet chemicals or by the dry etching method. Historically, this micromachining method was applied on silicon substrate to realize various physical or inertial sensors (pressure sensors, gyroscopes, accelerometers, etc.). Currently, this field is not limited to only to the world of sensors. Rather, it is expanding its domain for RF, microwave, or even THz-frequency range. The feature of precise lithography along with batch fabrication facility makes this silicon micromachining a unique process for high frequency devices. Moreover, the repeatability of the process is quite promising in comparison to other existing methods in the market. Initially, waveguide structures are developed for THz passive devices, which act as a base-line architecture for all other passive building blocks.

In this chapter, we discuss about two specific rectangular waveguide structures, namely WR-2.2 and WR-3.4. The nomenclatures of the waveguide are defined by the EIA (Electronic Industries Alliance). As per the EIA-standard, WR-3.4 has a cross-sectional dimension of 864×432 μm^2, which is suitable for 220–330 GHz applications, whereas WR-2.2 has a standard dimension of 560×280 μm^2, which is best for the 325 to 550 GHz band. This work comprises design and simulation of several passive components based upon rectangular waveguide structures.

These are:

a. Power divider (1:2 and 1:4).
b. Directional coupler (coupling coefficients of 7, 10 and 13 dB).
c. Multi-hole branch-line coupler.
d. Antenna.

Schematic diagrams of various popularly used passive components are shown in Figure 13.6 to Figure 13.12 [12–14]. All of these structures are designed based upon the high-aspect ratio process (HARP) called DRIE (deep

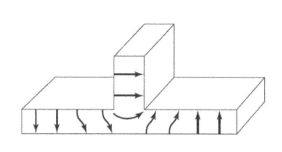
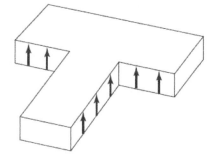

(a) Waveguide based E-plane T-junction   **(b)** Waveguide based H-plane T-Junction

FIGURE 13.6   Various T-junction power dividers [12].

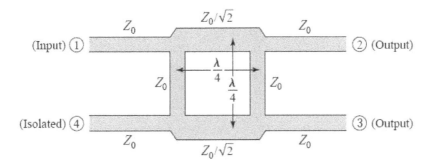

FIGURE 13.7   Geometry of quadrature hybrid [12].

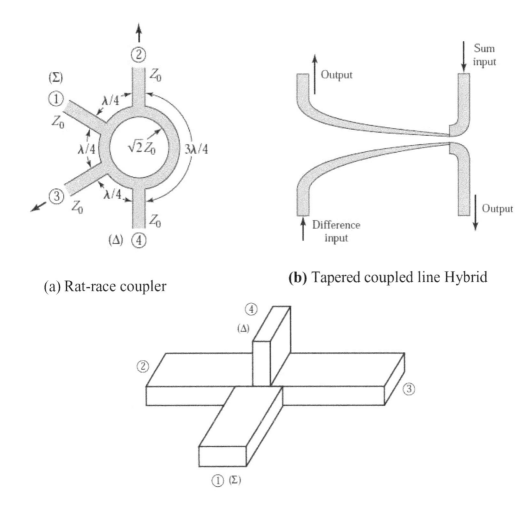

(a) Rat-race coupler

(b) Tapered coupled line Hybrid

(c) A waveguide 180°-Hybrid junction/ Magic Tee

FIGURE 13.8   Various 180°-hybrid junctions [12].

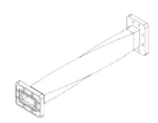

(a)Waveguide twist with flange

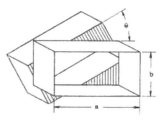

(b) Step twist geometry [9]

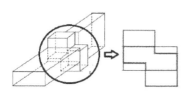

(c)Twisting by double square cut [13]
channel [14]

(d) Twisting by gradual change in

FIGURE 13.9 Waveguide twist with different designing approach.

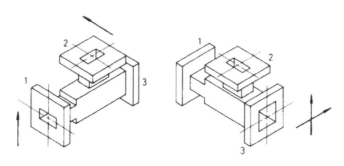

(a) Simple OMT

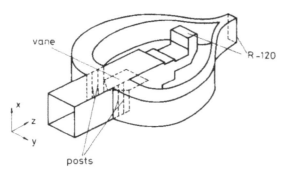

(b) OMT structure having side arms split into E-plane bends

FIGURE 13.10 OMT structures [15].

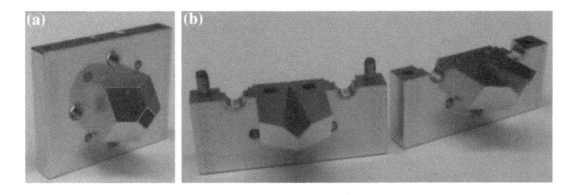

FIGURE 13.11 Split block horn with WR-1.5 input from Virginia Diodes: (a) assembled block and (b) upper and lower blocks [16].

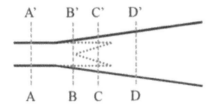

(a)Side view for diagonal horn structure

A-A'   B-B'   C-C'   D-D'

(b) Cross-section through blocks at various planes

FIGURE 13.12 Diagonal horn antenna geometry [14].

reactive ion etching). A generic two wafer process steps are shown in Figure 13.13, which is applicable for realizing all the previously mentioned passive components. The steps can be described as follows:

i. It is two wafer processes. Initially, two double-side polished silicon wafers are taken and given requisite chemical cleaning (RCA + SPM) to eliminate all possible organic and inorganic contaminants. After that, on the first wafer, the following process steps are performed:
   a. A thick oxide layer is deposited on its front side.
   b. The oxide layer is patterned to form a channel through DRIE etching.
   c. DRIE process is applied up to the depth of the lateral dimension of the rectangular waveguide.
   d. The oxide layer is removed.
   e. A conformal gold layer is evaporated to metalize the trench.
ii. On the second wafer, gold is deposited on the front-side by E-beam evaporation.
iii. The second wafer is flipped to make the metal layer on its bottom side, and then both of the wafers are bonded together adopting eutectic bonding at 363°C. Thus, the waveguide channel is formed.

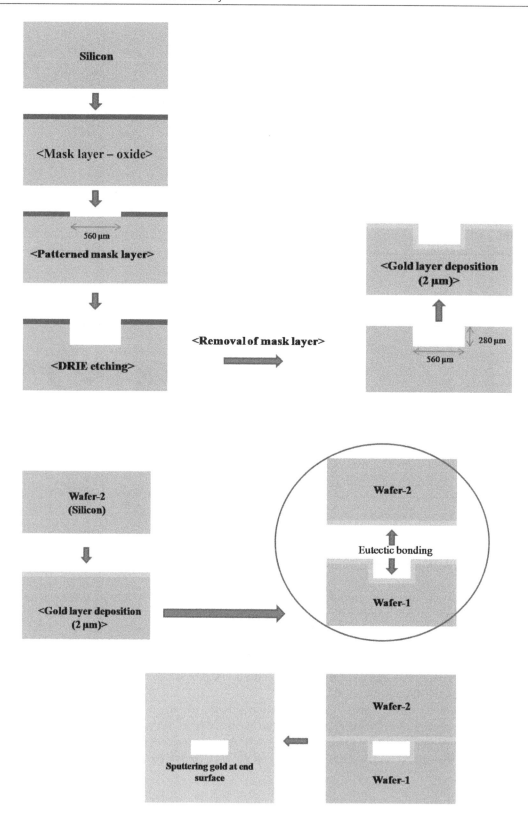

FIGURE 13.13 Generic fabrication process flow of silicon micro-machined passive device.

13.6 STUDY ON THE BASIC WAVEGUIDE MODEL FOR 325 TO 500 GHZ BAND

Waveguide is a media that guides an electromagnetic wave/energy from one point to another with minimal loss of energy. Rectangular waveguide configuration is chosen as the back-bone for all the designs in this research work. WR-2.2 is the preferred candidate of choice for this band. Traditionally, an air-filled metallic walled waveguide structure is used for lower frequency bands. Similarly, with the help of the DRIE (deep reactive ion etching) technique, the waveguide profile is etched within the silicon substrate, followed by conformal thin film deposition of suitable metals and sealing of the cavity. Geometry of the basic waveguide structure is shown in Figure 13.14, which defines properly the values of a and b. During the silicon micromachining, there is a chance of variation of vertical as well as horizontal dimensions of waveguide structure (a and b parameters of the rectangular waveguide shown in Figure 13.14) because of some process constraints. Figure 13.15 shows the effect of variation of lateral dimension of waveguide on its cut-off frequencies. Three basic waveguide modes (TE_{10}, TE_{11}, and TE_{20}) were studied. It has been observed that the variation of lateral dimension affects the cut-off frequencies at a rate of 0.5 GHz/μm, 0.2 GHz/μm, and 0.3 GHz/μm for TE_{10}, TE_{11}, and TE_{20} modes, respectively. TE_{10} being the dominant mode, tight tolerances are essential to maintain this mode's desired property. Basically, the vertical dimension of the waveguide is responsible for deciding power-handling capability and attenuation optimization. Usually, it is kept as half of the lateral dimension.

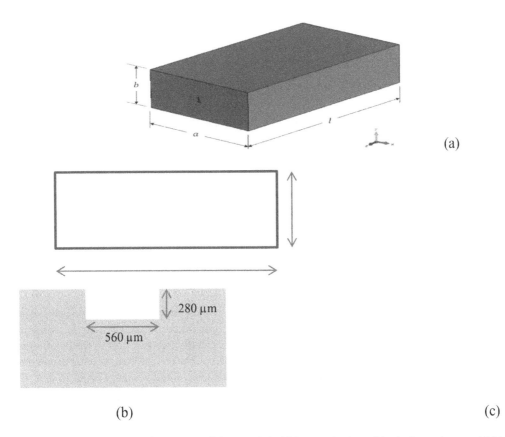

(a)

(b)   (c)

FIGURE 13.14 Geometry and dimension of the WR-2.2: (a) Isometric view, (b) c/s dimension, and (c) trench opening in silicon substrate

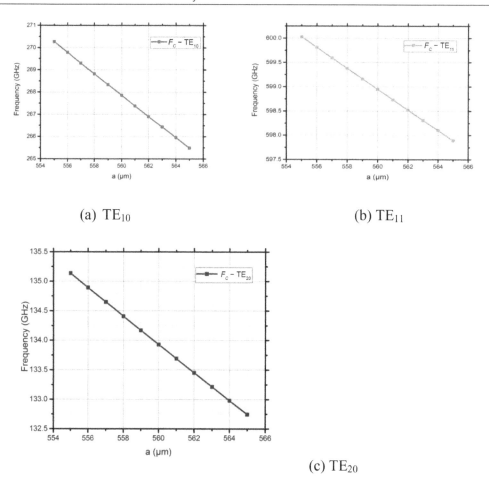

(a) TE_{10} (b) TE_{11}

(c) TE_{20}

FIGURE 13.15   Variation of cut-off frequencies for TE_{10}, TE_{11}, and TE_{20} modes with waveguide lateral dimension.

13.7 STUDY ON WAVEGUIDE BEND STRUCTURE AND ITS DESIGN

Frequently, a waveguide-based complete system realization requires bend structures. This is essential to mitigate the extra losses associated with the device or its interfacing circuit. It can affect the phase response of the whole assembly along with degradation of attenuation characteristics. Hence, it should be modeled well, considering all fabrication intricacies. In this work, single as well multiple bend structures were designed (Figure 13.16), and simulated results were compared with the right-angled straight wall waveguide configuration. In the right-angled bend structure, surface area is nearer to the bending region, which basically increases the capacitance of transmission line; hence the characteristics impedance is reduced. To maintain the line impedance as constant, surface area should be kept equal to that of other portions of the waveguide, and this is done by making the chamfered bend with a radius equal to the lateral dimension of the waveguide. The result of the analysis is depicted in Figure 13.17. It is noticed that the matching profile of the waveguide structure is improved significantly in the bending structure if the chamfering technique is used. Here, IL signifies the insertion loss and RL denotes the return loss parameters.

The following designs have been carried out using bend structures.

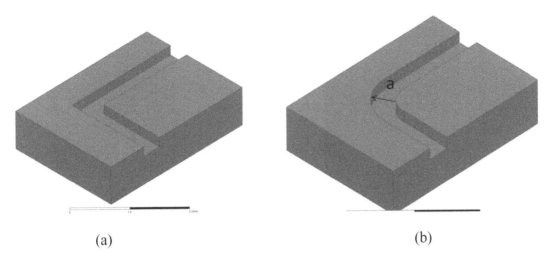

(a)   (b)

FIGURE 13.16 Geometry of the (a) right-angled bend and (b) chamfered bend.

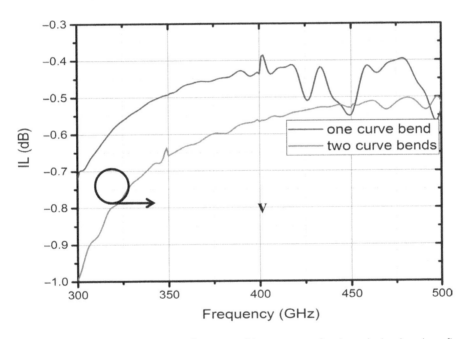

FIGURE 13.17 Insertion loss performance of a waveguide structure after introducing bend configuration in a waveguide structure.

13.7.1 Waveguide-Based Power Divider/Combiner

A power divider/combiner is the basic building block for any subsystem realization. Whether it is antenna feed or any routing path for an internal subsystem, it is necessary to split or combine the signals with desired amplitude and phase ratio. Here, in this research we have targeted a 1:2 power divider (as shown in Figure 13.18) with the targeted specifications, as shown in Table 13.3. Theory of waveguide bend optimization is applied here to obtain the optimum result. The main critical section of designing such a divider is the T-junction, which basically acts as an impedance transformer structure. Simulated RF performance of the power splitter is depicted in Figure 13.19. It is observed that the designed structure offers an insertion

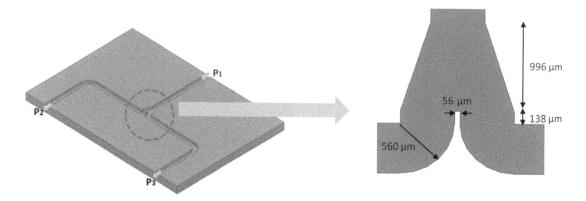

FIGURE 13.18   Layout of the waveguide-based power divider structure.

TABLE 13.3   Targeted Specs. of Power Divider (1:2)

PARAMETER	SPECIFICATION
Frequency range	370–470 GHz
Amplitude imbalance	±1 dB
Return loss	Better than 10 dB
Phase error	< ±10°

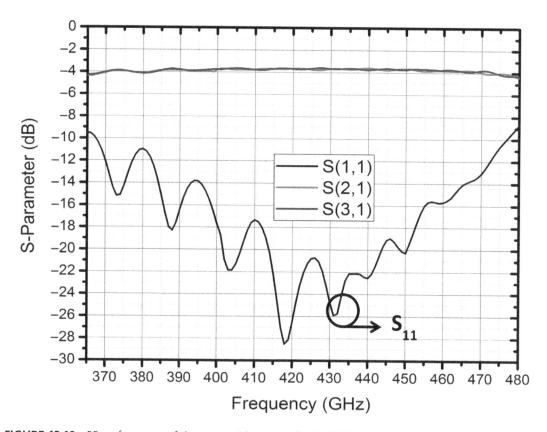

FIGURE 13.19   RF performance of the waveguide power divider (1:2).

loss of around 4.5 dB (max.) with a return loss of better than 12 dB for the entire frequency range. Phase imbalance of around ±2° is observed during FEM analysis between two output ports.

This power divider (1:2) topology is repeated twice to make the 1:4 configurations (Figure 13.20). The designed structure offers an insertion loss of around 7.5 dB (max.), with a return loss of better than 12 dB for the entire frequency range, as depicted in Figure 13.21. Phase imbalance of around ±2° is observed during FEM analysis between two output ports.

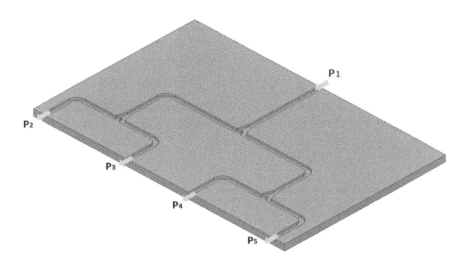

FIGURE 13.20   Four-way (1:4) waveguide-base power divider.

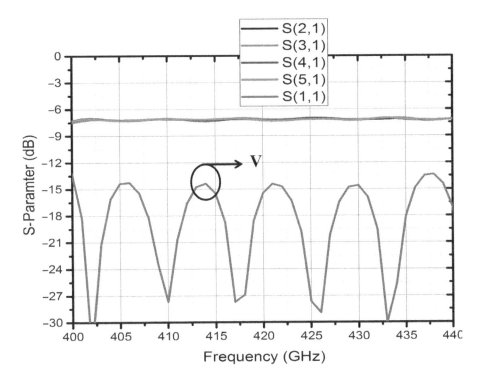

FIGURE 13.21   Simulated S-parameters for the power divider structure.

13.7.2 Directional Coupler

This component is used widely in various measurement instruments, and it is used to test the coupling between two transmission lines. Here in this work, we have targeted a directional coupler with three different coupling coefficients, viz. 7, 10, and 13 dB. An iris-based configuration is implemented to control the flow of EM energy in between primary and secondary waveguides. The base-line of the design is a rectangular waveguide with WR-2.2 configuration. Table 13.4 describes the compiled targeted specifications of the three directional couplers:

The isometric views of three directional couplers are shown in Figure 13.22.

Internal structure details have been outlined in Figure 13.23, where the dimensions of the parameters have been optimized to get the best results (Table 13.5). RF performances of the three coupler structures have been shown in Figure 13.24.

13.7.2.1 Measurement Topologies of the Coupler

Three performance metrics have to be measured for the directional coupler. For the ease of testing, three topologies have been proposed, which have been shown in Figure 13.25.

TABLE 13.4 Targeted Specifications of the Directional Coupler

PARAMETERS	VALUES
Operating frequency band (GHz)	300 to 500
Coupling (dB)	7, 10, and 13 dB
Reflection loss (dB)	Better than 20

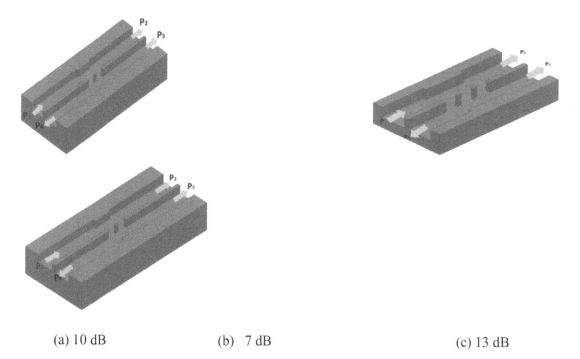

(a) 10 dB   (b) 7 dB   (c) 13 dB

FIGURE 13.22 Isometric view of directional couplers with various coupling coefficients.

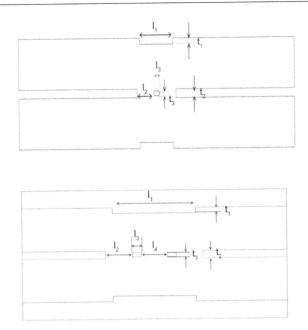

FIGURE 13.23   Internal structure of the directional coupler with various design parameters.

TABLE 13.5   Parameter Values of the Directional coupler

PARAMETER(S)	C = 7 DB (IN µM)	C = 10 DB (IN µM)	C = 13 DB (IN µM)
l_1	940	600	550
l_2	300	300	280
l_3	100	100	100
l_4	300	N.A.	N.A.
t_1	70	70	70
t_2	100	100	100
t_3	50	50	50

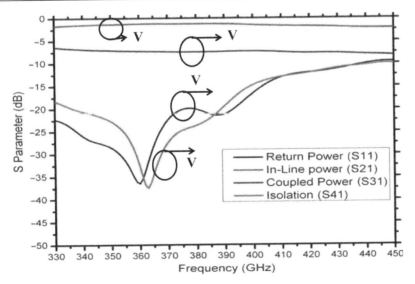

FIGURE 13.24   Simulated RF performances of directional couplers with various coupling coefficients.

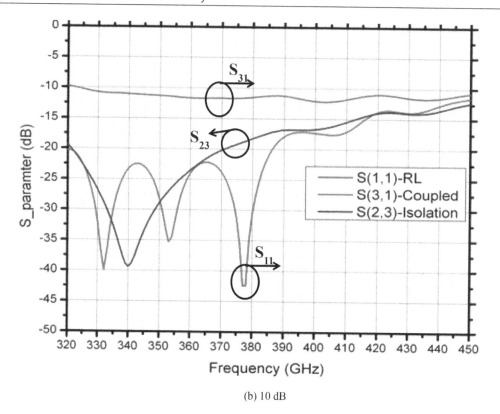

(b) 10 dB

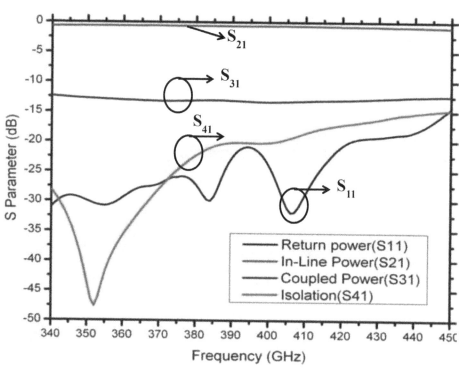

(c) 13 dB

FIGURE 13.24   (Continued)

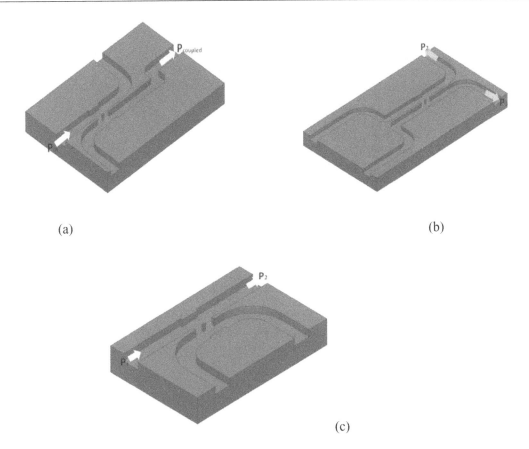

(a)

(b)

(c)

FIGURE 13.25 Measurement topologies for (a) coupling, (b) isolation, and (c) through port loss evaluation.

13.7.3 Multi-Hole Branch Line Coupler

This kind of multi-hole branch-line coupler is used where broadband applications are essential. A riblet-type [17] coupling mechanism is implemented in this structure. Geometry of the coupler is shown in Figure 13.26. Multiple iris structures have been placed to enhance the interaction of wave or coupling between two primary and secondary waveguides. Distances between two holes are usually kept a quarter-wavelength apart.

13.7.4 H-Plane Sectoral Horn

A horn antenna is traditionally popular because of its stable radiation pattern and broadband characteristics. It finds wide applications in antenna measurement and many other various applications. Here in this work, the rectangular waveguide structure is flared in a calculated proportion to get the maximum efficient antenna structure as per the norms given by the 'universal directivity curve' [18]. The fabrication process of the horn antenna has been kept aligned with the rest of the passive devices targeted here. Geometry of the antenna is shown in Figure 13.27. Targeted specification of the antenna is depicted in Table 13.6. Simulated antenna gain is shown in Figure 13.28, which describes the directive nature of the antenna with a gain of 13 dBi.

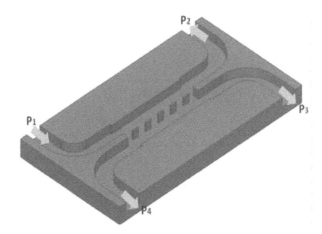

FIGURE 13.26   Multi-hole branch-line coupler.

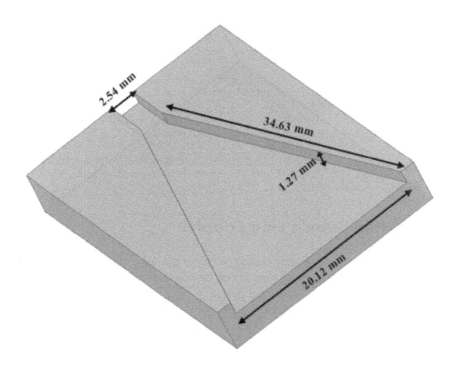

FIGURE 13.27   Cross-sectional view of sectoral horn antenna.

TABLE 13.6   Targeted Specifications of the Sectoral Horn Antenna

SPECIFICATION	VALUES
Frequency band (GHz)	75–110
Feeding element	Waveguide WR10 (2.54 × 1.27 mm²)
Gain (dBi)	13

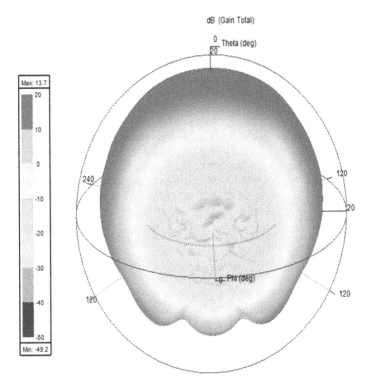

FIGURE 13.28   Simulated 3D-radiation pattern of the horn antenna at 100 GHz.

13.8 CONCLUSION

This chapter emphasizes the potentiality of silicon micromachining technology for the development of submmW and THz applications. The impetus of such high frequency technology in the modern communication world as well as in daily human life has also been outlined. The status of various research institutes at the national level (Indian scenario) along with their work initiative has been highlighted here. The inherent merits of silicon micromachining are implemented to develop such precise and tiny passive devices for submmW and THz applications. Various process-related issues have also been addressed with analytical simulations data, which clearly depicts the implication of fabrication tolerances. Generic process flow of such devices has also been proposed with two wafer approaches. Finally, four types of passive devices have been designed with this method. Fabrication of such devices is under progress.

ACKNOWLEDGMENT

The authors are very much thankful to Sh. Vaibhav Adhikar of COMSOL Multiphysics software for his great help in doing various simulations in this entire research work.

REFERENCES

[1] Z. D. Tayloretal, "THz Medical Imaging: In Vivo Hydration Sensing", *IEEE Transactions on Terahertz Science and Technology* 1.1 (2011), pp. 201–219.

[2] D. M. Mittleman et al., "Recent Advances in Terahertz Imaging", *Applied Physics B* 68.6 (1999), pp. 1085–1094.

[3] K. Fukunaga et al., "Terahertz Spectroscopy Applied to the Analysis of Artists' Materials", *Applied Physics A* 100 (Sept. 2010), pp. 591–597.

[4] C. M. Armstrong, "The Truth about Terahertz", *IEEE Spectrum* 49.9 (2012), pp. 36–41.

[5] T. G. Phillips and J. Keene, "Sub-millimeter Astronomy (Heterodyne Spectroscopy)", *Proceedings of the IEEE* 80.11 (1992), pp. 1662–1678.

[6] M. Rahaman et al., "THz Communication Technology in India Present and Future", *Defense Science Journal* 69.5 (Sept. 2019), pp. 510–516.

[7] B. Huang and Q. Jia, "Accurate Modeling of Conductor Rough Surfaces in Waveguide Devices", *Electronics* 8.269 (2019). https://doi.org/10.3390/electronics8030269.

[8] E. O. Hammerstad and F. Bekkadal, *Microstrip Handbook*, Norwegian Institute of Technology: Trondheim, Norway, pp. 4–8, 1975.

[9] P. G. Huray, O. Oluwafemi, J. Loyer, E. Bogatin, and X. Ye, "Impact of Copper Surface Texture on Loss: A Model that Works", in *Proceedings of the Design Conference*, Santa Clara, CA, USA, 1–4, pp. 462–483, February 2010.

[10] P. G. Huray, *The Foundations of Signal Integrity*, John Wiley & Sons, Inc., Hoboken, NJ, 2009.

[11] P. G. Huray, S. Hall, S. Pytel, F. Oluwafemi, R. Mellitz, D. Hua, and Ye. Peng, "Fundamentals of a 3-D 'Snowball' Model for Surface Roughness Power Losses", in *Signal Propagation on Interconnects*, SPI 2007. IEEE Workshop, pp. 121, 124, 13–16 May 2007. https://doi.org/10.1109/SPI.2007.4512227.

[12] D. M. Pozar, *Microwave Engineering*. 4th Edition, John Wiley Sons, Inc, NY, 2012. ISBN: 978-0-470-63155-3.

[13] A. A. Kirilenko, D. Y. Kulik, and L. A. Rud, "Compact 90° Twist Formed by a Double-Corner-Cut Square Waveguide Section", *IEEE Transactions on Microwave Theory and Techniques* 56.7 (2008), pp. 1633–1637.

[14] J. F. Johansson and N. D. Whyborn, "The Diagonal Horn as a Sub-millimeter Wave Antenna", *IEEE Transactions on Microwave Theory and Techniques* 40.5 (1992), pp. 795–800.

[15] A. M. Boifot, E. Lier, and T. Schaug-Pettersen, "Simple and Broadband Ortho-mode Transducer (Antenna Feed)", *IEE Proceedings H – Microwaves, Antennas and Propagation* 137.6 (1990), pp. 396–400.

[16] Goutam Chattopadhyay et al., "Terahertz Antennas and Feeds", in *Aperture Anten- nas for Millimeter and Sub-Millimeter Wave Applications*. Ed. by Artem Boriskin and Ronan Sauleau. Springer International Publishing: Cham, pp. 335–386, 2018.

[17] J. A. Ruiz-Cruz, J. R. Montejo-Garai, J. M. Rebollar, A. I. Daganzo, and I. Hidalgo-Carpintero, "Design of Riblet-type Couplers for Ka Band Applications", in *2007 IEEE Antennas and Propagation Society International Symposium*, Honolulu, HI, 2007, pp. 4276–4279. https://doi.org/10.1109/APS.2007.4396486.

[18] Lan Jun, "A Universal Method for Directivity Synthesis", *IEEE Transactions on Antennas and Propagation* 35.11 (Nov. 1987), pp. 1199–1205. https://doi.org/10.1109/TAP.1987.1144014.

Photodetectors for Security Application

14

Shonak Bansal, Krishna Parkash, Prince Jain, Sanjeev Kumar, Neena Gupta, and Arun K. Singh

Contents

DOI: 10.1201/9781003126645-14

14.1 INTRODUCTION

Over the past few decades, there is a massive increase in different technological applications that have significantly influenced the way of living. Many of those could never have been imagined over 20 years ago. The rapid growth in miniaturizing the integrated circuits (ICs) to nanometer (nm) sizes have opened up an enormous opportunity space for affordable consumer electronics, which are all around us nowadays. Miniaturization is key to the improvements in speed and power consumption of ICs. This technological revolution was predicted by R. P. Feynman in his legendary seminar, stating that "There's plenty of room at the bottom" [1], and by G. E. Moore, who stated that "In every two years, the number of transistors that are placed inexpensively on an integrated circuit (IC) doubles nearly" [2]. For every semiconductor industry, Moore's law became the road map.

ICs have continued scaling down for the past decades, and in that respect, it can be stated that Moore's law lies at the basis of the field of nanotechnology that has been growing exponentially since the 1980s. With the ever-improving performance of device processing technologies, several new research fields were born in the area of nanotechnology. Nanotechnology is one of the world-wide key technologies as its processes and products have a massive cost-effective potential for future markets. The increase in more efficient, faster, and smaller nano-technological devices with a tolerable price-to-performance ratio has become a crucial successful reason in the international competition for many industrial divisions. The technical ability in the nanotechnology field will become a significant condition in order to successfully compete with improved techniques and products in future high technology markets. Because of its interdisciplinary cross-section character, nanotechnology finds its wide use in the areas of life sciences, medicine, information technology, electronics, chemistry, materials, energy and environmental engineering, space, data storage, automotive manufacturing, and communications in addition to optics and precision engineering in numerous aspects [3], [4].

Moore's law stated that as transistor sizes shrink and communications related to the Internet grow, the conventional copper-based technology has been pushed to its limit and the increasing demands for large bandwidth and high-speed communications applications cannot be met [2]. For this reason, the existence of photonics with low noise and outstanding high-speed transmission properties is intended to have a dominant role in present communication applications. Recent advances are proving that photonics can provide high-performance solutions for the communication industries, such as chip-to-chip, fiber-to-the-premise, and even intra-chip applications at low cost [5].

Long-distance communication can be carried out in the 1,300–1,550 nm spectral regime (or region), which corresponds to the least optical loss window for silica doped optical fibers [5]. The optical devices working at this spectral regime can be aggressively pursued and connected directly to external servers without any wavelength conversion. Up to now, several optical devices such as optical modulators [6], light-emitting sources [6], and other passive components [6], [7] have been developed in this spectral regime.

Although much effort has been made in all-optical fiber communication systems, in optical fiber communications, system signals are not processed in the optical form and require an efficient conversion to electrical signals to be processed by electronic circuitry.

The theory of optoelectronic and photonic devices first began in the late 1980s. Conversely, in the last few decades, it has become one of the most intensive research areas driven by the rapid growth of communications technology [8]. Among photonic devices, a photodetector is a receiver unit that is employed at the end of the optical path to convert an optical signal into an electrical signal. The quality of the conversion will greatly affect the complete performance of the communication system. Photodetectors that hold a critical position in photonic devices have found a wide variety of consumer and scientific applications and have a dominant role in the growth of modern physics over the past century. Therefore, this chapter provides the basic mechanism behind photodetection, its various types, different materials used for photodetection, performance parameters, and their various potential applications.

14.2 SEMICONDUCTOR PHOTODETECTORS

To detect photons, that is, optical radiation in a particular spectral regime, both the internal and external photoemission effects of electrons are employed. Although external photoemission devices such as vacuum photodiodes and photomultiplier tubes meet certain performance standards, they are very heavy and need high voltages for their operation. Conversely, internal photoemission devices such as semiconductor photodetectors with or without internal gain offer better compatibility and performance. These photodetectors are designed with a growing number of semiconductors materials, for example, silicon (Si) [9], germanium (Ge) [10], III–V alloys [11], II–VI alloys [12], and numerous novel two-dimensional (2D) nanomaterials [13], all of which meet many photodetection necessities in numerous ways. Consequently, they have significant applications in all the major existing fiber optics communication systems [9].

The internal photoemission effect can occur in both the intrinsic and extrinsic kind of semiconductors. In intrinsic absorption, the incident photons excite electrons from the valence band in the semiconductor to the conduction band. On the other hand, the extrinsic absorption comprises impurity centers created in the material. Moreover, for efficient photons absorption with rapid response, the intrinsic absorption phenomenon is chosen. Generally, all photodetectors utilize intrinsic photodetection for optical fiber communications.

A basic photodiode is a reverse-biased p–n junction optoelectronic device with a junction exposed to light. Photodiodes meet most of the requirements and hence are widely utilized as photodetectors. A photodetector absorbs the incident optical radiation and converts it into the electrical signal, called photocurrent, which is amplified before additional processing. Thus, when taking into account signal attenuation associated with the link, the performance of the system is estimated at the photodetector. The enhancement in the photodetector performance thereby enables the installation of fewer repeater stations and reduces the operating costs and capital investment. Under equilibrium conditions, a potential barrier exists across the depleted areas on either side of the p–n junction. Generally, the photodetectors are sufficiently reverse-biased, meaning that there is almost no current flow without incident light; the intrinsic area has been completely depleted by carriers. The photodetection process involves the following basic steps [14]:

i. Absorption of energy from optical source and generation of charge carriers.
ii. Transportation of carriers across the depletion region.
iii. Carrier collection and flow of net photocurrent through an external circuit.

Such photodetectors are extensively utilized in optical communication systems, where the detector receives the transmitted optical pulses and converts them into electrical pulses. Therefore, the detector should be highly sensitive to optical signals with low noise and high bandwidth and be cost-effective [9]. They are used in numerous applications ranging from directing automatic lighting in superstores to sensing radiation from the outer galaxy as well as security-related applications. Moreover, photodetectors may vary from simple devices that repeatedly open supermarket doors to receivers on the television remote controls.

14.2.1 Basic Requirements for Photodetector

A photodetector must meet severe requirements in performance and compatibility with optical communication systems when converting an optical signal to an electrical signal. The optical signal becomes weakened and distorted after emerging from the fiber end, so the photodetector must satisfy the following stringent performance requirements [9], [15]:

i. Smaller size, that is, the photodetector must be well-matched with the physical sizes of the optical fiber.
ii. Better sensitivity (responsivity) in the desired spectral regime, that is, wavelength selectivity.

iii. High fidelity, that is, the photodetector response should be linear for a wide range of the optical signal.

iv. Efficient conversion of photons to electrons, that is, high external quantum efficiency.

v. Low noise (i.e., low dark currents, leakage currents, and shunt conductance) and high gain.

vi. Fast response time, that is, large bandwidth.

vii. Be insensitive to the temperature variations.

viii. Low voltages, that is, the photodetector should not require excessive currents or bias voltages.

ix. Long operating life and reasonable cost compared to other system components.

x. High reliability, that is, the photodetector must perform continuous stable room temperature operation for several years.

xi. High detection speed.

14.2.2 Materials Used for Photodetectors

Among photonic devices, photodetector holds a critical position and founds a variety of consumer and scientific applications. The major issue of a photodetector is its compatibility with CMOS technology to provide low power dissipation, good noise immunity, high performance, and better reliability. To attain high-performance and CMOS-compatible photodetection, a suitable semiconductor photosensitive material should be selected. The semiconductor materials used in a photodetector define many of its critical properties. Both the light wavelength to which it responds and the noise level are crucial parameters that depend on the photosensitive material utilized in the photodetector. The wavelength sensitivity of various materials happens because only photons having enough energy to excite an electron across the energy bandgap of the semiconductor material will generate significant energy in producing the current from the photodetector.

Si-based photodetectors [9], [16–18] show their better performance in the ultraviolet (UV) and visible spectral regime with sufficient speed, low dark current, negligible conductance, and long-term stability. The use of a Si-based photodetector is now commercially available and is restricted to the first-generation spectral regime due to an indirect energy bandgap of Si (~1.12 eV), which gives a response loss above 1,100 nm. Therefore, in the longer spectral regime, 1,100 to 1,600 nm (second-generation systems) research is dedicated to the search for semiconductor photosensitive materials with narrow energy bandgaps. This motivates the integration of Si with Ge [10], III–V [11], [16], [19], [20], and II–VI [12] compound alloys that provide a better response in the wider spectral regime. The performance of such devices has been further enhanced significantly in recent years, and a broad selection of photodetectors is now commercially available.

In addition to the fabrication of sophisticated photodetector architectures developed from III–V semiconductor materials for operation at 1,300 and 1,550 nm wavelengths, similar material systems are under investigation for longer wavelengths operations required for infrared (IR) transmission (700 to 30,000 nm). In terms of material properties, mercury cadmium telluride (MCT: $Hg_{1-x}Cd_xTe$) is one of the utmost favorable materials for IR photodetection due to its adjustable energy bandgap, high photon absorption coefficient, high lattice matching to give high-quality crystal growth, and moderate dielectric permittivity that confirms small device capacitance and a low thermal expansion coefficient to give device stability [15], [21–26]. Other semiconductor photodetector types, namely the heterojunction photoconductive detector and phototransistor, can also be developed from III–V alloy semiconductor materials. Particularly the photoconductive detector is suitable as a potential photodetector in the spectral regime of 1,100 to 16,00 nm.

Table 14.1 enlists the common photosensitive materials used for photodetectors along with their energy bandgap in eV and wavelength (λ) range at room temperature [13], [27–31]. However, the primary operating spectral regimes at present are 800 to 900, 1,300, and 1,550 nm, with the key device configurations as the p–n junction, p–i–n, and avalanche photodetectors. Although several types of photodetectors, by using conventional semiconductor material, demonstrate high specific detectivity of the order of 10^{10}–10^{13} cmHz$^{1/2}$/W, they suffer from the limited spectral regime, inflexibility, and cryogenic temperature operation [32].

Recently emerging 2D nanomaterials such as graphene (Gr: 0–0.43 eV), black phosphorus (BP: 0.3–2.0 eV), transition metal dichalcogenides (TMDs), for example molybdenum disulfide (MoS$_2$: 1.9–2.1 eV), molybdenum diselenide (MoSe$_2$: 1.55–1.58 eV), tungsten disulfide (WS$_2$: 1.98–2.05 eV), and tungsten diselenide (WSe$_2$: 0.9–1.66 eV), and hexagonal boron nitride (h-BN: 6.0 eV), etc., as shown in Figure 14.1 (a) [13], have gain intensive attentions in the photodetectors [34], [35]. These 2D nanomaterials demonstrate atomic layered thick configurations and weak van der Waals interactions among the layers. The

TABLE 14.1 Photosensitive Materials Used for Photodetectors Along with Their Energy Bandgap in eV.

MATERIAL	ENERGY BANDGAP (eV)	ELECTROMAGNETIC SPECTRUM WAVELENGTH RANGE (nm)
Si	1.12	320–1,100
Ge	0.67	800–1,800
ZnO	3.37	190–400
GaN	3.44	200–365
GaP	2.24	190–550
GaAs	1.424	650–870
GaAsP	1.98	280–680
In$_{0.53}$Ga$_{0.47}$As	0.75	1000–1,700
InGaAsP	0.75–1.35	1000–1,700
PbS	0.37	1000–3,500
GeSn	0–0.8	1000–2,500
Hg$_{1-x}$Cd$_x$Te	0–1.5	700–30,000
BP	0.3–2.0	310–8,050
Graphene	0–0.43	190–30,000

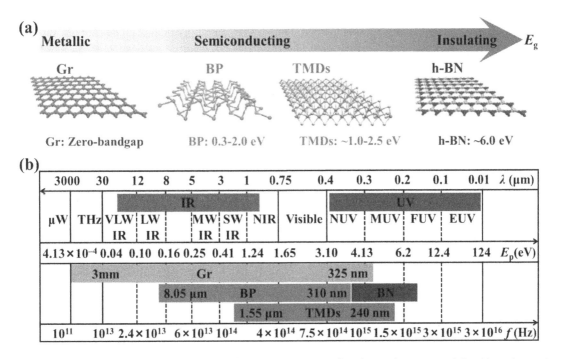

FIGURE 14.1   (a) The energy bandgap and atomic structures of 2D layered nanomaterials with an increasing bandgap from left to right [13], [33]. (b) Electromagnetic spectrum regime from THz to UV [13].

development of high-performance photodetectors is possible due to their broadband detection range, good flexibility, high transparency, sensitivity, strong light-matter interactions, ease of processing, bandgap engineering via a varying number of layers, and high speed, etc. [32], [34], [36].

The spectral regime of photodetector ranges from the UV (10–400 nm) – to the visible (400–750 nm) – to the IR (750 nm-1000 μm) – to the terahertz (THz: 0.1–10 THz, λ: 30–3000 μm) [13], [37]. The UV spectral regime is usually divided into the extreme ultraviolet (EUV: 10–100 nm), far-ultraviolet (FUV: 100–200 nm), mid-ultraviolet (MUV: 200–300 nm), and near-ultraviolet (NUV: 300–400 nm) (Figure 14.1b) [13]. The IR detection regime is often characterized into the near-IR (NIR: 0.75–1.1 μm), short-wave IR (SWIR: 1–3 μm), mid-wave IR (MWIR: 3–5 μm), long-wave IR (LWIR: 8–12 μm), very long-wave IR (VLWIR: 12–30 μm), and far IR (FIR: 30–1000 μm) ranges [13], [21].

As shown in Figure 14.1b, graphene is the only zero bandgap material that enables an enormously broadband light absorption from UV to IR and even up to THz [13]. Its absorption can be further improved by engineering graphene into other nanostructures or by using multilayers [34]. Moreover, its bandgap can be tuned either chemically or electrically [38] and is regarded as semimetal. The excellent response of graphene and its heterostructures with suitable material can demonstrate broadband photodetection and high-frequency operation with low parasitic capacitance [13], [39].

14.3 MECHANISM OF PHOTODETECTION

The photodetectors, also designated as photosensors, are the sensor of electromagnetic radiations or incident light. The operation of photodetectors is usually dependent on the photoelectric and thermal processes in photosensitive materials. The photoelectric effect governing the photodetection mechanism is demonstrated in Figure 14.2 [9]. If the energy of the incoming photons $E_p = hf = hc/\lambda$ (where h corresponds to Planck's constant; f is the optical frequency of incoming photon; c denotes the velocity of light; and λ represents wavelength) exceeds the energy bandgap (E_g) of the material, one electron is promoted to the conduction band for each photon absorbed. Due to this, an empty hole is created in the valence band and is recognized as the photogeneration of the carrier (an electron-hole pair), as illustrated in Figure 14.2a. In Figure 14.2, E_F denotes Fermi-level energy; E_{VB} and E_{CB} signify valence and conduction band energies, respectively; V corresponds to the applied bias; and R_L represents the load resistance. By applying an external biasing, the photogenerated carriers are collected in an external circuit and optical energy is converted into electric energy, as shown in Figure 14.2b. Photogeneration and the separation of an electron-hole pair in the depletion region of the reverse-biased p–n junction are demonstrated in Figure 14.2d [9].

The depletion region width should be large enough to allow a large number of incident photons to be absorbed so that maximum carrier pair generation can be achieved. Meanwhile, large carrier drift times in the depletion layer affect the operational speed of the photodetector; therefore, it is essential to limit the depletion region width. Therefore, a balance must be maintained among the response speed and the number of absorbed photons [9].

The photoelectric effect of photodetection can be categorized as the photovoltaic effect (PVE) and the photoconductive effect (PCE) based on different device architecture and material characteristics.

The PVE is based on the separation of photogenerated carriers in opposite directions under the influence of the built-in electric field $\left(\overrightarrow{E_{field}} \right)$ at the p–n junction (homojunction, heterojunction, and Schottky junction) without external bias voltage (Figure 14.3a). As a result, the photovoltaic photodetectors work with almost zero dark current [34]. The net photocurrent shifts the current-voltage characteristic curves, as depicted in Figure 14.3c. The built-in $\overrightarrow{E_{field}}$ in 2D nanomaterials like graphene can be introduced either by the electrostatically split gates and doping or by using the work function difference among the contacts and graphene [40]. A similar effect can be achieved by applying an external electric field with a

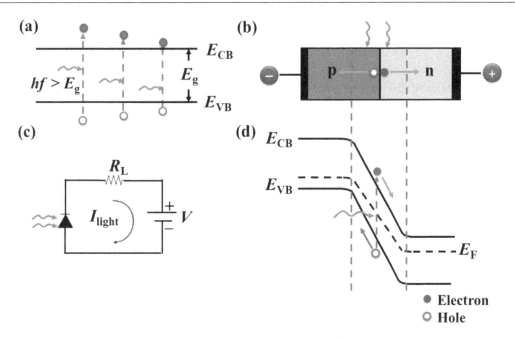

FIGURE 14.2   General photodetection mechanism: (a) Energy bandgap diagram. (b) Reverse-biased p–n junction under illumination demonstrating the drift of carrier in the depletion region. (c) Electrical circuit. (d) Energy bandgap diagram of the reverse-biased p–n junction illustrating the photogeneration and consequent carrier separation.

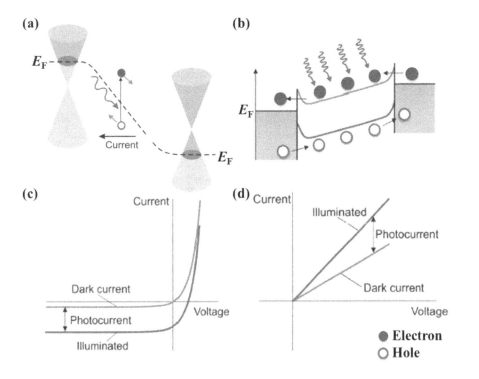

FIGURE 14.3   Schematic of the photocurrent generation through (a) PVE and (b) PCE. Typical current-voltage characteristics under dark and illumination conditions through (c) PVE and (d) PCE [40].

bias voltage that aligns the net electric field toward its direction, and hence photogenerated carriers are collected by the reverse-biased electrodes resulting in total photocurrent to the external circuit, as shown in Figures 14.3b and 14.3d.

A photodetector operating in photoconductive (or photovoltaic) mode is designated as a current source with (or without) the bias voltage. The photovoltaic mode has only the advantage of the lowest dark current density, which in turn improves the specific detectivity. Alternatively, the photoconductive mode operation offers the following advantages [41]:

i. Increased response speed due to the reduction in the junction capacitance.
ii. Weak signal detection capability due to the lesser recombination probability of the photogenerated carriers under the reverse bias.
iii. Increased quantum efficiency due to the increased depletion region width under the reverse bias.

The recent approach also utilizes the thermal effects to produce an electric field due to electron diffusion into metal contacts. The thermal effects include photothermoelectric and photobolometric effects. These thermal effects play a dominant role in the photocurrent generation in graphene and graphene-related 2D materials [40].

The photothermoelectric effect (PTE) is a thermal effect caused via the heating effect in the channel due to the nonuniform incident light and is also known as the Seebeck effect. When the spot of incident light is smaller than the channel length, a temperature gradient at the semiconducting channel leads to a temperature difference ΔT at the two ends, as demonstrated in Figure 14.4a [40]. A photothermoelectric voltage difference ΔV_{PTE} proportional to ΔT generated by the PTE is given as $\Delta V_{PTE} = \Delta T(S_2 - S_1)$, where S_1

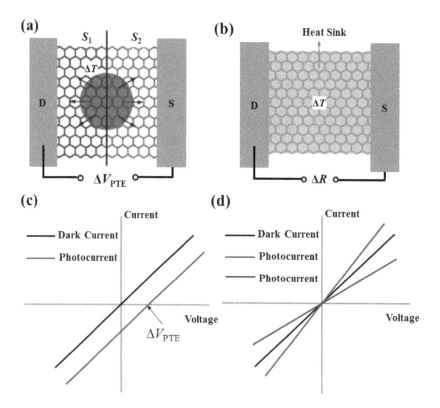

FIGURE 14.4 (a) Schematic of the photocurrent generation through (a) PTE and (b) PBE. The red portion in (a) indicates the incident light beam spot, S_1 and S_2 represent the Seebeck coefficients of the graphene with different doping. Current-voltage characteristics under dark and illumination conditions through (c) PTE and (d) PBE.

and S_2 represent the Seebeck coefficients of the material in V/K with different doping [13]. Generally, the value of ΔV_{PTE} is small and ranges from μV to mV. The typical current-voltage characteristics through PTE under dark and illumination conditions are illustrated in Figure 14.4c.

On the other hand, in the photobolometric effect (PBE), the resistivity of the temperature-sensitive material changes due to the uniform light heating effect caused by the absorption of the photon, as depicted in Figure 14.4b. Under the thermal radiation, the resistance of the material can be decreased or increased, resulting in an increased or decreased current across the device, as shown in Figure 14.4d [13].

The major difference between PTE and PBE is that the photocurrent in PTE is self-driven, as in photovoltaic devices, whereas no photocurrent generates for PBE without an external bias, as shown in Figure 14.4b. The photocurrent in PTE depends on the Seebeck coefficients. However, in PBE, the photocurrent depends on the change of resistivity, which is a function of temperature [13].

14.4 TYPES OF SEMICONDUCTOR PHOTODETECTORS

A semiconductor photodetector is a device that absorbs the incident, that is, incoming photons, and converts it into electrical signals. This conversion mechanism is outlined in previous sections. To understand the development of the photodetectors, it is now essential to elaborate upon the major types of semiconductor photodetectors.

14.4.1 The Semiconductor p–n Single-Junction and p–i–n Dual-Junction Photodetectors

The semiconductor p–n junction and p–i–n photodetectors are the semiconductor photodetectors without an internal gain mechanism, where a single electron-hole pair (EHP) is generated per absorbed photon.

Figure 14.5a illustrates a reverse-biased highly p-type doped (p$^+$) and lightly n-type doped (n$^-$) p$^+$-n$^-$ single-junction photodetector with diffusion and depletion regions. The depletion layer is created by immobile negatively charged acceptor ions in the p-type material and immobile positively charged donor atoms in the n-type material when mobile carriers are swept to their majority sides under the built-in $\overrightarrow{E_{field}}$. Therefore, the width of depletion region depends upon the doping density levels for an applied reverse bias voltage (i.e., the lower the doping level, the wider the depletion region) [9]. As illustrated by the absorption region in Figure 14.5a, photons are absorbed in both the diffusion and depletion regions. The position and width of the absorption region depend upon the energy of incoming photons and on fabricated material for the photodetector. Therefore, during the weak absorption of photons, the absorption region can be fully extended across the device. Thus, carriers are produced in both the diffusion and depletion regions. In the depletion region, the carriers are separated and drifted under the built-in $\overrightarrow{E_{field}}$, while outside this region, the hole diffuses near to the depletion region to be collected. Since most of the incident photons are absorbed in the depletion region, the width of depletion region should be made as large as possible by reducing the doping in n-region.

The homojunction photodetectors suffer from low quantum efficiency due to their gradual doping profile. Except for homojunction, the photodetectors with heterojunction are preferred, as they exhibit better quantum efficiency due to a more efficient built-in $\overrightarrow{E_{field}}$. [32], [34].

To allow the operation in a longer spectral regime where light enters deeper into the photosensitive material, a broader depletion width is required. To accomplish this, the semiconductor material is lightly doped (usually of n-type) so that it may be treated as intrinsic (i), and to create a low resistance contact a highly n-doped (n$^+$) thin layer is employed. This makes the p$^+$-i-n$^+$ architecture the place where all the absorption occurs in the depletion layer, as depicted in Figure 14.5b [9]. The p$^+$-i-n$^+$ photodetector

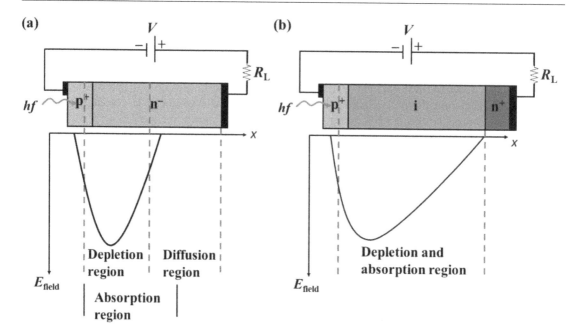

FIGURE 14.5   The schematic and electric field profile for (a) the p⁺-n⁻ junction and (b) the p⁺-i-n⁺ photodetectors showing the depletion, diffusion, and absorption regions.

performance can be significantly enhanced by employing a dual-heterostructure architecture where the central i-region is sandwiched among the wider bandgap p^{+-} and n^{+}-type layers of different energy bandgap semiconductors, such that the incident photon is absorbed mainly in the i-region [42]. The p^{+}-region is termed as a cladding or transparent window at long wavelengths so that photon absorption occurs in the i-region of the device.

14.4.2 Avalanche Photodetector

Another major type of photodetector with an internal gain mechanism is an avalanche photodetector (APD), as shown in Figure 14.6a, and has a more refined architecture than the p–i–n photodetector to produce an extremely high electric field region (~10^{5} V/cm). Therefore, in addition to the depletion region where the photons are absorbed and the carrier pairs generated, a high electric field region exists where electrons and holes can attain enough energy to excite newly generated carriers. This phenomenon is recognized as impact ionization or carrier multiplication (CM) and is the process that results in avalanche breakdown in conventional reverse-biased diodes. Normally, this needs a higher reverse bias voltage so that the newly generated carriers via impact ionization can produce more carriers with a similar phenomenon, as shown in Figure 14.6b [9]. More than 10^{4} of carrier multiplication can be achieved in APDs. Thus, APDs internally multiply the primary photocurrent before entering the input circuit of the amplifier. As the photocurrent is multiplied before facing the thermal noise associated with the receiver circuit, the sensitivity of the receiver increases.

A commonly used architecture for attaining carrier multiplication with slight excess noise is composed of a high resistive, that is, lightly p-type doped (p^{-}) material deposited on a p^{+} substrate. A p^{-}-type ion implant or diffusion is then made in the high resistive material, followed by n^{+}-layer. The i-region is intrinsic material that has some p-type doping. This configuration is termed as p^{+}-i-p^{-}-n^{+} reach through the avalanche photodetector (RAPD) and is shown in Figure 14.6a. The term reach-through arises from the photodetector operation. The high electric field region is also known as the avalanche multiplication

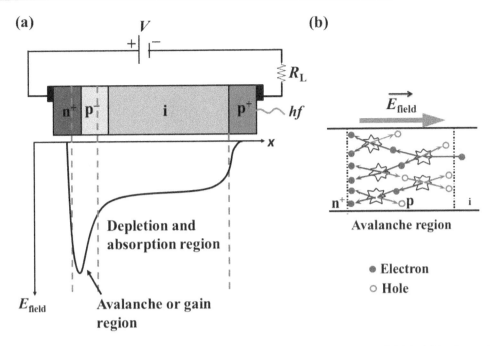

FIGURE 14.6   (a) Reach-through avalanche photodetector and the high electric field (gain) in the avalanche and depletion region. (b) The carrier multiplication process in the gain region of an avalanche photodetector.

gain region, which is comparatively narrow and is centered on the p^--n^+ junction. Therefore, under low reverse bias, mostly the voltage is dropped across the p^--n^+ junction. As the bias voltage increases, the depletion region extends to the p^--region till it "reaches through" to the almost i-region [6].

The photons are incident on the device through the p^+-region and are absorbed in the i-region, acting as the collection region for the photogenerated carriers. After being absorbed, the photons gives up their energy, thus generating EHP, and are separated by high $\overrightarrow{E_{field}}$ in the i-region. The photogenerated carriers drift through the i-region in the p^--n^+ junction. In the high field, i-region carrier multiplication takes place [6].

14.4.3 Metal-Semiconductor-Metal Photodetector

The metal-semiconductor-metal (MSM) photodetector is regarded as an interdigitated photoconductor with the metal-semiconductor and semiconductor-metal junctions as Schottky barriers. The MSM photodetector is monolithic integration and is fabricated using processing steps similar to those needed for the construction of field-effect transistors (FETs) [43]. Normally, the MSM photodetector can be realized utilizing either back-to-back ohmic contacts during photoconductive operation or back-to-back Schottky contacts during photovoltaic operation. However, neither of these device architectures can work under self-powered mode, that is, 0 V bias [44].

A typical device schematic of the MSM photodetector is depicted in Figure 14.7 [27] that comprises two metal electrodes on the top of the semiconductor absorption layer that absorbs a significant amount of incident photon over the desired spectral regime [9]. The low-resistance contacts are created to the conducting layer via the interdigitated electrodes with a finger width (1–7 µm) and finger spacing (1.5–5 µm), as demonstrated in Figure 14.7 [9], [27], [45], [46]. This MSM structure resulted in a planar architecture with an inherently low parasitic capacitance that allowed high-speed operation (up to 300 GHz)

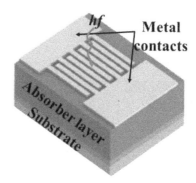

FIGURE 14.7　The schematic of MSM photodetector.

of MSM photodetectors. Also, the interdigitated metal finger contacts are made to maximize the coupling of light into the absorption layer by reducing their obstruction of the active area. This will reduce the distance that photogenerated carriers have to travel before being collected at one of the electrodes. The quantum efficiency of the MSM photodetector is reduced due to the shadowing effect caused by the interdigitated metal electrodes if the photon is incident from the electrode side. The quantum efficiency can be enhanced with the use of an antireflection coating to the surface of the photodetector receiving the optical input. The problem can also be avoided by back illumination if the substrate is transparent to the incident photon [47].

14.4.4　Quantum Wells and Dots Photodetector

The miniaturized physical sizes of nanomaterials provide various nanostructured photodetectors such as quantum wells (QWs) and quantum dots (QDs) with unique photodetection properties [48–50]. The photodetection ability of the photodetector can be enhanced by simply changing the corresponding material properties [34].

The QWs- and QDs-based photodetectors are the alternative solutions to replace the $Hg_{1-x}Cd_xTe$ photodetectors in the IR and THz spectral regime with the high-temperature operation and high-density chip integration [34]. The quantum wells and dots are usually fabricated by inserting a well or dot layer in energy barriers via epitaxial molecular beam epitaxy (MBE). The wide barrier isolates different layers resulting in increased material absorption in the active region, reducing tunneling effects, and thus reduced dark current [9]. The quantum confinement in such nanostructure is generally achieved through the materials with mismatched valence or conduction band. The schematic of the conduction band alignment in QW and QD photodetectors is shown in Figure 14.8a. The barrier potential at the interface due to the energy band mismatch creates confinement to the free charge carriers in the well (or dot) layer. Upon light absorption, the photoexcited electrons from the confined conduction band well (or dot) layer are swept out to the electrodes. The intraband absorption in QWs and QDs permits IR photodetection far beyond the intrinsic bandgap of materials. To further enhance the absorption capability, multilayers of QWs (or QDs) are employed in the active region of the photodetector device. The resultant photodetector device is also designated as the dots-in-well (DWELL), as shown in Figure 14.8b [9], [34].

For the efficient coupling of incident light to the active region, an optical coupling layer (corrugated, gratings layers, etc.) is included in the photodetector device. The inclusion of such a layer is important in QWs-based photodetectors, as 2D QWs do not absorb the incident photons directly from normal incidence. However, no such limitation exists for the zero-dimensional (0D) QDs due to their better three-dimensional quantum confinements, thus allowing the QDs based IR photodetectors in device integration [34].

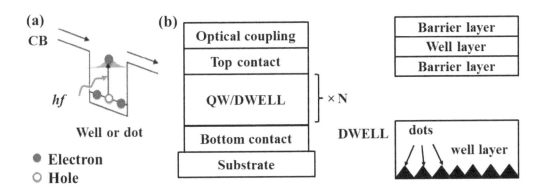

FIGURE 14.8   (a) The schematic of the conduction band (CB) in a QW and QD photodetector. (b) The schematic structure of a QW- and QD-based IR photodetector including the *N* numbers of repeated well and dot layers. The schematic structures of sandwiched QWs and DWELL are shown on the right.

The spectral response of QWs-based photodetectors can be tuned in IR regime by simply adjusting the width of the barrier layer. The DWELL photodetectors show the additional feature of multiwavelength operation due to the presence of multiple electron energy levels in the well, dots, and barrier layers. This tunable photoresponse in QWs and QDs based photodetectors allow the development of multi-color sensitive IR photodetectors in imaging applications [9], [34].

However, for the efficient IR intraband absorption in QWs and QDs, the constituting photosensitive materials are often heavily doped to create enough carriers in the valence band. As a result, QWs and QDs photodetectors often suffer from the high dark current [34].

14.4.5 Barrier Photodetectors

Among different device architecture and device design concepts, barrier photodetectors have been most extensively used, which was first developed by White [51]. Normally, an n-type dual-heterojunction unipolar barrier photodetector comprises a narrow bandgap contact layer, a thin wide bandgap barrier layer (B), and a narrow bandgap active (or absorber) layer, resulting in an nBn photodetector, as shown in Figure 14.9a. The majority carriers, that is, electrons are blocked in such devices due to the absence of valence band discontinuity, which allows the passage of photogenerated minority carriers, that is, holes. Due to a large conduction band offset and a nearly zero valence band offset, the majority carrier current between two electrodes is blocked, as illustrated in Figure 14.9b. This unipolar photodetector operates as a minority carrier photoconductor. Generally, an nBn photodetector eliminates/suppresses the noise level and dark-current by reducing surface-related leakage current and depletion layer-related thermal generation-recombination current, resulting in better uniformity in contrast with an ordinary p–n junction photodetector [52], [53]. The nBn photodetectors have been effectively developed by employing $Hg_{1-x}Cd_xTe$ for IR regime [54], InAsSb [55], [56], $InAs/InAs_{1-x}Sb_x/AlAs_{1-x}Sb_x$ based dual-band IR photodetector [57], and InGaAsSb/GaSb [58].

Compared to the III–V based semiconductor heterostructures, $Hg_{1-x}Cd_xTe$ does not show a near-zero valance band offset, thus requiring high bias voltage to collect the photogenerated carriers [59]. To resolve this issue, a proper p-doped barrier is used to increase the conduction band offset and reduce the offset in the valence band. Moreover, due to existence of the barrier, the n-type contact layer in nBn device can be replaced with the p-type contact layer, resulting in a pBn device, as shown in Figure 14.9a. Therefore, a p–n junction is located between the barrier and the active layer.

A wide variety of barrier photodetectors have been developed that can be categorized as CB_nn and CB_pp photodetectors [59], [60]. In the first category, both the barrier and active layers are of n-type,

(a)

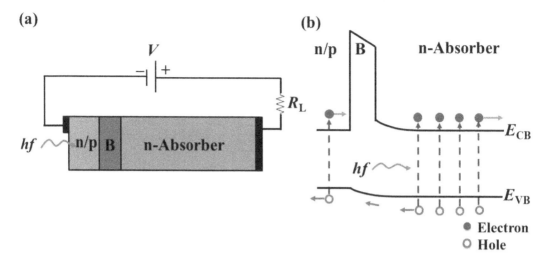

(b)

FIGURE 14.9 (a) The schematic of barrier photodetector. (b) The energy bandgap diagram of the reverse-biased barrier photodetector with zero valence band offset. The barrier (B) layer can be of n-type (B_n) or p-type (B_p).

making a B_nn structural unit, whereas the contact layer (C) can be of n- or p-type. Accordingly, they were named as C_nB_nn (i.e., nBn) and CpB_nn [59–61]. In the second category, both the barrier and active layers are of p-type, making a B_pp structural unit, while the contact layer can be of n- or p-type. As a result, two devices are C_nB_pp and CpB_pp (i.e., pBp) [60–62].

The unipolar nBn photodetectors avoid additional processing steps required for p-type doping via ion-implantation or by MBE. It also eliminates the processing steps for p–n heterojunction formation using expensive equipment [54], [63]. At low temperatures, a unipolar nBn photodetector offers higher-temperature operation than the conventional photodetector [60].

14.5 CHARACTERISTICS PARAMETERS OF THE PHOTODETECTORS

To better compare and quantify the photodetector performance, a set of fundamental characteristic parameters or figures-of-merit need to be considered. This section briefly explains the characteristics parameters of photodetectors.

14.5.1 Electrical Parameters

14.5.1.1 Current Density and Resistance Area Product

In photodetectors, the electrical characteristics depend on many parameters, such as doping profiles, device length, material type, and defects. The dark current density (J_{dark}) is an important characteristic parameter of photodetectors for estimation of operating temperature and bias voltage. The total dark current density J_{dark} is a combination of diffusion current density (J_{diff}), drift current density (J_{gr}), and the tunneling current density (J_{TUN}). J_{TUN} comprises trap-assisted tunneling (TAT) and band-to-band (BTB) tunneling [64]. The high reverse bias leads to bending in the energy band to tunnel the electrons from the valence to the conduction band. This results in the BTB tunneling current density mechanism in photodetectors [64], [65].

The effective or net resistance area product $(RA)_{NET}$ arises due to these different current densities components, and can be written as:

$$\frac{1}{(RA)_{NET}} = \frac{1}{(RA)_{diff}} + \frac{1}{(RA)_{gr}} + \frac{1}{(RA)_{TAT}} + \frac{1}{(RA)_{BTB}} \tag{14.1}$$

where $(RA)_x = \left(\dfrac{dJ_x}{dV}\right)^{-1}$. The subscript x corresponds to the different components of current density.

When the device is illuminated with an incident power density (P_{in}), carriers are produced in the junction. Such photogenerated carriers get separated or drifted away due to the presence of a built-in E_{field} and contribute to the net photogenerated current through an external circuit.

14.5.1.2 Photosensitivity and Linear Dynamic Range

The photosensitivity is defined as the ratio of the photocurrent density J_{light} at a desired wavelength to the dark current density J_{dark}. The generated J_{light} and photosensitivity of the device must be larger to achieve high-performance photodetector.

Linear dynamic range (LDR) defines the input illumination intensity range within which the photocurrent of a photodetector varies linearly with the light intensity and is approximated in a logarithmic scale as [66]:

$$\left(LDR(dB) = 20\log\left(\frac{J_{light}}{J_{dark}}\right)\right) \tag{14.2}$$

A high value of the LDR is preferable to detect both weak and strong light.

14.5.1.3 Photoswitching Time

The photoswitching time measures the response speed of a photodetector and is characterized by the rise time and fall time. The rise time (or response time, τ_r) and fall time (or recovery time, τ_f) are measured at 10% and 90% of maximum J_{light} and are strongly influenced by the charge carrier transport and collection mechanism during the photoresponse. A rapid photoswitching response, that is, smaller values of rise and fall times, is desirable for the high-performance photodetector.

14.5.2 Optical Parameters

14.5.2.1 Spectral Photoresponse

For any given photodetectors, they can only respond to a specific wavelength range. When designing a photodetector, the first consideration is choosing the proper photosensitive material that can respond to the photo signal. To detect the photo signal, the bandgap of the selected material should be smaller or equal to the incident photon energy.

14.5.2.2 Quantum Efficiency

Quantum efficiency (QE) determines how efficiently the semiconductor material converts light into current, that is, it measures the sensitivity of the photodetector. It is the percentage of the photogenerated charge that contributes to the photocurrent and, mathematically, is the ratio of number of electron-hole pairs (N_e) generated to the number of incident photons (N_p) [9], i.e.,

$$QE = \frac{N_e}{N_p}$$

(14.3)

The QE of the device is wavelength-dependent due to the semiconductor energy bandgap and wavelength-dependence of optical absorption [13]. The QE may decrease due to the reflection, absorption, and electrical properties of the semiconductor device. The two types of quantum efficiencies, that is, external QE (QE_{ext}) and internal QE (QE_{int}) are taken into account for a photodetector. QE_{ext} is a measure of the number of photogenerated carriers collected by the electrodes due to excitation by incident photons. The QE_{int} of internal photoemission (IPE) process is the number of carriers emitted to semiconductor material per absorbed photon. If the lifetime of photogenerated carriers exceeds the transit time, a QE of more than 100% is possible in the photodetectors [34].

14.5.2.3 Photocurrent Responsivity

The photocurrent responsivity (R_i) is a ratio of photocurrent to the incident light power and is given by

$$R_i = \frac{J_{light}}{P_{in}} \, \text{A} \, / \, \text{W}$$

(14.4)

The QE_{int}, which gives the internal photocurrent responsivity $\left(R_i^{int} \right)$, and the QE_{ext}, which determines the external photocurrent responsivity $\left(R_i^{ext} \right)$, are calculated using the following relations [66–69]:

$$R_i^{int} = QE_{int} \left(\frac{q\lambda}{hc} \right) = QE_{int} \left(\frac{\lambda}{1.24} \right) \text{A} \, / \, \text{W}$$

(14.5)

$$R_i^{ext} = QE_{ext} \left(\frac{\lambda}{1.24} \right) \text{A/W}$$

(14.6)

14.5.2.4 Specific Detectivity

The specific detectivity (D^*) determines the smallest signal that a photodetector can detect and is calculated by using the relation

$$D^* = \frac{R_i^{ext}}{2} \sqrt{\frac{(R_0 A)_{NET}}{kT}} \, \text{cmHz}^{1/2} \, / \, \text{W}$$

(14.7)

where $(R_0 A)_{NET} = \left(\frac{dJ}{dV} \right)_{V=0}^{-1}$ is the resistance area product at zero biased, that is, under the self-powered mode. The high value of $(R_0 A)_{NET}$ is desirable to result in high specific detectivity.

14.5.2.5 Noise Equivalent Power

The noise equivalent power (*NEP*) is the minimum incident photon signal that a photodetector can decide from noise and is calculated using the following relations [66–69]:

$$NEP = \frac{\sqrt{\Delta f A}}{D^*} \text{ W / Hz}^{1/2} \tag{14.8}$$

where A and Δf denote the active area and bandwidth, respectively.

14.5.3 Noise

Noise refers to spontaneous fluctuations in the current or voltage passing through the semiconductor devices. The shot (quantum) and Johnson-Nyquist (thermal) noise [69] are major sources of noise in the photodetector. Shot noise occurs due to the discrete behavior of charge carriers that contribute to current flow in the devices. On the other hand, thermal noise occurs due to the random thermal motion of the charge carriers. It is also known as white noise due to the existence of the same level at all frequencies [65]. Therefore, the shot (i_s) and Johnson-Nyquist (i_j) noise currents are estimated by [64].

$$i_s = \sqrt{2q\left(J_{light} + J_{dark}\right) A \Delta f} \tag{14.9}$$

$$i_j = \sqrt{\frac{4kT\Delta f}{(RA)_{NET}}} \tag{14.10}$$

The total noise current (i_n) [69] and signal-to-noise ratio (SNR) [9], [64] are given by

$$i_n = i_s + i_j \tag{14.11}$$

$$SNR = 10\log\left(\frac{\frac{1}{2}\left(Q_{ext}P_{in}\lambda\big/1.24\right)}{i_n}\right)\text{dB} \tag{14.12}$$

In general, an ideal high-performance photodetector should possess low dark current density, high sensitivity, high quantum efficiency, high detection speed, high detectivity, high responsivity, low noise equivalent power, small standby power consumption, low noise level, and low applied voltage-bias requirements [15].

14.6 APPLICATIONS

The development in photodetectors has been largely focused on pursuing miniaturized photodetectors with low leakage current, low noise, short response time, high responsivity, large bandwidth, and high gain-bandwidth [15], [31]. Photodetection is possible due to the photoelectric effect in photosensitive materials and large-scale fabrication and integration technologies [31], [70]. Photodetection in the different spectral regimes has gained significant research interest due to their several scientific applications in life, the military, industry, and agriculture, etc.

Depending on the applications, different spectral regimes are frequently used for photodetection, as shown in Figure 14.10 [34]. UV photodetectors found their potential applications in environmental monitoring, space science, military investigation, sterilization, territory intrusions, etc., whereas visible photodetectors have been used for video imaging, displays, industrial safety, optoelectronic storage, etc. [31], [71]. The IR photodetectors have applications in various fields, such as thermal and biomedical

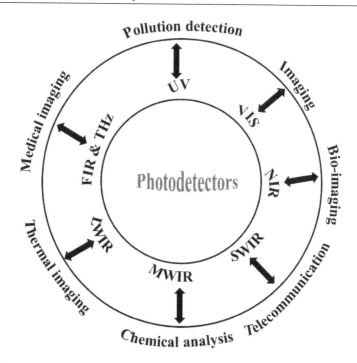

FIGURE 14.10 The potential applications of the photodetector in different spectral regimes covering from UV to IR and THz.

imaging, remote sensing, missile guidance, motion detection, telecommunications, security, free-space communication, chemical analysis, and so on [23], [34], [72–74].

As stated earlier, the tunable optoelectronic properties of $Hg_{1-x}Cd_xTe$ semiconductor material makes it a promising choice for the development of IR broadband photodetectors [21], [53], [59], [74], [75]. Several $Hg_{1-x}Cd_xTe$-based high performance IR photodetectors with different device architectures, such as p–n [22], [68], p–i–n [74], APD [62], [76], and dual-band IR photodetector [23], [77] are being reported at cryogenic temperatures. $Hg_{1-x}Cd_xTe$ technology development continues to be primarily for military- and civilian-related security or safety applications for the detection of a condition or an object [78]. Another most common application is IR-sensitive motion detection for home-security systems. On the other hand, the ultrahigh mobility and higher temperature operation due to the lower thermo-generation rate of graphene led to the development of graphene/$Hg_{1-x}Cd_xTe$-based p–n [39], [79], p–i–n [42], and barrier type [62], [63] IR photodetector architectures with superior performance that can be utilized for the detection of weak signal and security-related applications.

14.7 CONCLUSION

In summary, this chapter presented an introduction to photodetectors, their various photodetection mechanisms, different types, materials used for photodetection, characteristics parameters, and their various applications. Photodetectors are the optoelectronic receiver component that is used to detect incident radiations. Depending on the amount of light absorbed by the photosensitive material in a specific spectral regime, photodetectors can found their potential applications in UV, visible, and IR spectral regimes. In terms of material properties, $Hg_{1-x}Cd_xTe$ is one of the most widely used semiconductor material for IR photodetection due to its adjustable optoelectronic properties by tuning the Cd composition. Particularly, $Hg_{1-x}Cd_xTe$ is most widely used for the development of military-related security applications. But the operation of $Hg_{1-x}Cd_xTe$ based photodetectors is limited to cryogenic temperatures. To deal with such an issue, the heterojunction of graphene with $Hg_{1-x}Cd_xTe$ can be the prominent solution to the production of high-performance photodetectors at near-room temperature.

REFERENCES

[1] R. P. Feynman, "There's plenty of room at the bottom," *Engineering and Science*, vol. 23, no. 5, pp. 22–36, 1959.

[2] G. E. Moore, "Cramming more components onto integrated circuits," *Proceedings of the IEEE*, vol. 86, no. 1, pp. 82–85, 1998.

[3] A. L. Rogach, A. Eychmüller, S. G. Hickey, and S. V. Kershaw, "Infrared-emitting colloidal nanocrystals: Synthesis, assembly, spectroscopy, and applications," *Small*, vol. 3, no. 4, pp. 536–557, 2007.

[4] L. Merhari, *Hybrid Nanocomposites for Nanotechnology*. Boston, MA: Springer, 2009.

[5] J. Wang and S. Lee, "Ge-photodetectors for Si-based optoelectronic integration," *Sensors*, vol. 11, no. 1, pp. 696–718, 2011.

[6] G. Keiser, *Optical Fiber Communications*, Fourth Edition. New Delhi: Tata McGraw-Hill Publishing Company Limited, 2008.

[7] B. Jalali, S. Yegnanarayanan, T. Yoon, T. Yoshimoto, I. Rendina, and F. Coppinger, "Advances in silicon-on-insulator optoelectronics," *IEEE Journal on Selected Topics in Quantum Electronics*, vol. 4, no. 6, pp. 938–947, 1998.

[8] B. Jalali, M. Paniccia, and G. Reed, "Silicon photonics," *IEEE Microwave Magazine*, vol. 7, no. 3, pp. 58–68, 2006.

[9] J. M. Senior, *Optical Fiber Communications Principles and Practice*, Third Edition. England: Pearson Education Limited, 2009.

[10] S. J. Koester, S. Member, J. D. Schaub, G. Dehlinger, and J. O. Chu, "Germanium-on-SOI infrared detectors for integrated photonic applications," *IEEE Journal of Quantum Electronics*, vol. 12, no. 6, pp. 1489–1502, 2006.

[11] Y. Kang, P. Mages, A. R. Clawson, P. K. L. Yu, M. Bitter, Z. Pan, A. Pauchard, S. Hummel, and Y. H. Lo, "Fused InGaAs-Si avalanche photodiodes with low-noise performances," *IEEE Photonics Technology Letters*, vol. 14, no. 11, pp. 1593–1595, 2002.

[12] X. Zhang, D. Wu, and H. Geng, "Heterojunctions based on II-VI compound semiconductor one-dimensional nanostructures and their optoelectronic applications," *Crystals*, vol. 7, no. 10, pp. 1–24, 2017.

[13] M. Long, P. Wang, H. Fang, and W. Hu, "Progress, challenges, and opportunities for 2D material based photodetectors," *Advanced Functional Materials*, vol. 29, no. 19, pp. 1803807-1–1803807-28, 2018.

[14] P. Bhattacharya, *Semiconductor Optoelectronic Devices*, Second Edition. New Delhi: Prentice-Hall India, 2004.

[15] P. C. Eng, S. Song, and B. Ping, "State-of the art photodetectors for optoelectronic integration at telecommunication wavelength," *Nanophotonics*, vol. 4, no. 3, pp. 277–302, 2015.

[16] J. Brouckaert, G. Roelkens, D. Van Thourhout, and R. Baets, "Thin-film III-V photodetectors integrated on silicon-on-insulator photonic ICs," *Journal of Lightwave Technology*, vol. 25, no. 4, pp. 1053–1060, 2007.

[17] L. Shi and S. Nihtianov, "Comparative study of silicon-based ultraviolet photodetectors," *IEEE Sensors Journal*, vol. 12, no. 7, pp. 2453–2459, 2012.

[18] M. Casalino, "Recent advances in silicon photodetectors based on the internal photoemission effect," in *New Research on Silicon – Structure, Properties, Technology*, V. I. Talanin (Eds.). Rijeka, Croatia: InTech, 2017.

[19] A. Beling and J. C. Campbell, "InP-based high-speed photodetectors," *Journal of Lightwave Technology*, vol. 27, no. 3, pp. 343–355, 2009.

[20] Y. Chen and B. Chen, "Design of InP-based high-speed photodiode for 2-μm wavelength application," *IEEE Journal of Quantum Electronics*, vol. 55, no. 1, pp. 1–8, 2019.

[21] A. Rogalski, "HgCdTe infrared detector material: history, status and outlook," *Reports on Progress in Physics*, vol. 68, no. 10, pp. 2267–2336, 2005.

[22] P. K. Saxena and P. Chakrabarti, "Computer modeling of MWIR single heterojunction photodetector based on mercury cadmium telluride," *Infrared Physics and Technology*, vol. 52, no. 5, pp. 196–203, 2009.

[23] P. K. Saxena, "Numerical study of dual band (MW/LW) IR detector for performance improvement," *Defence Science Journal*, vol. 67, no. 2, pp. 141–148, 2017.

[24] S. Bansal, K. Sharma, N. Gupta, and A. K. Singh, "Simulation and optimization of Hg1-xCdxTe based mid-wavelength IR photodetector," in *2016 IEEE Uttar Pradesh Section International Conference on Electrical, Computer and Electronics Engineering (UPCON)*, Varanasi, 2016, pp. 422–425.

[25] S. Bansal, K. Sharma, K. Soni, N. Gupta, K. Ghosh, and A. K. Singh, "Hg1−xCdxTe based *p−i−n* IR photodetector for free space optical communication," in *2017 Progress in Electromagnetics Research Symposium-Spring (PIERS)*, St Petersburg, Russia, 2017, pp. 981–983.

[26] S. Bansal, K. Sharma, P. Jain, N. Gupta, and A. K. Singh, "Atlas simulation of a long-infrared P+-N homo-junction photodiode," in *2018 6th Edition of International Conference on Wireless Networks & Embedded Systems (WECON)*, Rajpura (near Chandigarh), India, 2018, pp. 19–22.

[27] B. Tekcan, C. Ozgit-Akgun, S. Bolat, N. Biyikli, and A. K. Okyay, "Metal-semiconductor-metal ultraviolet photodetectors based on gallium nitride grown by atomic layer deposition at low temperatures," *Optical Engineering*, vol. 53, no. 10, pp. 107106-1–107106-4, 2014.

[28] A. Bablich, S. Kataria, and M. Lemme, "Graphene and two-dimensional materials for optoelectronic applications," *Electronics*, vol. 5, no. 1, pp. 1–16, 2016.

[29] W. Tian, H. Sun, L. Chen, P. Wangyang, X. Chen, J. Xiong, and L. Li, "Low-dimensional nanomaterial/Si heterostructure-based photodetectors," *InfoMat*, vol. 1, no. 2, pp. 140–163, 2019.

[30] S. Tyagi, P. K. Saxena, and R. Kumar, "Numerical simulation of $In_xGa_{1-x}As$ InP PIN photodetector for optimum performance at 298 K," *Optical and Quantum Electronics*, vol. 52, no. 374, pp. 1–12, 2020.

[31] J. Yao and G. Yang, "2D material broadband photodetectors," *Nanoscale*, vol. 12, no. 2, pp. 454–476, 2020.

[32] J. Wang, J. Han, X. Chen, and X. Wang, "Design strategies for two-dimensional material photodetectors to enhance device performance," *InfoMat*, vol. 1, no. 1, pp. 33–53, 2019.

[33] F. Xia, H. Wang, D. Xiao, M. Dubey, and A. Ramasubramaniam, "Two-dimensional material nanophotonics," *Nature Photonics*, vol. 8, no. 12, pp. 899–907, 2014.

[34] F. Zhuge, Z. Zheng, P. Luo, L. Lv, Y. Huang, H. Li, and T. Zhai, "Nanostructured materials and architectures for advanced infrared photodetection," *Advanced Materials Technologies*, vol. 2, no. 8, pp. 1700005-1–1700005-26, 2017.

[35] J. Gusakova, X. Wang, L. L. Shiau, A. Krivosheeva, V. Shaposhnikov, V. Borisenko, V. Gusakov, and B. K. Tay, "Electronic properties of bulk and monolayer TMDs: Theoretical study within DFT framework (GVJ-2e method)," *Physica Status Solidi (A)*, vol. 214, no. 12, pp. 1700218-1–1700218-7, 2017.

[36] C. Xie, C. Mak, X. Tao, and F. Yan, "Photodetectors based on two-dimensional layered materials beyond graphene," *Advanced Functional Materials*, vol. 27, no. 19, pp. 1603886-1–1603886-41, 2017.

[37] H. Chen, H. Liu, Z. Zhang, K. Hu, and X. Fang, "Nanostructured photodetectors: From ultraviolet to terahertz," *Advanced Materials*, vol. 28, no. 3, pp. 403–433, 2016.

[38] V. Ryzhii, M. Ryzhii, N. Ryabova, V. Mitin, and T. Otsuji, "Terahertz and infrared detectors based on graphene structures," *Infrared Physics & Technology*, vol. 54, no. 3, pp. 302–305, 2011.

[39] S. Bansal, K. Sharma, P. Jain, N. Sardana, S. Kumar, N. Gupta, and A. K. Singh, "Bilayer graphene/HgCdTe based very long infrared photodetector with superior external quantum efficiency, responsivity, and detectivity," *RSC Advances*, vol. 8, no. 69, pp. 39579–39592, 2018.

[40] A. Rogalski, M. Kopytko, and P. Martyniuk, "Two-dimensional infrared and terahertz detectors: Outlook and status," *Applied Physics Reviews*, vol. 6, no. 2, pp. 021316-1–021316-23, 2019.

[41] X. Wei, F. G. Yan, C. Shen, Q. S. Lv, and K. Y. Wang, "Photodetectors based on junctions of two-dimensional transition metal dichalcogenides," *Chinese Physics B*, vol. 26, no. 3, pp. 038504-1–038504-15, 2017.

[42] S. Bansal, A. Das, P. Jain, K. Prakash, K. Sharma, N. Kumar, N. Sardana, N. Gupta, S. Kumar, and A. K. Singh, "Enhanced optoelectronic properties of bilayer graphene/HgCdTe based single- and dual-junction photodetectors in long infrared regime," *IEEE Transactions on Nanotechnology*, vol. 18, pp. 781–789, 2019.

[43] A. Rogalski, *Infrared Detectors*, Second Edition. Boca Raton: CRC Press, 2010.

[44] X. Sun, D. Li, Z. Li, H. Song, H. Jiang, Y. Chen, G. Miao, and Z. Zhang, "High spectral response of self-driven GaN-based detectors by controlling the contact barrier height," *Scientific Reports*, vol. 5, no. 16819, pp. 1–7, 2015.

[45] X. Chen, B. Nabet, A. Cola, F. Quaranta, and M. Currie, "An AlGaAs – GaAs-based RCE MSM photodetector with delta modulation doping," *IEEE Electron Device Letters*, vol. 24, no. 5, pp. 312–314, 2003.

[46] C.-K. Wang, Y.-Z. Chiou, S.-J. Chang, W.-C. Lai, S.-P. Chang, C.-H. Yen, and C.-C. Hung, "GaN MSM UV photodetector with sputtered AlN nucleation layer," *IEEE Sensors Journal*, vol. 15, no. 9, pp. 4743–4748, 2015.

[47] G. P. Agrawal, *Fiber-optic Communication Systems*, Fourth Edition. Hoboken, NJ: John Wiley & Sons, 2010.

[48] D. Shin and S.-H. Choi, "Graphene-based semiconductor heterostructures for photodetectors," *Micromachines*, vol. 9, no. 7, pp. 1–29, 2018.

[49] A. Asgari and S. Razi, "High performances III-nitride quantum dot infrared photodetector operating at room temperature," *Optics express*, vol. 18, no. 14, pp. 14604–14615, 2010.

[50] M. R. Hao, Y. Yang, S. Zhang, W. Z. Shen, H. Schneider, and H. C. Liu, "Near-room-temperature photon-noise-limited quantum well infrared photodetector," *Laser and Photonics Reviews*, vol. 8, no. 2. pp. 297–302, 2014.

[51] A. M. White, "Infra red detectors," U.S. Patent US 4679063, 1987.

[52] N. D. Akhavan, G. Jolley, G. A. Umana-Membreno, J. Antoszewski, and L. Faraone, "Design of band engineered HgCdTe nBn detectors for MWIR and LWIR applications," *IEEE Transactions on Electron Devices*, vol. 62, no. 3, pp. 722–728, 2015.

[53] N. D. Akhavan, G. A. Umana-Membreno, R. Gu, M. Asadnia, J. Antoszewski, and L. Faraone, "Superlattice barrier HgCdTe nBn infrared photodetectors: Validation of the effective mass approximation," *IEEE Transactions on Electron Devices*, vol. 63, no. 12, pp. 4811–4818, 2016.

[54] A. M. Itsuno, J. D. Phillips, and S. Velicu, "Mid-wave infrared HgCdTe nBn photodetector," *Applied Physics Letters*, vol. 100, no. 16, pp. 161102-1–161102-3, 2012.

[55] A. P. Craig, A. R. J. Marshall, Z. B. Tian, S. Krishna, and A. Krier, "Mid-infrared $InAs_{0.79}Sb_{0.21}$-based nBn photodetectors with $Al_{0.9}Ga_{0.2}As_{0.1}Sb_{0.9}$ barrier layers, and comparisons with $InAs_{0.87}Sb_{0.13}$ p–i–n diodes, both grown on GaAs using interfacial misfit arrays," *Applied Physics Letters*, vol. 103, no. 25, pp. 253502-1–253502-4, 2013.

[56] A. P. Craig, M. D. Thompson, Z.-B. Tian, S. Krishna, A. Krier, and A. R. J. Marshall, "InAsSb-based nBn photodetectors: Lattice mismatched growth on GaAs and low- frequency noise performance," *Semiconductor Science and Technology*, vol. 30, no. 10, pp. 105011-1–105011-7, 2015.

[57] A. Haddadi, A. Dehzangi, R. Chevallier, S. Adhikary, and M. Razeghi, "Bias-selectable nBn dual-band long-/very long-wavelength infrared photodetectors based on $InAs/InAs_{1-x}Sb_x/AlAs_{1-x}Sb_x$ type-II superlattices," *Scientific Reports*, vol. 7, no. 3339. pp. 1–7, 2017.

[58] T. D. Nguyen, J. O. Kim, Y. H. Kim, E. T. Kim, Q. L. Nguyen, and S. J. Lee, "Dual-color short-wavelength infrared photodetector based on InGaAsSb/GaSb heterostructure," *AIP Advances*, vol. 8, no. 2, pp. 025015-1–025015-7, 2018.

[59] M. Kopytko, A. Keblowski, W. Gawron, A. Kowalewski, and A. Rogalski, "MOCVD grown HgCdTe barrier structures for hot conditions (July 2014)," *IEEE Transactions on Electron Devices*, vol. 61, no. 11, pp. 3803–3807, 2014.

[60] P. Martyniuk, M. Kopytko, and A. Rogalski, "Barrier infrared detectors," *Opto-Electronics Review*, vol. 22, no. 2, pp. 127–146, 2014.

[61] P. Martyniuk, "HOT mid-wave HgCdTe nBn and pBp infrared detectors," *Optical and Quantum Electronics*, vol. 47, no. 6, pp. 1311–1318, 2015.

[62] J. He, Q. LI, P. Wang, F. Wang, Y. Gu, C. Shen, M. Luo, C. Yu, L. Chen, X. Chen, W. Lu, and W. Hu, "Design of a bandgap-engineered barrier-blocking HOT HgCdTe long-wavelength infrared avalanche photodiode," *Optics Express*, vol. 28, no. 22, pp. 33556–33563, 2020

[63] S. Bansal, K. Prakash, N. Sardana, S. Kumar, K. Sharma, P. Jain, N. Gupta, and A. K. Singh, "Bilayer graphene/HgCdTe based self-powered mid-wave IR nBn photodetector," in *2019 IEEE 14th Nanotechnology Materials and Devices Conference (NMDC)*, Stockholm, Sweden, 2019, pp. 1–4.

[64] P. K. Saxena and P. Chakrabarti, "Analytical simulation of HgCdTe photovoltaic detector for long wavelength infrared (LWIR) applications," *Optoelectronics and Advanced Materials, Rapid Communications*, vol. 2, no. 3, pp. 140–147, 2008.

[65] S. M. Sze and K. K. Ng, *Physics of Semiconductor Devices*, Third Edition. Hoboken, NJ: John Wiley & Sons, 2006.

[66] X. Wan, Y. Xu, H. Guo, K. Shehzad, A. Ali, Y. Liu, J. Yang, D. Dai, C.-T. Lin, L. Liu, H.-C. Cheng, F. Wang, X. Wang, H. Lu, W. Hu, X. Pi, Y. Dan, J. Luo, T. Hasan, X. Duan, X. Li, J. Xu, D. Yang, T. Ren, and B. Yu, "A self-powered high-performance graphene/silicon ultraviolet photodetector with ultra-shallow junction: breaking the limit of silicon?," *NPJ 2D Materials and Applications*, vol. 1, no. 4, pp. 1–8, 2017.

[67] L.-H. Zeng, M.-Z. Wang, H. Hu, B. Nie, Y.-Q. Yu, C.-Y. Wu, L. Wang, J.-G. Hu, C. Xie, F.-X. Liang, and L.-B. Luo, "Monolayer graphene/germanium Schottky junction as high-performance self-driven infrared light photodetector," *ACS Applied Materials and Interfaces*, vol. 5, no. 19, pp. 9362–9366, 2013.

[68] A. D. D. Dwivedi, "Analytical modeling and numerical simulation of P$^+$-Hg$_{0.69}$Cd$_{0.31}$Te/n-Hg$_{0.78}$Cd$_{0.22}$Te/CdZnTe heterojunction photodetector for a long-wavelength infrared free space optical communication system," *Journal of Applied Physics*, vol. 110, no. 4, pp. 043101-1–043101-10, 2011.

[69] M. Casalino, U. Sassi, I. Goykhman, A. Eiden, E. Lidorikis, S. Milana, D. De Fazio, F. Tomarchio, M. Iodice, G. Coppola, and A. C. Ferrari, "Vertically illuminated, resonant cavity enhanced, graphene-Silicon Schottky photodetectors," *ACS Nano*, vol. 11, no. 11, pp. 10955–10963, 2017.

[70] C. Li, W. Huang, L. Gao, H. Wang, L. Hu, T. Chen, and H. Zhang, "Recent advances in solution-processed photodetectors based on inorganic and hybrid photo-active materials," *Nanoscale*, vol. 12, no. 4, pp. 2201–2227, 2020.

[71] S. Bansal, K. Prakash, K. Sharma, N. Sardana, S. Kumar, N. Gupta, and A. K. Singh, "A highly efficient bilayer graphene/ZnO/silicon nanowire based heterojunction photodetector with broadband spectral response," *Nanotechnology*, vol. 31, no. 40, pp. 405205-1–405205-10, 2020.

[72] A. Rogalski, "Infrared detectors: Status and trends," *Progress in Quantum Electronics*, vol. 27, no. 2–3, pp. 59–210, Jan. 2003.

[73] V. Ryzhii and M. Ryzhii, "Graphene bilayer field-effect phototransistor for terahertz and infrared detection," *Physical Review B-Condensed Matter and Materials Physics*, vol. 79, no. 24, pp. 245311-1–245311-8, 2009.

[74] P. K. Saxena, "Modeling and simulation of HgCdTe based p$^+$-n-n$^+$ LWIR photodetector," *Infrared Physics & Technology*, vol. 54, no. 1, pp. 25–33, 2011.

[75] N. D. Akhavan, G. A. Umana-membreno, R. Gu, J. Antoszewski, and L. Faraone, "Optimization of super-lattice barrier HgCdTe nBn infrared photodetectors based on an NEGF approach," *IEEE Transactions on Electron Devices*, vol. 65, no. 2, pp. 591–598, 2018.

[76] A. Singh, A. K. Shukla, and R. Pal, "Performance of graded bandgap HgCdTe avalanche photodiode," *IEEE Transactions on Electron Devices*, vol. 64, no. 3, pp. 1146–1152, 2017.

[77] E. Bellotti and D. D'Orsogna, "Numerical analysis of HgCdTe simultaneous two-color photovoltaic infrared detectors," *IEEE Journal of Quantum Electronics*, vol. 42, no. 4, pp. 418–426, 2006.

[78] A. Rogalski, M. Kopytko, P. Martyniuk, and W. Hu, "Comparison of performance limits of HOT HgCdTe photodiodes with 2D material infrared photodetectors," *Opto-Electronics Review*, vol. 28, no. 2, pp. 82–92, 2020.

[79] S. Bansal, P. Jain, N. Gupta, A. K. Singh, N. Kumar, S. Kumar, and N. Sardana, "A highly efficient bilayer graphene HgCdTe heterojunction based p+-n photodetector for long wavelength infrared (LWIR)," in *2018 IEEE 13th Nanotechnology Materials and Devices Conference (NMDC)*, Portland, OR, 2018, pp. 1–4.

Nano-Rectifier: A Review, Current Status, and Future Scope

15

Ankur Garg, Neelu Jain, Sanjeev Kumar, and Arun K. Singh

Contents

15.1 INTRODUCTION

The technical advancement in semiconductor technology enables the integration of billions of transistors on a single chip. In 1965, G.E Moore, cofounder of Intel, predicted that the number of transistors on an IC will double exponentially in almost 18 to 24 months, shrinking the device size by scaling factor 0.7 [1], [2]. The semiconductor industry adopted Moore's observation for solid state devices, which in turn improved

DOI: 10.1201/9781003126645-15

the device performance at reduced cost [2]. In 1981, the first computer by IBM having an Intel 8088 microprocessor with 4.77 MHz speed and 16 KB of memory space cost $1,600. As a comparison, today's computer is available at less than a third of the cost and contains an Intel core 2 Duo processor with 2.66 GHz speed and 4 GB of memory [2]. The scaling of semiconductor technology offers devices of smaller size, better performance, and less power consumption. In 2002, the device size was reduced down to 100 nm, as compared to an expected reduction to 10 nm by 2020, which can further saturate, as per the 2015 ITRS report, as shown in Figure 15.1 [1], [3]–[5]. The continuous advancement of silicon-based devices offered complementary metal-oxide semiconductor (CMOS) technology with less power consumption of the order of 20–30 W in response to 100 W at the input side. The James D. Meindl model estimated that trillions of transistors would be fabricated on a chip in 2020, and the number of transistors on a single chip (N_{tran}) can be calculated using $N_{\mathrm{tran}} = F^{-2} \times D^2 \times PE$, where F is the feature size, D is the square root of chip area, and PE is the packing efficiency of transistors [6]. Despite such predictions, the lithography tool cost of a scalable device is increasing linearly from the last 60 years [4]. This scalability saturates the reduced device dimension and is not following, as expected, Moore's observation [4], [5]. The reduction of gate length in CMOS devices creates gate current leakage that causes the short barrier height between source and channel. This phenomenon is known as the short channel effect (SCE). This undesired phenomenon can be suppressed by selecting novel geometries with different working principles [5].

Device designers are not only facing problems in further reduction beyond 40 nm, but also it is difficult to overcome other effects, such as thermal noise due to integration density problems of these devices [1]. The doped bulk materials change their properties when the device size shrinks beyond a limit, including avalanche breakdown, which may occur on applying the high electric field and increases the probability of device damage. The increased density of transistors is also responsible for thermal heat dissipation and noise, which results in a wrong output due to malfunctioning [1], [7]. As a result, researchers had explored new types of novel devices, including new materials and new geometries to operate at high frequencies.

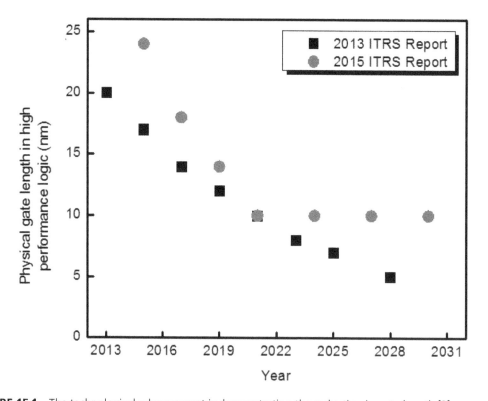

FIGURE 15.1 The technological advancement is demonstrating the reduction in gate length [3].

15.2 HIGH SPEED MATERIALS

The requirement of high speed electronic components and devices in communication, medicine, and security is increasing day by day, which results in the requirement of novel materials. The progress in material science and engineering enable the growth of new materials with high carrier concentration and mobility, for example, III–V materials and graphene.

15.2.1 III–V Based Materials

High speed devices have been fabricated using III–V material based quantum well heterostructure. The III–V material contains different compound materials such as InP, AlGaAs, InGaAs, and GaN [8], [9]. High mobility enables the heterostructure-based devices to operate at very high frequency, up to several GHz, and at room temperature. These semiconductor structures can be grown with materials having almost the same lattice constant but with a different bandgap.

The combination of different compound materials generates the two-dimensional electron gas (2DEG) in the channel where charge carriers are free to move only in two directions. As a result, these charge carriers provide better control to design the devices [10]. The fabrication process of heterostructure is similar than that of bulk materials, which consist of different layers such as substrate, buffer, channel, spacer, barrier, and cap layer [9]. These materials can be grown using molecular beam epitaxy (MBE) [11] and molecular organic chemical vapor deposition (MOCVD) [12] techniques.

15.2.2 Novel Materials

Novel materials including graphene have been exposed with very high carrier mobility. Graphene was first demonstrated at the University of Manchester, UK, in 2004 [13]. The graphene is a basic unit of graphite having a two-dimensional (2D) honeycomb lattice structure with zero bandgap [14], [15]. Graphene has excellent properties such as very high carrier mobility, high thermal conductivity, mechanical strength, flexibility, transparency, etc. Therefore, graphene has attracted many industries and academia to utilize it for various applications [15], [16]. High quality monolayer graphene and few layer graphene can be grown by mechanical exfoliation [13], [17]; however, other techniques such as chemical vapor deposition (CVD) [18], [19], chemical exfoliation [20], [21], and ultrasonic exfoliation [22] can also be used to grow the multilayer graphene. The graphene has mobility up to 200,000 cm^2/Vs at room temperature [23]. The mean free path for graphene has been estimated to be in the μm range, at the mobility of 10,000 cm^2/Vs and carrier concentration of 5×10^{12} cm^{-2} [13]. Such a large mean free path allows the fabrication of electronics devices requiring ballistic transport.

Several research groups are now concentrating on development of new classes of 2D materials such as dichalcogenides (MoS_2 and WS_2), 2D ZnO, h-BN, and black phosphorus [16], [24]. These materials have structure similar to that of graphene. The nanofabrication facilities create prospects to combine the graphene with other 2D materials such as a graphene/MoS_2 heterostructure to take the advantage of both the material properties. The graphene/h-BN-based heterostructure is expected to provide high carrier mobility up to 200,000 cm^2/Vs [25], [26].

15.3 NANOELECTRONIC DEVICES

The limitation on further reduction in dimension encourages creating novel devices that are small enough resulting in high speed and less heat dissipation. At the same time, these devices have been realized at room temperature. In this direction, several devices have been proposed with different working principles, as given in Table 15.1.

TABLE 15.1   Nanoelectronic devices with different working principles

DEVICE	REFERENCE
Quantum point contact (QPC)	[27]–[31]
Single electron transistor (SET)	[32]–[37]
Self-switching diode (SSD)	[38]–[51]
Three terminal junction (TTJ)	[52]–[87]
Four terminal ballistic rectifier (FTBR)	[77]–[79], [88]–[107]
Ballistic deflection transistor (BDT)	[108]–[121]
Artificial nanomaterials	[88], [90], [122]

15.3.1 Quantum Point Contact (QPC)

The QPC is a narrow constriction region with comparable wavelength to the charge carriers. The constriction can be formed by breaking the conductor into two parts so it can generate the 2DEG easily in the narrow region. The quantization of conductance can be realized in the QPC using GaAs/AlGaAs-based heterostructure [27], [31].

15.3.2 Single Electron Transistor (SET)

The behavior of conventional semiconductor devices is different from submicron devices. The conventional transistor becomes ON in response to electrons, in contrast to submicron/nanometer transistors, which are ON and OFF every time in response to electrons. This behavior of small transistors is due to charge quantization and coulomb integration between charge carriers [32], [37]. The submicron-sized SET was first invented by Fulton et al. in 1987 [35]. The SET contains two gates, in comparison to conventional metal-oxide semiconductor field effect transistors (MOSFETs). When a positive voltage is applied to the upper gate terminal, keeping the bottom gate at a negative potential, the electrons are accumulated in the gap of bottom gate which is of the order of 700 A°, and 2DEG is formed in SET where electrons are confined in one direction and free to move in the other two directions [32], [37]. This electron gas separated from lead by tunnel or barrier junction results in conductance oscillation with respect to charge density. The potential applications of SET are at high frequency operation and multilevel logic [32].

15.3.3 Self Switching Diode (SSD)

The conventional diode is a part of various electronic circuits for many decades. The counterpart of this diode at high frequency is SSD [38]. The I–V characteristic of SSD is similar to a diode without any need of doping and/or potential barrier for getting the nonlinear effects. The channel becomes open and closed in response to applied voltage between trenches. A wide variety of turn-on voltage can be tuned in the SSD by changing the width and length of the channel [43], [44], [46]. The SSDs can be designed using III–V heterostructures [38] and graphene [123]. These devices have also been demonstrated at 1.5 THz and room temperature [41], [48]. The rectifier circuit and digital logic can also be realized using multiple SSDs [38], [42], [51].

The classical physics-based mathematical models are unable to explain the working of novel devices [57], [101]. Hence, semiclassical-based models are being used for estimating the nonlinear ballistic transport. Such devices include the TTJ [52]–[79], [82]–[87] and FTBR [77]–[79], [88]–[99], [101]–[107].

15.3.4 Three Terminal Junction (TTJ)

The YBJ/TTJ contains three terminals named left, right, and center terminals. In a Y-branch junction (YBJ), the angle between the left and right terminal is kept at 60° as shown in Figure 15.2 [66]. The left and right terminals are directed 180° from each other in the case of TTJ [57]. The defined configuration is also known as the T-branch junction (TBJ). The AC signal is given in either push-fixed and/or push-pull configuration to the input terminals, which provides the DC output signal at the central terminal of the device [56], [57], [79], [87]. The working principle and outcome of a YBJ junction is similar to that of TTJ. Both the configurations have been fabricated using Si [55], III–V group-based heterostructure (GaAs/AlGaAs [62], InAlAs/InGaAs [78], [79]) and novel material such as graphene [81], [124]–[136]. The TTJ can be used to get the 0 and 1 binary logic in various digital circuits such as AND gate [84], NOR gate [83], NAND gate [61], half-adder [59], SR latch [75], frequency mixer [73], and differential amplifier [84].

15.3.5 Four Terminal Ballistic Rectifier (FTBR)

The FTBR is a nonlinear nanoelectronic device that converts the AC signal into the DC using asymmetric device geometry. A triangular antidot at the central part of the device directs the carriers in a particular direction for getting the DC response [101], [105], [107]. The ballistic rectifier works in the nonlinear regime; therefore, it can be characterized using the extended Landauer–Büttiker approach [90], [105]. The size of the input terminals can be less as compared to the size of output terminal for the deflection of the maximum number of electrons toward the lower terminal of the device. The FTBR is expected to generate a non-zero response when a small signal is applied between the input terminals. The working of ballistic rectifier is similar to that of a bridge rectifier without any requirement of doping junction and/or Schottky barrier, as shown in Figure 15.3 [98], [101]. The active region of the device is smaller than or comparable to the mean free path of charge carriers, and as a result, the charge carrier moves like a billiard ball without encountering any impurity scattering [90]. Therefore, scattering takes place only at the boundary of the device structure. Being a planar device, it offers negligible parasitic capacitance with a minimal impedance mismatch with external hardware circuitry that allows operating at several THz frequencies [98], [103]. The ballistic rectifier has been designed using III–V materials (GaAs-AlGaAs [101], InGaAs/InP [91], InAs/AlGaSb [96], [102], InGaAs/InAlAs [78], [79], [89], [95]) and graphene [25], [26], [137]. These devices can be used for frequency multiplication [103], THz detection [138], energy harvesting [137], etc.

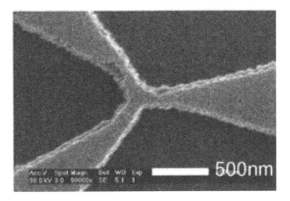

FIGURE 15.2 SEM image of YBJ/TTJ illustrating the input (L and R) and output (C) terminals. In YBJ input branches are rotated by an angle of 60° as compared to TTJ with 180° rotation. The rectified response is obtained at the C terminal while applying the AC input signal between the L and R terminals [66].

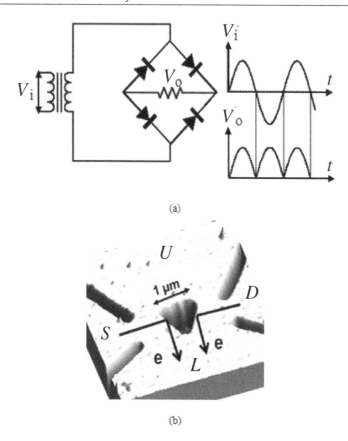

FIGURE 15.3 (a) Schematic of a bridge rectifier made of four conventional PN or Schottky diodes. The rectified output (V_o) is obtained across load while applying AC input signal (Vi) irrespective of positive or negative amplitude. (b) An atomic force micrograph (AFM) image of FTBR is indicating typical flow of electrons ejected from source (S) or drain (D) terminals indicating rectified voltage between lower (L) and upper (U) terminal [98].

15.3.6 Ballistic Deflection Transistor (BDT)

Recently, the ballistic rectifier geometry has been modified as BDT, which was based on ballistic deflection of electron and steering [112], [114], [121]. Two gate terminals (left gate and right gate) were introduced with respect to the source terminal to control the electron movement in an east or west direction by actuating it from the south direction [108], [114], [118]. The BDT can be used as a NAND gate [119], flip-flop [109], AND gate [110], or OR gate [110].

15.3.7 Graphene Based Nano-Rectifiers

Due to excellent extensive properties, graphene has been explored to design the futuristic devices that can be used for many potential applications. A graphene-based multiplier was demonstrated by Wang et al. in 2009 in which a graphene-based single transistor was used as frequency multiplier by rectifying the input signal. It was demonstrated to be a high frequency operation with more efficiency when compared to conventional diode or field effect transistor (FET) based multipliers [139]. Theoretical work on graphene-based SSD demonstrated the higher carrier mobility and rectification ratio allow operation at higher frequency and room temperature [140]. The first graphene-based three terminal junction (G-TTJ) rectifier was reported by A. Jacobsen et al. in 2010, as shown in Figure 15.4a [136]. The rectification in this device

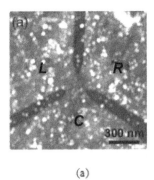
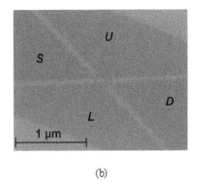

(a)   (b)

FIGURE 15.4   (a) Scanning force microscope (SFM) image of G-TTJ [136] and (b) SEM image of G-FTBR [26], illustrate the input (*L* and *R*) and output (*C*) terminals for G-TTJ, and input (*S* and *D*), and output (*L* and *U*) terminals for G-FTBR, respectively. Both rectifiers used graphene as a channel, where charge carriers can be tuned with back gate voltage.

exhibits the change in sign by reversing the gate voltage at room temperature [136]. Similarly, graphene-based four terminal rectifiers (G-FTBR) were first reported by A.K. Singh et al. in 2015, as demonstrated in Figure 15.4b, which confirms the rectified outcome using Si/SiO_2 substrate [26].

The TTJ and FTBR were explored experimentally and theoretically using III–V materials. However, graphene-based rectifiers are required to be explored more concerning their theoretical and experimental perspective. As predicted, these devices can also be demonstrated at very high microwave frequencies by carefully selecting a substrate that induces low parasitic capacitance.

15.4 BACKGROUND

These days, a number of nanoelectronic devices are available with dimensions smaller than the mean free path of charge carriers. The ballistic transport of charge carriers is dominant in the active region of the device. The ballistic transport governs the elastic travel of carriers within the channel region of the device without losing any energy. In contrast, scattering takes place only at the device boundary, which results in heat dissipation through the electrical contacts [141]. During the last two decades, ballistic transport has been well explored in various nanoelectronic devices at room temperature and high frequency.

The phase coherence conductance in a four terminal conductor was carried out by Büttiker in 1986 [142]. The electrochemical potential and hence transmission probability and resistance between leads of the conductor has been used to formulate the current in the output leads. In contrast to previous studies, in which the voltage drop has been studied over a distance comparable to the mean free path of carriers without considering the phase coherence, in [142], four terminal resistance of conductors has been derived to relate the symmetry between voltage and current at the input and output leads. Similarly, experiments have been performed on QPC to study the quantum phenomenon [30]. The QPC shows the conductance plateaus at lower temperature due to comparable mean free path, and with the increase of temperature and magnetic field, the quantized conductance become invisible. Hirayama and Tarucha in 1993 reported the room temperature quantum ballistic transport in the four terminal device fabricated from AIGaAs/InGaAs/GaAs materials in Figure 15.5 [143]. The bending resistance of cross-junction confirmed the negative characteristic obtained in response to the applied magnetic field at different temperatures ranging from 1.5 K to 290 K. The result clearly demonstrates higher carrier mobility at low temperature (6.5×10^4 cm²/Vs at below 35 K) as compared to the value attained at high temperature (7.8×10^3 cm²/Vs at above 70 K), which suggests that the mean free path is more abundant at lower temperatures.

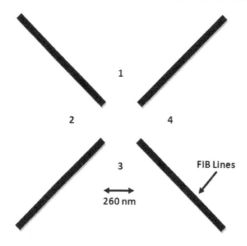

FIGURE 15.5 Four terminal cross-junction structure fabricated by focused ion beam (FIB) with 260 nm gap spreading between opposite junctions [143].

15.4.1 Four Terminal Ballistic Rectifier (FTBR)

In particular, an FTBR proposed by Song et al. in 1998, as shown in the inset of Figure 15.6, works easily with different principle as compared to the conventionally used diodes [101]. The device was first demonstrated on a GaAs-AlGaAs-based heterostructure exhibiting the nonlinear behavior using an artificial scatterer (antidot) at the center of the geometry. The FTBR characteristics can be explained using the extended Landauer–Büttiker approach in the nonlinear regime of electron transport [101], [105]. The mathematical models have been derived to correlate the theoretical and experimental studies [105]. The applied voltage increases the charge carrier's velocity in the direction of input terminals without changing in output terminals. As a result, electrons trying to cross the junction are deflected by the central antidot toward the lower junction. The small ejection angle accelerates the electrons and collimates at the output terminal as compared to the de-collimate with the large ejection angles. This process can change the transmission probability of electrons from one terminal to the other and is responsible for nonlinear rectified response, as demonstrated in Figure 15.6 [101], [105]. The FTBR can easily be fabricated using single-step electron beam lithography and wet and/or dry etching process. The ballistic rectifier does not have any doping junction, intrinsic current, and potential barrier when compared to a semiconductor diode based on PN or a Schottky junction [101].

The ballistic rectifiers start working from a non-zero input current or voltage irrespective to the direction of input signal [101], [107]. Some asymmetry may occur in the rectified outcome due to the presence of fabrication defects. The rectifier demonstrates the increase in output at 4.2 K as compared to 77 K due to the increase in the mean free path and mobility of charge carriers at a lower temperature. The device is planar in nature, it offers less parasitic capacitance, which allows operating at higher frequencies. The devices have been demonstrated to confirm output response up to 2 GHz frequency without any decrease of rectification efficiency [101].

The FTBR has been demonstrated to work at low input signal of the order of 0.5 mV at 4.2 K up to the frequency of 10 GHz to verify the zero intrinsic threshold property of the device. The rectification efficiency was found to be independent of temperature and never affect the ballistic transport. The device rectification efficiency was reported to be around 1% at 1 GHz input AC signal. To improve the rectification efficiency, the geometry of the rectifier may be optimized with the top gate electrode to tune the Fermi levels in the channel region enabling, the ballistic transport of electrons [107].

The InGaAs/InP-based quantum well structure has been used to design the functional nanomaterial rectifier, which was based on the similar working principle of the ballistic rectifiers [88], [122]. The

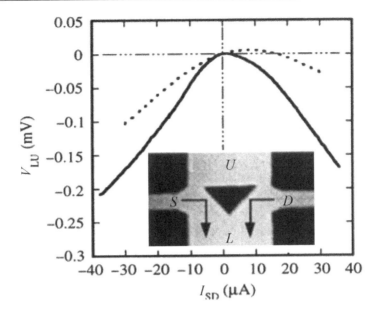

FIGURE 15.6 AFM image of the FTBR (inset) with triangular antidot in the center and *I–V* characteristic between input-output terminal of rectifier demonstrate nonlinear output at 4.2 K (solid) and 77 K (dashed) [101].

Indium-based heterostructure provides the high mobility electrons, as compared to GaAs-based heterostructure. Also, the InGaAs/InP based quantum well structure offered less depletion region, which was approximately equal to 10–30 nm [91]. In artificial nonmaterial, the number of nanometer-size triangular elements were arranged to get the nonlinear characteristic suitable to work at high frequency up to 50 GHz. Further, the improvement in arrangement of a triangular micro-junction can be made to get the lesser size comparable to the mean free path [88], [122].

The InGaAs/InP heterostructure-based FTBR with different dimensions has been fabricated to check the device rectification. The rectification efficiency of the device has been enhanced using InGaAs/InP, which offered less depletion region as compared to the GaAs-AlGaAs [90], [91]. The rectified response demonstrated that the smaller devices have high efficiency of about 23% in comparison to the value obtained from the bigger device (10%) at 108 K. However, the efficiency of smaller devices gets saturated at large applied voltage due to electron heating. Furthermore, instead of applying a DC signal, a 1 KHz AC signal was applied between the input terminals that results in 14% rectification efficiency for the smaller device at room temperature. The device performance has also been demonstrated up to 50 GHz frequency and is expected to work well up to THz frequency [88], [91].

Similarly, a ballistic rectifier was reported [88] in which gates were considered in four terminal devices as compared to chemical doping. Instead of using an antidot, the mirror symmetry of the device was broken by shifting the input and output regions such that a higher number of charge carriers were deflected toward the output terminal. The device demonstrated the nonlinear rectified response at a different value of gate voltages [88].

In contrast to studies by Fleischmann et al. [92], a quantum phenomenon was explained by Löfgren et al. [106] to justify the previous studies [101], [105]. At lower temperature, the mean free path of charge carriers is comparable to the Fermi wavelength, and quantum effect dominates in the channel region. Also, as compared to the ballistic rectifier, the space between the lower and upper output terminal is almost identical in artificial nanomaterial, but still, output reversal and oscillation has been seen in the rectified response. These oscillations were explained using lateral confinement modes, which were increased by applying the input across the device. Due to the variations of these modes, oscillations have occurred in the output of the ballistic rectifier and artificial nanomaterials [90], [106].

The study on ballistic rectifier has been further extended by considering the tunable rectification using gate control Fermi levels [93], [94]. The rectification output reversed in the positive direction with the increase of gate voltage. The rectification was quasi-classical for wide channel region rectifier as compared to quantum rectification for a narrow channel region rectifier. The temperature-dependent sign-reversal of rectification characteristics has been explained in [95], considering the effect of variation in the channel conductance. The electrons were accumulated in the narrow or wide channel regions, which fluctuates the average conductance of the device.

Similarly, the rectification effects have been studied for non-centrosymmetric ballistic rectifier for different branch angles (30°, 45°, 60°, and 90°) between input and output terminals. The decrease in the angle increases the rectifier output while voltage at the upper terminal is independent of change in angle. In addition, variation in the gate voltage made the opposite effect on overall rectification output, and upper terminal voltage [77], [104].

The rectification in an InAs/AlGaSb heterostructure-based FTBR with square antidot has been studied [96], [102], as shown in Figure 15.7 [96]. The magneto-resistance property of ballistic rectifier demonstrated the Shubnikov–de Hass oscillations at magnetic field above ±1T [96]. In addition, the bending resistance was characterized, which represents the electron focusing peak at the positive or negative side of the weak magnetic field and again depends on used input configuration [102]. The proposed heterostructure can be used to extend the research even at higher temperatures [96].

The electro-thermal phenomenon of the ballistic rectifier has been studied by Matthews et al. [97]. Instead of applying a voltage between input terminals, thermal effects have been used to get the rectified output. The InP/GaInAs heterostructure was used to fabricate such a device. A temperature difference has been applied across the input terminals, which generated the thermal voltage at the output terminals. An output reversal similar to the previous discussion has also been observed in the thermal study of the ballistic rectifier. This work suggests the possibility of designing the electronic thermal rectifiers and thermoelectric energy converters [97]. Furthermore, the electro-thermal phenomenon has been studied in cross-junction in which a QPC was used in the output terminal to create a potential barrier [99].

The device performance is typically affected by the noise. The low frequency noise of the ballistic rectifier has been studied and formulated with the channel and antidot dimensions. The zero intrinsic threshold property of the rectifier eliminates the low frequency flicker noise. In addition, the speed of the device has been checked using the noise equivalent power (NEP) and responsivity characteristic of the rectifier at different frequencies [98].

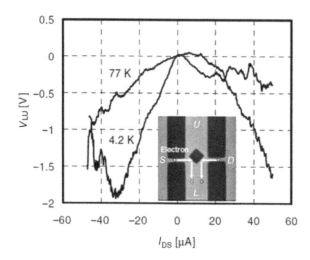

FIGURE 15.7 The ballistic rectifier with square central antidot is demonstrated in the inset with rectified characteristic at 77 K and 4 K [96].

15.4.2 Three Terminal Junction (TTJ)

Similar to the ballistic rectifier, another device called the TTJ based on ballistic effect was proposed by H. Q. Xu in 2001, with two input terminals (left and right) and one central output terminal [52]. In TTJ, one of the output terminals of FTBR has been omitted to study the rectification effects. Also, theoretical formulation relates the output current at the central terminal with the transmission probability of all three terminals, which can further depend on electrochemical potentials. The device can be realized in both symmetric and asymmetric configurations, and output has been predicted to be varying quadratically with the applied input voltage [52].

In another experiment, the rectified output of GaAs/AlGa-As heterostructure-based YBJ has been studied for different gate voltages at 4.2 K [62]. The TTJ was fabricated on a high mobility GaInAs/InP heterostructure, as shown in Figure 15.8a [57], which provided the rectified output and transmission coefficients up to 200 K, which was further validated by mathematical modeling [52]. The rectified output of the device increased with the decrease in gate voltage and demonstrated a good correlation with the quadratic curve fitting, as demonstrated in Figure 15.8b [57].

The TTJ has operating capability to work up to high frequency (10 GHz) and room temperature [82]. The parasitic capacitance limits the high frequency operation of TTJ, which can be further improved ten times by separating the junctions to decrease in the capacitance. The TTJ has been demonstrated up to 1 GHz with the second harmonic generation up to 300 GHz [58].

Furthermore, the TTJ can also be combined with devices of its class such as FET to amplify the rectified response [87]. The designed device can easily be used as an AND logic gate and frequency doubler [82], [87]. The more complex digital operations, for example, half-adder, have been demonstrated using rectified effects in YBJ [59], [60]. In addition, TTJ can also be worked as a diode and triode depending upon the voltage applied between the input terminals [61]. The digital operation such as the SR latch has been demonstrated in TTJ, utilizing two TTJs and in-plane gate terminals to measure the output. The single step lithography process was used to design the latch to comply with the CMOS processes [75].

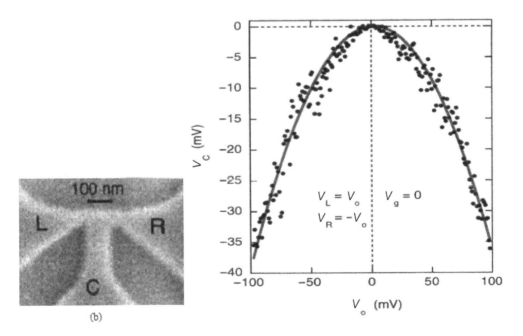

(b)

FIGURE 15.8   The (a) SEM image of TTJ with input terminal (*L* and *R*) and (b) quadratic output response of TTJ at output terminal (*C*) [57].

As compared to the conventional PN diode, the mentioned applications can work up to THz frequencies due to the ballistic nature of the device. The TBJ demonstrates a nonlinear rectified characteristic, but the device geometry also shows rectified response for long branches exhibiting diffusive transport. At low bias conditions, the device works in the parabolic region in contrast to the linear region for high biases [56]. The high frequency operation of YBJ up to 40 GHz has been demonstrated utilizing two YBJ, which minimized the input impedance [63]. The carbon nanotube (CNT) has also been used to design the TBJ to generate the rectified effects [65].

The high frequency RF response operation of YBJ has been simulated at room temperature using the ATLAS simulator [53]. The device demonstrates the decrease in the rectified response with the increase in frequency from 2 to 20 GHz [53]. The HF to DC response of double YBJ was also demonstrated up to 50 GHz at 77 K [66]. The degradation in the operation at higher frequencies was observed due to interconnect feeding losses. Similarly, the high frequency response of YBJ was theoretically reported up to 94 GHz by Bednarz et al. [85]. Two YBJs were connected in parallel to minimize the impedance of the complete device. Furthermore, the transistor characteristic of TTJ was studied [64], where the central output terminal was used as a gate terminal. The current and voltage gain of 60 and −30 with a cutoff frequency of 4 GHz was reported for signal processing and sensing application [64].

The nonlinear rectified response of TTJ has also been observed for larger device size with low mobility materials [54]. The TTJ with Schottky wrap gates was used to control the two input nanowires. The wrap gates controlled the nonlinearity in the output and changed the curvature of rectified response to get the transistor-like characteristics [74], [76]. The InAs-based nanowires were also utilized as TTJ, showing a tunable nonlinear rectified electrical characteristic with frequency mixing and phase detection operation [71].

The NOR gate digital output has been produced using three TTJs [83]; one of the TTJ was used as a load, and the other was used as the input to the circuit. Other applications such as frequency mixing and phase detection have also been demonstrated [73]. The frequency mixing operation can be useful in the modulation process, where the carrier signal and message signal can be mixed to transmit the modulated signal [85]. Further, a phase detection process is quite useful in telecommunications, where it works as an essential part of the phase locked loop. As predicted in [85], double TTJ has been operated up to 94 GHz practically and can be used as a multiplexer or demultiplexer. Besides, a Monte-Carlo simulation predicted TTJ operation up to THz frequency. Generally, the THz performance of a TTJ rectifier is limited due to the presence of intrinsic impedance [72].

The physical dimensions, temperature, and mean free path of charge carriers largely affect the rectified response of TTJ. Ballistic effects are responsible for rectification in small size TTJ's, but as the device dimensions increase when compared to the mean free path, the quasi-ballistic effects become dominating. The shifting from ballistic to the quasi-ballistic or diffusive region does not affect the rectifier output characteristics due to its novel geometry [70].

The high mobility material InAS/AlGaSb was further utilized to fabricate the TTJ to work at 300 and 77 K. Due to a large mean free path (430 nm to 1.15 μm from 300 to 77 K), the device generated the rectified nonlinear response [86]. Further, the transistor-like characteristic was generated from TTJ with maximum transconductance of 100 μA/V and power gain up to 1.5 GHz.

The fabrication facilities for nano-devices are expensive; therefore, it is better to use a simulation tool for designing and optimization of the nanostructures before real fabrication. Different III–V material based structure such as TTJ and YBJ in addition to FTBR has been simulated using Monte-Carlo and semiclassical simulations methods [53], [78], [79], [85], [89]. Monte-Carlo simulations consider the semiclassical nature of the device rather than the quantum behavior [78], [79]. These devices were simulated in 2D by performing the separate simulation for front view and top view. The front view of the device contains all the layers of the heterostructure, which is used to evaluate the carrier density and mobility of electrons, and the top view contains the air-channel (air-InGaAs) combination of the device. For compensating the effect of other layers, a net background doping and surface charge density have been assigned to the air-channel region. The simulated results confirm the THz response of TTJ and FTBR at different temperatures [79], [85], [89]. However, to consider the real device structure, there is a need to consider the 3D geometry of the nano-rectifiers [79], [89].

After the invention of graphene with the number of novel properties, many efforts have been made during the last few years to design the novel nano-rectifiers such as G-TTJ and G-FTBR.

15.4.3 Graphene Based Nano-Rectifiers (G-TTJ and G-FTBR)

The first successful demonstration of G-TTJ was done by Jacobsen et al. in 2010 [136]. The rectifier was designed on graphene, which was prepared by mechanical exfoliation on a Si/SiO$_2$ substrate with 285 nm dielectric thickness. The device exhibits rectification efficiency around 4% at 4 K as compared to 1% at 300 K. The rectified response of the G-TTJ can be tuned using a back gate electrode. Furthermore, the graphene TTJ has been designed on silicon carbide (SiC) using epitaxial growth. The device shows rectified response for both push-pull and push-fixed input configurations. The rectification efficiency of 25% at an input voltage of 2 V has been achieved [130]. As compared to Jacobsen et al., a chemical doped (polyethyleneimine) exfoliated G-TTJ has been designed on a Si/SiO$_2$ substrate with a thickness of 300 nm [127], [134]. The chemical doping provides the passivation to the graphene layer, which results into higher mobility. The device demonstrated the rectification efficiency of 15% at the input voltage of 100 mV. Another similar G-TTJ has been designed on exfoliated graphene having a 300 nm Si/SiO$_2$ substrate demonstrating rectification of 8% for the input signal of 500 mV [126]. The rectification was attributed to charge transfer between metal contact and graphene layer between two input terminals.

The finite element simulation has been used to design the G-TTJ on a Si/SiO$_2$ substrate [131]. The device demonstrated the rectification efficiency of 1% at the input voltage of 100 mV. Recently, the drift-diffusion modeling and noise analysis of G-TTJ has been reported in [81]. Similarly, TTJ has been fabricated on a passivated Si/SiO2 substrate with the dielectric thickness of 90 nm. The device shows a voltage gain of 0.013, which need to be improved for practical application. The designed device confirmed 15% rectification efficiency at an input voltage of 50 mV [135]. Furthermore, the SiC-based G-TTJ was proposed [133] with rectification efficiency of 30% for an input signal of 2 V. The curvature of the rectified output can be changed with the branch width and by changing the device shape from TTJ to YBJ [133].

TABLE 15.2 Summary of FTBR and TTJ to be designed with different materials for different applications

DEVICE	MATERIAL	REFERENCE	APPLICATION	REFERENCE
FTBR	GaAs-AlGaAs	[77], [88], [93], [94], [99], [101], [104], [107]	Signal detection	[77], [98], [101]
	InGaAs/InP	[88], [90], [91], [97], [106], [122]	Weak signal mixing	[101]
	InAs/AlGaSb	[96], [102]	Frequency multiplication	[103]
	InGaAs/InAlAs	[78], [79], [89], [95], [98]	THz detection	[138]
	Graphene	[25], [26], [100], [137]	Energy harvesting	[137]
TTJ	Si	[55]	Digital gate	[59], [61], [82]–[84]
	GaAs/AlGaAs	[59], [60], [62], [64], [69], [74], [76], [77], [82]–[84]	Half-adder	[59], [60]
	GaInAs/InP	[54], [57], [58], [61], [73], [87]	SR latch	[75]
	InAlAs/InGaAs	[53], [56], [63], [66], [70], [72], [78], [79], [85]	D flip-flop	[84]
	InAs	[71]	Frequency doubler	[53], [58], [82], [87]
	InAS/AlGaSb	[86]	Differential amplifier	[84]
	Carbon nanotube	[65]	Frequency mixing and phase detection	[71], [73]
	Graphene	[81], [124]–[136]	Multiplexer and demultiplexer	[72]

In the same manner, G-TTJ has been exfoliated grown on 280 nm Si/SiO_2 with rectification efficiency of 8% at input potential of 400 mV [129].

Recently, a G-FTBR has been demonstrated first time to work at room temperature [26]. Back-gated voltage was used to control the type of charge carriers, that is, electrons or holes resulting into the negative or positive DC output voltage. The G-FTBR demonstrates the nonlinear characteristic at −15 V back-gate voltage. The suppression of flicker noise in G-FTBR is obtained, as graphene has no bandgap and zero threshold voltage. The graphene provides high carrier mobility that improves its speed up to THz frequencies for imaging, detection, communication, and microwave/THz applications. Later, G-FTBR was reported replacing SiO_2 with an h-BN substrate to enable very high mobility carriers up to 200,000 cm²/Vs [23], [26]. The G-FTBR on h-BN substrate provided a smooth surface to the charge carriers, enabling the highest mobility for ballistic movement of charge carriers [25], [137]. The device demonstrated the change in the sign of output voltage with the change in the back gate voltage and shows flicker noise at a lower frequency, which can be taken over by thermal noise at higher frequencies. Similarly, the G-FTBR has been demonstrated with drift-diffusion modeling to analyze the 3D and 2D rectification behavior of rectifier [100]. Besides, G-FTBR can also be used in imaging applications [138].

The summary of FTBR and TTJ structure design with different materials and application is shown in Table 15.2.

15.5 CONCLUSIONS AND FUTURE SCOPE

15.5.1 Conclusions

The scaling issues in conventional electronics devices create the opportunity for novel devices to process the high speed data. The FTBR and TTJ are the types of novel nano-rectifiers analogous to a bridge rectifier. The high speed materials such as III–V heterostructures and graphene can generate the device response at THz frequencies. The rectifiers can be used as voltage inverter and frequency doubler at room temperature in addition to other applications such as THz imaging, RFID.

15.5.2 Future Scope

Graphene-based rectifiers can be utilized as thermo-electric generators, which will respond according to the change in temperature at input terminals. These thermo-electric generators can use the heat generated in the electronics circuitry and convert it into useful electrical signals. Graphene based rectifiers can be integrated with antennas to perform the energy harvesting using electromagnetic radiations. The combination of rectifier and antenna is known as rectenna. The generated DC outcome can be used to charge the mobile batteries. This will save the wastage of electromagnetic signal that continually comes from the base station. These rectifiers can also be used in future to perform the digital operations such as logic gates, etc. Due to THz response, graphene-based rectifiers can be utilized in imaging applications.

REFERENCES

[1] L. B. Kish, "End of Moore's law: thermal (noise) death of integration in micro and nano electronics," *Phys. Lett. A*, vol. 305, no. 3–4, pp. 144–149, Dec. 2002.

[2] X. Sun, "Nanoscale bulk MOSFET design and process technology for reduced variability," University of California: Berkeley, Technical Report No. UCB/EECS-2010–80, 2010.

[3] R. Courtland, "Transistors could stop shrinking in 2021," *IEEE Spectr.*, vol. 53, no. 9, pp. 9–11, Sep. 2016.

[4] G. D. Hutcheson, "Moore's law, lithography, and how optics drive the semiconductor industry," in *Extreme Ultraviolet (EUV) Lithography IX, International Society for Optics and Photonics,* vol. 1058303, 2018, p. 501.

[5] S. E. Thompson and S. Parthasarathy, "Moore's law: the future of Si microelectronics," *Mater. Today*, vol. 9, no. 6, pp. 20–25, Jun. 2006.

[6] J. D. Meindl, "Beyond Moore's law: the interconnect era," *Comput. Sci. Eng.*, vol. 5, no. 1, pp. 20–24, Jan. 2003.

[7] D. Goldhaber-Gordon, M. S. Montemerlo, J. C. Love, G. J. Opiteck, and J. C. Ellenbogen, "Overview of nanoelectronic devices," *Proc. IEEE*, vol. 85, no. 4, pp. 521–540, Apr. 1997.

[8] W. R. Frensley, "Heterostructure and quantum well physics," in *Heterostructures and Quantum Devices*, Academic Press, New York, 1994, pp. 1–24.

[9] T. R. Lenka and A. K. Panda, "Characteristics study of 2DEG transport properties of AlGaN/GaN and AlGaAs/GaAs-based HEMT," *Semiconductors*, vol. 45, no. 5, pp. 650–656, May 2011.

[10] S. Das Sarma, S. Adam, E. H. Hwang, and E. Rossi, "Electronic transport in two-dimensional graphene," *Rev. Mod. Phys.*, vol. 83, no. 2, pp. 407–470, May 2011.

[11] R. J. Matyi, J. W. Lee, and H. F. Schaake, "Substrate orientation and processing effects on GaAs/Si misorientation in GaAs-on-Si grown by MBE," *J. Electron. Mater.*, vol. 17, no. 1, pp. 87–93, Jan. 1988.

[12] P. D. Dapkus, "A critical comparison of MOCVD and MBE for heterojunction devices," *J. Cryst. Growth*, vol. 68, no. 1, pp. 345–355, Sep. 1984.

[13] K. S. Novoselov *et al.*, "Electric field effect in atomically thin Carbon films," *Science*, vol. 306, no. 5696, pp. 666–669, Oct. 2004.

[14] K. S. Novoselov *et al.*, "Two-dimensional atomic crystals," *Proc. Natl. Acad. Sci. U. S. A.*, vol. 102, no. 30, pp. 10451–10453, 2005.

[15] K. S. Novoselov, V. I. Fal'ko, L. Colombo, P. R. Gellert, M. G. Schwab, and K. Kim, "A roadmap for graphene," *Nature*, vol. 490, no. 7419, pp. 192–200, Oct. 2012.

[16] J.-W. Jiang, "Graphene versus MoS2: a short review," *Front. Phys.*, vol. 10, no. 3, pp. 287–302, Jun. 2015.

[17] M. Yi and Z. Shen, "A review on mechanical exfoliation for the scalable production of graphene," *J. Mater. Chem. A*, vol. 3, no. 22, pp. 11700–11715, 2015.

[18] M. Losurdo, M. M. Giangregorio, P. Capezzuto, and G. Bruno, "Graphene CVD growth on copper and nickel: role of hydrogen in kinetics and structure," *Phys. Chem. Chem. Phys.*, vol. 13, no. 46, pp. 20836–20843, 2011.

[19] K. S. Kim *et al.*, "Large-scale pattern growth of graphene films for stretchable transparent electrodes," *Nature*, vol. 457, no. 7230, pp. 706–710, 2009.

[20] S. Stankovich *et al.*, "Synthesis of graphene-based nanosheets via chemical reduction of exfoliated graphite oxide," *Carbon N. Y.*, vol. 45, no. 7, pp. 1558–1565, 2007.

[21] L. Zhang, J. Liang, Y. Huang, Y. Ma, Y. Wang, and Y. Chen, "Size-controlled synthesis of graphene oxide sheets on a large scale using chemical exfoliation," *Carbon N. Y.*, vol. 47, no. 14, pp. 3365–3368, 2009.

[22] S. M. Notley, "Highly concentrated aqueous suspensions of graphene through ultrasonic exfoliation with continuous surfactant addition," *Langmuir*, vol. 28, no. 40, pp. 14110–14113, 2012.

[23] K. I. Bolotin *et al.*, "Ultrahigh electron mobility in suspended graphene," *Solid State Commun.*, vol. 146, no. 9–10, pp. 351–355, Jun. 2008.

[24] A. K. Geim and I. V. Grigorieva, "Van der Waals heterostructures," *Nature*, vol. 499, no. 7459, pp. 419–425, 2013.

[25] G. Auton *et al.*, "Graphene ballistic nano-rectifier with very high responsivity," *Nat. Commun.*, vol. 7, p. 11670, May 2016.

[26] A. K. Singh, G. Auton, E. Hill, and A. Song, "Graphene based ballistic rectifiers," *Carbon N. Y.*, vol. 84, pp. 124–129, Apr. 2015.

[27] T. Ando, "Quantum point contacts in magnetic fields," *Phys. Rev. B*, vol. 44, no. 15, pp. 8017–8027, 1991.

[28] H. Van Houten *et al.*, "Coherent electron focusing with quantum point contacts in a two-dimensional electron gas," *Phys. Rev. B*, vol. 39, no. 12, pp. 8556–8575, 1989.

[29] L. W. Molenkamp *et al.*, "Electron-beam collimation with a quantum point contact," *Phys. Rev. B*, vol. 41, no. 2, pp. 1274–1277, Jan. 1990.

[30] R. Taboryski, A. Kristensen, C. B. Sorensen, and P. E. Lindelof, "Conductance-quantization broadening mechanisms in quantum point contacts," *Phys. Rev. B*, vol. 51, no. 4, pp. 2282–2286, 1995.

[31] H. Lehmann *et al.*, "Spin-resolved conductance quantization in InAs," *Semicond. Sci. Technol.*, vol. 29, no. 7, p. 075010, July 2014.

[32] M. A. Kastner, "The single-electron transistor," *Rev. Mod. Phys.*, vol. 64, no. 3, pp. 849–858, 1992.

[33] K. Matsumoto, M. Ishii, K. Segawa, Y. Oka, B. J. Vartanian, and J. S. Harris, "Room temperature operation of single electron transistor made by STM nano-oxidation process," in *1995 53rd Annual Device Research Conference Digest, IEEE*, 1995, pp. 46–47.

[34] R. H. Chen, A. N. Korotkov, and K. K. Likharev, "Single-electron transistor logic," *Appl. Phys. Lett.*, vol. 68, no. 14, pp. 1954–1956, 1996.

[35] T. Fulton and G. Dolan, "Observation of single-electron charging effects in small tunnel junctions," *Phys. Rev. Lett.*, vol. 59, no. 1, pp. 109–112, 1987.

[36] U. F. Keyser, H. W. Schumacher, U. Zeitler, R. J. Haug, and K. Eberl, "Fabrication of a single-electron transistor by current-controlled local oxidation of a two-dimensional electron system," *Appl. Phys. Lett.*, vol. 76, no. 4, pp. 457–459, Jan. 2000.

[37] M. Kastner, R. Kwasnick, J. Licini, and D. Bishop, "Conductance fluctuations near the localized-to-extended transition in narrow Si metal-oxide-semiconductor field-effect transistors," *Phys. Rev. B*, vol. 36, no. 15, pp. 8015–8031, 1987.

[38] A. M. Song, M. Missous, P. Omling, A. R. Peaker, L. Samuelson, and W. Seifert, "Unidirectional electron flow in a nanometer-scale semiconductor channel: a self-switching device," *Appl. Phys. Lett.*, vol. 83, no. 9, pp. 1881–1883, 2003.

[39] K. Xu, W. Gang, and A. M. Song, "Electron transport in self-switching nano-diodes," *J. Comput. Electron.*, vol. 6, no. 1, pp. 59–62, 2007.

[40] A. Song, I. Maximov, M. Missous, and W. Seifert, "Diode-like characteristics of nanometer-scale semiconductor channels with a broken symmetry," *Phys. E Low-dimens. Syst. Nanostruct.*, vol. 21, no. 2–4, pp. 1116–1120, Mar. 2004.

[41] S. R. Kasjoo and A. M. Song, "Terahertz detection using nanorectifiers," *IEEE Electron Device Lett.*, vol. 34, no. 12, pp. 1554–1556, 2013.

[42] S. Garg, A. Garg, S. Bansal, A. Chaudhary, A. K. Singh, and S. R. Kasjoo, "Effect of filling dielectric in etched trenches of novel unipolar nanodiode," in *2016 International Conference on Microelectronics, Computing and Communications (MicroCom), IEEE*, 2016, pp. 1–3.

[43] I. Iñiguez-de-la-Torre, H. Rodilla, J. Mateos, D. Pardo, A. M. Song, and T. González, "Terahertz tunable detection in self-switching diodes based on high mobility semiconductors: InGaAs, InAs and InSb," *J. Phys. Conf. Ser.*, vol. 193, no. 1, p. 012082, 2009.

[44] I. Iñiguez-De-La-Torre, J. Mateos, D. Pardo, and T. González, "Microscopic analysis of noise in self-switching diodes," *AIP Conf. Proc.*, vol. 922, no. 1, pp. 317–320, 2007.

[45] Z. M. Chen, Z. Y. Zheng, K. Y. Xu, and G. Wang, "Ballistic transport in nanoscale self-switching devices," *Chinese Sci. Bull.*, vol. 56, no. 21, pp. 2206–2209, 2011.

[46] I. Iñiguez-De-La-Torre, J. Mateos, D. Pardo, and T. González, "Monte Carlo analysis of noise spectra in self-switching nanodiodes," *J. Appl. Phys.*, vol. 103, no. 2, p. 024502, 2008.

[47] K. Y. Xu, G. Wang, and A. M. Song, "Gunn oscillations in a self-switching nanodiode," *Appl. Phys. Lett.*, vol. 93, no. 23, p. 233506, 2008.

[48] C. Balocco, S. R. Kasjoo, L. Q. Zhang, Y. Alimi, S. Winnerl, and A. M. Song, "Planar terahertz nanodevices," in *2011 41st European Microwave Conference, IEEE*, 2011, pp. 1146–1149.

[49] J. Mateos, B. G. Vasallo, D. Pardo, and T. González, "Operation and high-frequency performance of nanoscale unipolar rectifying diodes," *Appl. Phys. Lett.*, vol. 86, no. 21, p. 212103, 2005.

[50] C. Balocco, S. R. Kasjoo, L. Q. Zhang, Y. Alimi, and A. M. Song, "Low-frequency noise of unipolar nanorectifiers," *Appl. Phys. Lett.*, vol. 99, no. 11, p. 113511, 2011.

[51] A. M. Song, M. Missous, P. Omling, I. Maximov, W. Seifert, and L. Samuelson, "Nanometer-scale two-terminal semiconductor memory operating at room temperature," *Appl. Phys. Lett.*, vol. 86, no. 4, p. 042106, 2005.

[52] H. Q. Xu, "Electrical properties of three-terminal ballistic junctions," *Appl. Phys. Lett.*, vol. 78, no. 14, pp. 2064–2066, Apr. 2001.

[53] Rashmi, L. Bednarz, B. Hackens, G. Farhi, V. Bayot, and I. Huynen, "Nonlinear electron transport properties of InAlAs/InGaAs based Y-branch junctions for microwave rectification at room temperature," *Solid State Commun.*, vol. 134, no. 3, pp. 217–222, 2005.

[54] D. Wallin, I. Shorubalko, H. Q. Xu, and A. Cappy, "Nonlinear electrical properties of three-terminal junctions," *Appl. Phys. Lett.*, vol. 89, no. 9, p. 092124, 2006.

[55] F. Meng *et al.*, "Nonlinear electrical properties of Si three-terminal junction devices," *Appl. Phys. Lett.*, vol. 97, no. 24, p. 242106, 2010.

[56] J. Mateos *et al.*, "Nonlinear effects in T-branch junctions," *IEEE Electron Device Lett.*, vol. 25, no. 5, pp. 235–237, May 2004.

[57] I. Shorubalko, H. Q. Xu, I. Maximov, P. Omling, L. Samuelson, and W. Seifert, "Nonlinear operation of GaInAs/InP-based three-terminal ballistic junctions," *Appl. Phys. Lett.*, vol. 79, no. 9, pp. 1384–1386, Aug. 2001.

[58] R. Lewén, I. Maximov, I. Shorubalko, L. Samuelson, L. Thylén, and H. Q. Xu, "High frequency characterization of a GaInAs/InP electronic waveguide T-branch switch," *J. Appl. Phys.*, vol. 91, no. 4, pp. 2398–2402, 2002.

[59] S. Reitzenstein, L. Worschech, and A. Forchel, "Room temperature operation of an in-plane half-adder based on ballistic Y – junctions," *IEEE Electron Device Lett.*, vol. 25, no. 7, pp. 462–464, Jul. 2004.

[60] S. Reitzenstein, L. Worschech, and A. Forchel, "A novel half-adder circuit based on nanometric ballistic Y-branched junctions," *IEEE Electron Device Lett.*, vol. 24, no. 10, pp. 625–627, 2003.

[61] H. Q. Xu *et al.*, "Novel nanoelectronic triodes and logic devices with TBJs," *IEEE Electron Device Lett.*, vol. 25, no. 4, pp. 164–166, Apr. 2004.

[62] L. Worschech, H. Q. Xu, A. Forchel, and L. Samuelson, "Bias-voltage-induced asymmetry in nanoelectronic Y-branches," *Appl. Phys. Lett.*, vol. 79, no. 20, pp. 3287–3289, 2001.

[63] L. Bednarz, Rashmi, B. Hackens, G. Farhi, V. Bayot, and I. Huynen, "Broad-band frequency characterization of double Y-branch nanojunction operating as room-temperature RF to dc rectifier," *IEEE Trans. Nanotechnol.*, vol. 4, no. 5, pp. 576–580, 2005.

[64] C. R. Muller, L. Worschech, D. Spanheimer, and A. Forchel, "Current and voltage gain in a monolithic GaAs/AlGaAs TTJ at room temperature," *IEEE Electron Device Lett.*, vol. 27, no. 4, pp. 208–210, Apr. 2006.

[65] P. R. Bandaru, C. Daraio, S. Jin, and A. M. Rao, "Novel electrical switching behaviour and logic in carbon nanotube Y-junctions," *Nat. Mater.*, vol. 4, no. 9, pp. 663–666, 2005.

[66] L. Bednarz *et al.*, "Low and room temperature studies of RF to DC rectifiers based on ballistic transport," *Microelectron. Eng.*, vol. 81, no. 2–4, pp. 194–200, Aug. 2005.

[67] S. Reitzenstein, L. Worschech, P. Hartmann, and A. Forchel, "Logic AND/NAND gates based on three-terminal ballistic junctions," *Electron. Lett.*, vol. 38, no. 17, p. 951, 2002.

[68] I. Iñiguez-de-la-Torre *et al.*, "Three-terminal junctions operating as mixers, frequency doublers and detectors: a broad-band frequency numerical and experimental study at room temperature," *Semicond. Sci. Technol.*, vol. 25, no. 12, p. 125013, Dec. 2010.

[69] D. Spanheimer, C. R. Müller, J. Heinrich, S. Höfling, L. Worschech, and A. Forchel, "Power gain up to gigahertz frequencies in three-terminal nanojunctions at room temperature," *Appl. Phys. Lett.*, vol. 95, no. 10, pp. 16–19, 2009.

[70] H. Irie, Q. Diduck, M. Margala, R. Sobolewski, and M. J. Feldman, "Nonlinear characteristics of T-branch junctions: transition from ballistic to diffusive regime," *Appl. Phys. Lett.*, vol. 93, no. 5, pp. 12–15, 2008.

[71] D. B. Suyatin *et al.*, "Electrical properties of self-assembled branched inas nanowire junctions," *Nano Lett.*, vol. 8, no. 4, pp. 1100–1104, 2008.

[72] S. Bollaert *et al.*, "Ballistic nano-devices for high frequency applications," *Thin Solid Films*, vol. 515, no. 10, pp. 4321–4326, Mar. 2007.

[73] J. Sun, D. Wallin, P. Brusheim, I. Maximov, Z. G. Wang, and H. Q. Xu, "Frequency mixing and phase detection functionalities of three-terminal ballistic junctions," *Nanotechnology*, vol. 18, no. 19, p. 195205, 2007.

[74] S. F. B. A. Rahman, D. Nakata, Y. Shiratori, and S. Kasai, "Boolean logic gates utilizing GaAs three-branch nanowire junctions controlled by schottky wrap gates," *Jpn. J. Appl. Phys.*, vol. 48, no. 6, p. 06FD01, Jun. 2009.

[75] J. Sun, D. Wallin, I. Maximov, and H. Q. Xu, "A novel SR latch device realized by integration of three-terminal ballistic junctions in InGaAs/InP," *IEEE Electron Device Lett.*, vol. 29, no. 6, pp. 540–542, Jun. 2008.

[76] T. Nakamura, S. Kasai, Y. Shiratori, and T. Hashizume, "Fabrication and characterization of a GaAs-based three-terminal nanowire junction device controlled by double Schottky wrap gates," *Appl. Phys. Lett.*, vol. 90, no. 10, pp. 9–12, 2007.

[77] U. Wieser, M. Knop, M. Richter, U. Kunze, D. Reuter, and A. Wieck, "Ballistic transport and rectification in mesoscopic GaAs/AlGaAs cross junctions," *Phase Transitions*, vol. 79, no. 9–10, pp. 755–764, 2006.

[78] J. Mateos *et al.*, "Ballistic nanodevices for terahertz data processing: Monte Carlo simulations," *Nanotechnology*, vol. 14, no. 2, p. 117, 2003.

[79] J. Mateos *et al.*, "Microscopic modeling of nonlinear transport in ballistic nanodevices," *IEEE Trans. Electron Devices*, vol. 50, no. 9, pp. 1897–1905, Sep. 2003.

[80] A. Garg, K. Prakash, N. Jain, N. Gupta, S. Kumar, and A. K. Singh, "III-V heterostructure based three terminal thermal rectifier," in *2017 Progress in Electromagnetics Research Symposium – Spring (PIERS), IEEE*, 2017, pp. 3681–3683.

[81] A. Garg, N. Jain, and A. K. Singh, "Modeling and simulation of a graphene-based three-terminal junction rectifier," *J. Comput. Electron.*, vol. 17, no. 2, pp. 562–570, Jun. 2018.

[82] L. Worschech, A. Schliemann, S. Reitzenstein, P. Hartmann, and A. Forchel, "Microwave rectification in ballistic nanojunctions at room temperature," *Microelectron. Eng.*, vol. 63, no. 1, pp. 217–221, 2002.

[83] C. R. Muller, L. Worschech, P. Hopfner, S. Hofling, and A. Forchel, "Monolithically integrated logic nor gate based on GaAs/AlGaAs three-terminal junctions," *IEEE Electron Device Lett.*, vol. 28, no. 10, pp. 859–861, Oct. 2007.

[84] L. Worschech, D. Hartmann, S. Reitzenstein, and A. Forchel, "Nonlinear properties of ballistic nanoelectronic devices," *J. Phys. Condens. Matter*, vol. 17, no. 29, p. R775, 2005.

[85] L. Bednarz *et al.*, "Theoretical and experimental characterization of Y-branch nanojunction rectifier up to 94 GHz," in *2005 European Microwave Conference, IEEE*, 2005, vol. 1, p. 4.

[86] M. Koyama, T. Inoue, N. Amano, T. Maemoto, S. Sasa, and M. Inoue, "Nonlinear electron transport in InAs/AlGaSb three-terminal ballistic junctions," *J. Phys. Conf. Ser.*, vol. 109, no. 1, p. 012023, 2008.

[87] I. Shorubalko *et al.*, "A novel frequency-multiplication device based on three-terminal ballistic junction," *IEEE Electron Device Lett.*, vol. 23, no. 7, pp. 377–379, Jul. 2002.

[88] A. M. Song, "Electron ratchet effect in semiconductor devices and artificial materials with broken centrosymmetry," *Appl. Phys. A*, vol. 75, no. 2, pp. 229–235, Aug. 2002.

[89] B. G. Vasallo, T. Gonz´Alez, D. Pardo, and J. Mateos, "Monte Carlo analysis of four-terminal ballistic rectifiers," *Nanotechnology*, vol. 15, no. 4, p. S250, 2004.

[90] A. M. Song, "Room-temperature ballistic nanodevices," *Encycl. Nanosci. Nanotechnol.*, vol. 9, no. 389, pp. 371–389, 2004.

[91] A. M. Song, P. Omling, L. Samuelson, W. Seifert, I. Shorubalko, and H. Zirath, "Operation of InGaAs/InP-based ballistic rectifiers at room temperature and frequencies up to 50 GHz," *Jpn. J. Appl. Phys.*, vol. 40, no. Part 2, No. 9A/B, pp. L909–L911, Sep. 2001.

[92] R. Fleischmann and T. Geisel, "Mesoscopic rectifiers based on ballistic transport," *Phys. Rev. Lett.*, vol. 89, no. 1, p. 016804, 2002.

[93] S. De Haan, A. Lorke, J. P. Kotthaus, W. Wegscheider, and M. Bichler, "Rectification in mesoscopic systems with broken symmetry: quasiclassical ballistic versus classical transport," *Phys. Rev. Lett.*, vol. 92, no. 5, p. 056806, 2004.

[94] S. De Haan, A. Lorke, J. Kotthaus, M. Bichler, and W. Wegscheider, "Quantized transport in ballistic rectifiers: sign reversal and step-like output," *Phys. E Low-dimens. Syst. Nanostruct.*, vol. 21, no. 2, pp. 916–920, 2004.

[95] B. Hackens *et al.*, "Sign reversal and tunable rectification in a ballistic nanojunction," *Appl. Phys. Lett.*, vol. 85, no. 19, pp. 4508–4510, 2004.

[96] T. Maemoto, M. Koyama, M. Furukawa, H. Takahashi, S. Sasa, and M. Inoue, "Electron transport in InAs/AlGaSb ballistic rectifiers," *J. Phys. Conf. Ser.*, vol. 38, pp. 112–115, May 2006.

[97] J. Matthews, D. Sánchez, M. Larsson, and H. Linke, "Thermally driven ballistic rectifier," *Phys. Rev. B*, vol. 85, no. 20, p. 205309, 2012.

[98] A. K. Singh, S. R. Kasjoo, and A. M. Song, "Low-frequency noise of a ballistic rectifier," *IEEE Trans. Nanotechnol.*, vol. 13, no. 3, pp. 527–531, May 2014.

[99] M. Wiemann, U. Wieser, U. Kunze, D. Reuter, and A. D. Wieck, "Full-wave rectification based upon hot-electron thermopower," *Appl. Phys. Lett.*, vol. 97, no. 6, p. 062112, Aug. 2010.

[100] A. Garg, N. Jain, S. Kumar, S. R. Kasjoo, and A. K. Singh, "Analysis of nonlinear characteristics of a graphene based four-terminal ballistic rectifier using a drift-diffusion model," *Nanoscale Adv.*, vol. 1, no. 10, pp. 4119–4127, 2019.

[101] A. M. Song, A. Lorke, A. Kriele, J. P. Kotthaus, W. Wegscheider, and M. Bichler, "Nonlinear electron transport in an asymmetric microjunction: a ballistic rectifier," *Phys. Rev. Lett.*, vol. 80, no. 17, pp. 3831–3834, Apr. 1998.

[102] M. Koyama, K. Fujiwara, N. Amano, T. Maemoto, S. Sasa, and M. Inoue, "Electron transport properties in InAs four-terminal ballistic junctions under weak magnetic fields," *Phys. status solidi*, vol. 6, no. 6, pp. 1501–1504, 2009.

[103] A. Garg, N. Jain, and A. K. Singh, "Drift-diffusion modeling and simulation of four terminal ballistic rectifier," in *2016 IEEE International Conference on Recent Trends in Electronics, Information & Communication Technology (RTEICT), IEEE*, 2016, pp. 1995–1998.

[104] M. Knop, U. Wieser, U. Kunze, D. Reuter, and A. D. Wieck, "Ballistic rectification in an asymmetric mesoscopic cross junction," *Appl. Phys. Lett.*, vol. 88, no. 8, p. 082110, Feb. 2006.

[105] A. M. Song, "Formalism of nonlinear transport in mesoscopic conductors," *Phys. Rev. B*, vol. 59, no. 15, pp. 9806–9809, Apr. 1999.

[106] A. Löfgren, I. Shorubalko, P. Omling, and A. M. Song, "Quantum behavior in nanoscale ballistic rectifiers and artificial materials," *Phys. Rev. B*, vol. 67, no. 19, p. 195309, 2003.

[107] A. M. Song *et al.*, "A nonlinear transport device with no intrinsic threshold," *Superlattices Microstruct.*, vol. 25, no. 1–2, pp. 269–272, Jan. 1999.

[108] V. Kaushal, I. Iñiguez-de-la-Torre, and M. Margala, "Nonlinear electron properties of an InGaAs/InAlAs-based ballistic deflection transistor: room temperature DC experiments and numerical simulations," *Solid. State. Electron.*, vol. 56, no. 1, pp. 120–129, Feb. 2011.

[109] M. Margala, H. Wu, and R. Sobolewski, "Ballistic deflection transistors and their application to THz amplification," *J. Phys. Conf. Ser.*, vol. 647, no. 1, p. 012020, 2015.

[110] V. Kaushal, I. Ifñiguez-De-la-torre, and M. Margala, "Room temperature nonlinear ballistic nanodevices for logic applications," in *68th Device Research Conference*, IEEE, 2010, pp. 115–116.

[111] V. Kaushal, I. Iñiguez-de-la-torre, and M. Margala, "Topology impact on the room temperature performance of THz-range ballistic deflection transistors," in *ACM Great Lakes Symposium on VLSI*, 2010, pp. 159–162. https://doi.org/10.1145/1785481.1785520

[112] D. Wolpert, H. Irie, R. Sobolewski, P. Ampadu, Q. Diduck, and M. Margala, "Ballistic deflection transistors and the emerging nanoscale era," in *2009 IEEE International Symposium on Circuits and Systems*, IEEE, 2009, pp. 61–64.

[113] Ravita, A. Garg, and A. Sharma, "Multiplexer based logic gates design using ballistic deflection transistors," in *2020 5th International Conference on Communication and Electronics Systems (ICCES)*, 2020, no. Icces, pp. 99–103.

[114] V. Kaushal *et al.*, "A study of geometry effects on the performance of ballistic deflection transistor," *IEEE Trans. Nanotechnol.*, vol. 9, no. 6, pp. 723–733, 2010.

[115] V. Kaushal, M. Margala, Q. Yu, P. Ampadu, G. Guarino, and R. Sobolewski, "Current transport modeling and experimental study of THz room temperature ballistic deflection transistors," *J. Phys. Conf. Ser.*, vol. 193, no. 1, p. 012092, 2009.

[116] Q. Diduck, H. Irie, and M. Margala, "A Room temperature ballistic deflection transistor for high performance applications," *Int. J. High Speed Electron. Syst.*, vol. 19, no. 01, pp. 23–31, Mar. 2009.

[117] P. Marthi, J.-F. Millithaler, I. Iniguez-de-la-Torre, J. Mateos, T. Gonzalez, and M. Margala, "Exploration of digital latch design using ballistic deflection transistors – Modeling and simulation," in *2015 IEEE Nanotechnology Materials and Devices Conference (NMDC)*, IEEE, 2015, pp. 1–4.

[118] V. Kaushal *et al.*, "Effects of a high-k dielectric on the performance of III-V ballistic deflection transistors," *IEEE Electron Device Lett.*, vol. 33, no. 8, pp. 1120–1122, 2012.

[119] D. Wolpert, Q. Diduck, and P. Ampadu, "NAND gate design for ballistic deflection transistors," *IEEE Trans. Nanotechnol.*, vol. 10, no. 1, pp. 150–154, 2009.

[120] V. Kaushal *et al.*, "A study of effects of deflector position variation on leakage currents in ballistic deflection transistors," in *2009 IEEE Nanotechnology Materials and Devices Conference*, IEEE, 2009, pp. 13–18.

[121] Q. Diduck, M. Margala, and M. Feldman, "A terahertz transistor based on geometrical deflection of ballistic current," in *2006 IEEE MTT-S International Microwave Symposium Digest*, 2006, pp. 345–347.

[122] A. M. Song, P. Omling, L. Samuelson, W. Seifert, I. Shorubalko, and H. Zirath, "Room-temperature and 50 GHz operation of a functional nanomaterial," *Appl. Phys. Lett.*, vol. 79, no. 9, pp. 1357–1359, Aug. 2001.

[123] F. Al-Dirini, F. M. Hossain, A. Nirmalathas, and E. Skafidas, "All-graphene planar self-switching MISFEDs, metal-insulator-semiconductor field-effect Diodes," *Sci. Rep.*, vol. 4, pp. 1–8, 2014.

[124] J. Pezoldt, R. Göckeritz, B. Hähnlein, B. Händel, and F. Schwierz, "T-and Y-branched three-terminal junction graphene devices," *Mater. Sci. Forum*, vol. 717, pp. 683–686, 2012.

[125] Z.-X. Xie, K.-M. Li, L.-M. Tang, C.-N. Pan, and K.-Q. Chen, "Nonlinear phonon transport and ballistic thermal rectification in asymmetric graphene-based three terminal junctions," *Appl. Phys. Lett.*, vol. 100, no. 18, p. 183110, Apr. 2012.

[126] W. Kim, P. Pasanen, J. Riikonen, and H. Lipsanen, "Nonlinear behavior of three-terminal graphene junctions at room temperature," *Nanotechnology*, vol. 23, no. 11, p. 115201, 2012.

[127] S. F. A. Rahman, A. M. Hashim, and S. Kasai, "Fabrication and transport performance of three-branch junction graphene nanostructure," in *2012 International Conference on Enabling Science and Nanotechnology*, IEEE, 2012, pp. 1–2.

[128] T. Ouyang *et al.*, "Ballistic thermal rectification in asymmetric three-terminal graphene nanojunctions," *Phys. Rev. B*, vol. 82, no. 24, p. 245403, Dec. 2010.

[129] R. J. Zhu, Y. Q. Huang, N. Kang, and H. Q. Xu, "Gate tunable nonlinear rectification effects in three-terminal graphene nanojunctions," *Nanoscale*, vol. 6, no. 9, pp. 4527–4531, 2014.

[130] R. Göckeritz, J. Pezoldt, and F. Schwierz, "Epitaxial graphene three-terminal junctions," *Appl. Phys. Lett.*, vol. 99, no. 17, p. 173111, 2011.

[131] P. Butti, I. Shorubalko, U. Sennhauser, and K. Ensslin, "Finite element simulations of graphene based three-terminal nanojunction rectifiers," *J. Appl. Phys.*, vol. 114, no. 3, p. 033710, Jul. 2013.

[132] W. Kim *et al.*, "All-graphene three-terminal-junction field-effect devices as rectifiers and inverters," *ACS Nano*, vol. 9, no. 6, pp. 5666–5674, 2015.

[133] B. Händel, B. Hähnlein, R. Göckeritz, F. Schwierz, and J. Pezoldt, "Electrical gating and rectification in graphene three-terminal junctions," *Appl. Surf. Sci.*, vol. 291, pp. 87–92, 2014.

[134] S. Fadzli Abd Rahman, S. Kasai, and A. Manaf Hashim, "Room temperature nonlinear operation of a graphene-based three-branch nanojunction device with chemical doping," *Appl. Phys. Lett.*, vol. 100, no. 19, p. 193116, 2012.

[135] X. Yin and S. Kasai, "Graphene-based three-branch nano-junction (TBJ) logic inverter," *Phys. Status Solidi*, vol. 10, no. 11, pp. 1485–1488, 2013.

[136] A. Jacobsen, I. Shorubalko, L. Maag, U. Sennhauser, and K. Ensslin, "Rectification in three-terminal graphene junctions," *Appl. Phys. Lett.*, vol. 97, no. 3, p. 032110, Jul. 2010.

[137] G. Auton, R. K. Kumar, E. Hill, and A. Song, "Graphene triangular ballistic rectifier: fabrication and characterisation," *J. Electron. Mater.*, vol. 46, no. 7, pp. 3942–3948, Jul. 2017.

[138] G. Auton *et al.*, "Terahertz detection and imaging using graphene ballistic rectifiers," *Nano Lett.*, vol. 17, no. 11, pp. 7015–7020, Nov. 2017.

[139] Han Wang, D. Nezich, Jing Kong, and T. Palacios, "Graphene frequency multipliers," *IEEE Electron Device Lett.*, vol. 30, no. 5, pp. 547–549, May 2009.

[140] F. Al-Dirini, E. Skafidas, and A. Nirmalathas, "Graphene self switching diodes with high rectification ratios," in *2013 13th IEEE International Conference on Nanotechnology (IEEE-NANO 2013)*, IEEE, 2013, pp. 698–701.

[141] S. Datta, *Electronic Transport in Mesoscopic Systems*. Cambridge: Cambridge University Press, 1995.

[142] M. Büttiker, "Four-terminal phase-coherent conductance," *Phys. Rev. Lett.*, vol. 57, no. 14, pp. 1761–1764, Oct. 1986.

[143] Y. Hirayama and S. Tarucha, "High temperature ballistic transport observed in AlGaAs/InGaAs/GaAs small four-terminal structures," *Appl. Phys. Lett.*, vol. 63, no. 17, pp. 2366–2368, Oct. 1993.

Graphene SymFET and SiNW FET for IP Piracy Prevention Security Systems

16

M. Arun Kumar

Contents

DOI: 10.1201/9781003126645-16

16.1 INTRODUCTION

For a decade, novelty-based device technologies have been investigated to design the sub-10-nm devices for a cybersecurity operating system. Tunneling devices fabricated by two-dimensional (2D) materials exposed good potential for the operation of low voltage. Hence it became a promising material for providing good energy to design the digital circuit design and systems. Some emerging tunneling devices like SymFETs and BiSFETs are attractive. The $I–V$ characteristics of the tunneling devices are different from those of MOSFETs.SymFET devices are especially attractive for security applications because they utilized low power, have controlled speed requirements, and establish improved efficient dynamic energy compared to CMOS logic circuits. In addition, CMOS technology does not support security applications. The larger leakage becomes an interesting challenge in SymFET circuits. The most required property for these devices is a large on-current to off-current ratio to permit minimum power and low-voltage operation. Due to its unique $I–V$ characteristics, it has been used for a perspective of new topologies in security-based circuit design. For cybersecurity protection schemes, these new methods are developing for the hardware infrastructure, and it is being changed to support security policies. Hence, the protection scheme at the system level will become more effective.

In this chapter, the 2D structure of graphene-based symmetric field-effect transistor (SymFET) will be explained along with its distributed potentials and the current flows internally through the device. A single-particle tunneling model was modeled to calculate the current-voltage characteristics of the device. Tunneling current and voltage distribution in the graphene electrodes will decide the device's behavior. If the tunneling current is small, due to either small coherence length or large tunneling thickness, zero variation will be achieved in the voltage distribution of the graphene electrodes, which was discussed by Zhao et al. [1]. If the tunneling current is large, due to either large coherence length or small tunneling thickness, the voltage distribution will be appreciable. The tunneling effects are not present in 1D (one-dimensional) SymFET, but it can be captured in the 2D graphene SymFET devices [2,3]. These effects can increase the $V–I$ characteristics of the device; hence, a graphene SymFET is more suitable for hardware security applications [27–35].

16.2 GRAPHENE

Graphene is a 2D material a single atom thick composed of Sp^2-bonded carbon atoms. It is attractive for its interesting honeycomb lattice structure. Each carbon atom is arranged in the structure with a space of about 1.42 Å. The 2s orbital is hybridized with the 2px and 2py orbital in the plane over trichotomy covalent bonds at the angle of 120°. A weak π-bond is arranged perpendicular to the p_z orbital in the plane. The layers in the graphite are arranged together with the van der Waals force of attraction, which is consequently weak, and hence it can be separated as a sheet very easily. In graphitic structures, the π-bond is assured for electronic conduction properties. Graphene is an eminent material that is taken from building blocks of graphite. The electronic band structure was measured by the tight-binding theory and explained by Wallace et al. [4]. The bandgap between the conduction band and valence band is zero. Hence, the valance band and conduction band are coincidences at the Brillouin zone point called the Dirac point. The structure of graphene was more attractive, and it was remembered several times in various circumstances (Ando et al.) [5]. The hexagonal structure of graphene is shown in Figure 16.1.

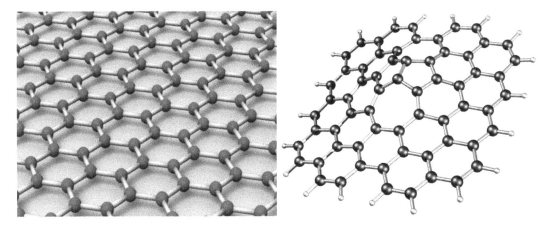

FIGURE 16.1   Hexagonal structure of graphene.

16.2.1 Electronic Properties of Graphene

The evolution of graphene makes scientists bring out a rigorous approach in the development of electronic devices for various sensor applications. Nonlinear short channel effects cause major issues in the electron device community. It was identified that 2D materials like graphene do not have short channel effects, even when the length of the channel is very small. Consequently, the 2D materials are more efficient than the Si material. The structure of graphene also a basic reason for its attractive property. The three sp^2 hybridized orbitals are distributed symmetrically with the nearest carbon atoms at an angle of 120°and forms the sigma bond, as shown in Figure 16.2a, b, and c. This sigma bond makes graphene the strongest material. Near the Dirac point, graphene shows a linear dispersion relation. The conduction band and the valance band met at the Dirac point where the energy band is symmetric and the band density at the Fermi level is zero. Graphene is one atom thick and massless; hence the motion of an electron is comparatively equal to light. Graphene has an interesting feature that it does not have backscattering like a carbon nanotube. It increases carrier mobility in graphene. Therefore, high mobility is applicable for pristine samples to reduce the scattering centers. Hence, carrier transport in the material is completely ballistic. High mobility and high velocity in graphene are promised for high frequency digital and analog circuits. Hence, graphene became a promising material for all the FET industry to apply in various applications.

16.3 GRAPHENE FET

After the discovery of graphene, researchers were hopeful, thinking that the graphene material became a potential applicant to replace the silicon material. However, the researchers were soon disappointed with its gapless energy band because the zero-bandgap does not make the FET switch-off. Hence, it causes more leakage current and excessive energy dissipation. Researchers took several trials to tempt the bandgap in graphene such as nanoribbon, surface functionalization, etc. All efforts failed to make the graphene appropriate for the digital logic design. Bandgap playing a major role in switching activity for low power devices; the absence of required bandgap in graphene makes it difficult to use in on/off switching ratios in low power dissipation. But graphene-based FETs are suitable for sensor applications [6]. Finally, researchers found an alternative way to design the graphene transistor architecture based on the quantum

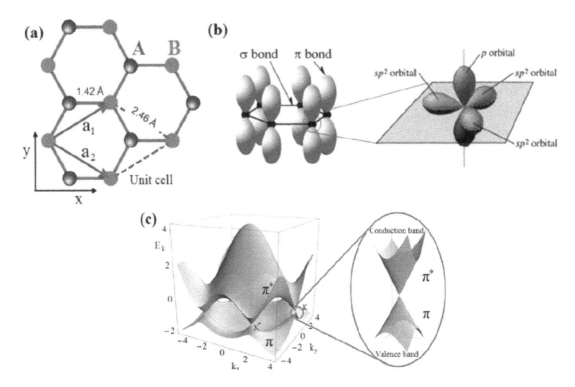

FIGURE 16.2 (a) Lattice structure of the graphene, (b) σ and π bonds in graphene, and (c) band structure of the graphene.

tunneling method by Britnell et al. [7]. This type of graphene transistors exposes high switching ratio and are suitable for high-frequency operation. It creates a new path to design the field-effect tunneling transistors for graphene-based nanoelectronics applications. Currently, Zhao et al. [1] designed the most significant graphene FET architecture to provide a high switching ratio, and it is called graphene symmetric FET or SymFET. Herein, the tunnel-based field-effect transistor was designed in such a way that the insulator layer was sandwiched between two layers of graphene.

16.4 GRAPHENE SYMMETRIC FET (SYMFET)

16.4.1 Fabrication of Graphene SymFET

The graphene h-BN graphene SymFET structure is shown in Figure 16.3. The graphene is obtained by the chemical vapor deposition (CVD) method and moved onto 90 nm-thick SiO_2 on a silicon substrate. After patterning the graphene, Ni/Au contacts were deposited on this bottom layer of graphene. A Ti layer was deposited as a protection layer using e-beam evaporation. Hexagonal boron nitride was grown on the Ni by the CVD method. Ammonia borane was heated to the substrate with the presence of $Ar:H_2$ gas and rapidly cooled for 10 minutes. It acts as an h-BN precursor. This process creates a multilayer (3–4 layers) h-BN, and it was identified by the TEM characterization. Then the graphene was formed on the wet-transferred h-BN. Ni/Au contacts were developed on the graphene layer as a top gate contact. The bottom gate manages the concentration of the bottom layer, and the gate controls the graphene layer.

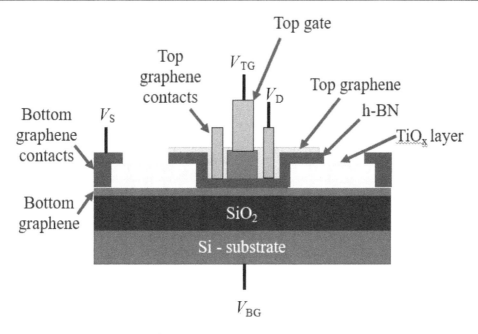

FIGURE 16.3 Cross-sectional view of the graphene SymFET structure.

16.4.2 Structure of the Device

The graphene SymFET device structure is shown in Figure 16.4a. The insulator layer is reserved to separate the two layers of graphene. Herein, the boron nitride (BN) acts as an insulating layer. Therefore, it creates a new structure graphene-insulator-graphene (GIG) structure sandwiched between a top and bottom gate. The 2D materials like boron nitride (BN) might be a good choice to decrease the interface trap density until it reduces the dangling bond. BN material will be present along with the graphene layer to align the structure. Stacking graphene and BN will provide a smooth surface; meanwhile, BN shares a similar hexagonal lattice structure with graphene. The graphene layers achieved ohmic contacts individually to represent the source (S) and drain (D). In this structure, the top electrode is considered to be the n-doped and the bottom electrode is considered to be the p-doped region [36–40]. The top gate voltage (V_{TG}) and bottom voltage (V_{BG}) are controlled by the quasi-Fermi level (μ_n and μ_p). In the graphene layer, the quasi-Fermi level is represented as ΔE; for an n-type region, the Fermi level is ΔE present above the Dirac point, and for the p-type region, it is present below the Dirac point; this was stated by Hasan et al., [8] In graphene, both the top and back gate are symmetric; hence the voltage $V_{TG} = -V_{GB} = V_G$ and the drain to source voltage is given as $V_{DS} = V_D - V_S$. The 2D representation of the device as well as internal current drifts are clearly shown in Figure 16.4b.

16.4.3 Operation of the Device

The operation of the graphene SymFET device is shown in Figure 16.5. In the Dirac point, if the top and bottom layer are skewed, then the solitary energy level will achieve the condition of concurrent energy and momentum conservation, therefore, the tunneling current is minor as shown in Figure 16.5a and b. If both layers are aligned equally at $V_{DS} = 2\Delta E/q$, then the electrons present at all the energies between the Fermi level (μ_n and μ_p) achieve energy and momentum conservation [8, 41–45]; hence the tunneling current is large under this condition, as shown in Figure 16.5 c.

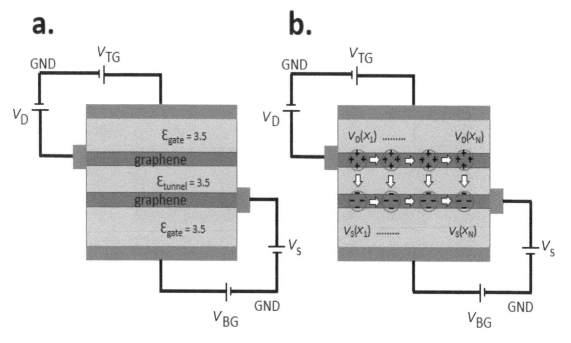

FIGURE 16.4   (a) Graphene SymFET structure. (b) 2D representation of SymFET device.

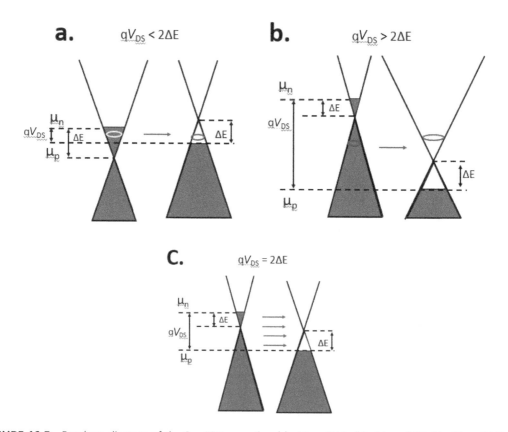

FIGURE 16.5   Bandgap diagram of the SymFET operation (a) $qV_{DS} < 2\Delta E$, (b) $qV_{DS} > 2\Delta E$, (c) $qV_{DS} = 2\Delta E$.

TABLE 16.1   List of Device Parameters

SYMBOL	PARAMETER	EMPLOYED VALUE
ΔE_{doping}	Chemical doping	0.1 eV
k_t	Decay constant for tunneling current in the barrier	6 nm^{-1}, 17 nm^{-1}
v_F	Fermi velocity	1×10^8 cms^{-1}
L	Coherence length	50 nm, 100 nm, 300 nm, 1,200 nm
t_t	Tunneling insulator thickness	0.33 nm, 0.66 nm, 0.99 nm, 1.32 nm
t_g	Gate insulator thickness	20 nm
ε_{BN}	Dielectric constant of boron nitride	3.5
μ	Mobility of electrons in graphene	10,000 cm^2/Vs
d	Typical inter-planner separation in graphite	0.33 nm

Graphene's high electron mobility can exceed up to 15,000 cm^2/Vs under room temperature conditions and endures high even with more carrier density (> 10^{12} cm^{-2}) for both chemically and electrically doped samples. If the extrinsic scattering is eliminated in graphene, its mobility will be increased due to the weak electron-phonon interactions. In the graphene SymFET, the Fermi degeneracy is changed over a huge energy gap, then the critical temperature for the excitonic condensate has been considered to be more than the room temperature. A macroscopic tunneling current will be provided by the excitonic condensate formation between the layers. This will be helpful to design the structure of the device for high-frequency applications. For designing the graphene SymFET structure, consider the list of device parameters given in Table 16.1 [8]. The tunneling insulator layer (BN) is fixed with the different thickness of 0.33 nm (~1 atomic layer), 0.66 nm (~ 2 atomic layers), 0.99 nm (~ 3 atomic layers), and 1.32 nm (~ 4 atomic layers). Whenever the tunneling insulator layer is very thin with a large variation of ΔE, the resonant peak produced by the tunneling current leads to substantial broadening. The insulator layer thickness (< 0.66 nm = ~ 2 atomic layers) and tunnel barrier thickness (~ 0.5 nm) are very important to achieve the electron momentum relaxation time in graphene.

In graphene, the charge carriers (electrons or holes) can be tempted by the electric field or chemical doping. The doping effect of the electric field is frequently achieved in graphene-based field-effect transistors (FET); the charge carriers are attracted by changing the electric potential between gate terminal and graphene, for example, a Si+/SiO$_2$ substrate [9,10]. The concentration and type of carriers can be tuned by varying the gate voltage, V_g. The sign of the applied gate voltage is opposite to the induced carrier's sign. A positive gate voltage (V_g) persuades electrons, while a negative gate voltage (V_g) encourages holes. Graphene interacts with the chemical species by chemical doping [11]. Chemical doping is of two types: one is surface transfer and another is substitutional doping. In such cases, the effect of doping occurs once some of the carbon atoms in the graphene lattice are swapped by other atoms with a dissimilar number of valence electrons. For example, boron, B$_s$, and nitrogen, N$_s$, substitutional atoms and lead top- and n-type conductivity, respectively [12,13, 46–49]. However, the electronic structure of the graphene can be modified by the integration of foreign atoms into the graphene lattice. Hence, it can be used for ultra-high frequency and THz applications in a high-security hardware device.

16.4.4 Electrical Characteristics of the Device

The electrical characteristics of the SymFET provide the information about conductivity distribution infinite along with the electrodes. The electrical characteristics of the device show how the tunnel current becomes small and the voltage will drop in the graphene layers are insignificant. The density of the tunnel current is large when the tunnel thickness becomes small, or if there is a huge coherence length, then the voltage drops in the graphene layer will be reduced, for example, if the I–V_{DS} characteristics of a SymFET were measured at K_t =17 nm^{-1}, t_t = 0.66 nm (~ 2 atomic layers of BN) with L = 50 nm, as shown in Figure 16.6. The results are found using a 1D approximation ($N = 0$) and a 2D model consisting of 20 distributed cells ($N = 20$). The

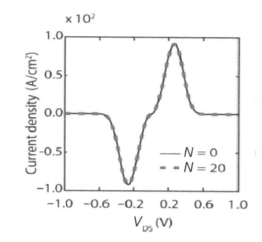

FIGURE 16.6 I–V_{DS} characteristics of the SymFET.

obtained results from both the methods are completely matched. The gate voltage V_g =−0.76 V was selected, which delivers symmetric I–V_{DS} characteristics. The tunnel thickness is decreased from 0.66 nm (~ 2 atomic layers of BN) to 0.33 nm (~ 1 atomic layer of BN) by maintaining the coherence length as L = 50 nm. Across the resistive elements, the potential drops that become sufficient will increase the current in the tunnel, and there is a deviation in the results from the 2D model to 1D approximation [8]. The electrical characteristics of the SymFET were studied and measured, as shown in Figure 16.6 by M. Hasan et al., [8].

16.5 PRACTICAL ASPECTS OF GRAPHENE SYMFET TECHNOLOGY

Nowadays, hardware security has become a big concern for IP piracy and hardware Trojans [14]. The Trojans have triggered the researchers to design the applications with circuit protection and spiteful logic detection from different design perspectives. In this chapter, emerging technologies are examined by their unique properties for hardware security domain applications. Circuit structures including power regulators, camouflaging gates, and polymorphic gates are measured to prove the high efficiency of silicon nanowire FETs and graphene SymFETs for circuit protection and IP piracy prevention applications. Simulation results of the devices designate that extremely efficient and secure circuit design structures can be attained via the custom of emerging technologies. Overall, the knowledge of using nanoelectronics like graphene symmetric FET and silicon nanowire FET structures for security applications will be a new platform and interesting path of research for both electronic and security researchers [50–52].

16.6 APPLICATIONS

16.6.1 Graphene SymFET-Based Circuit Protectors

In addition to the protection of the IP protocol, the devices can also stabilize circuit strength to counter different hardware assaults, such as fault infusion and analysis of the side-channel signal, with particularly low performance and small circuit redesign. For instance, cryptographic circuits are regularly susceptible to power flexibility-based issue infusions [15]. Increasing the register's setup time to switch into the accurate state causes power supply faults that affect paths of high capacitance, which provides the slowest path of the circuit.

16.6.1.1 Current-Based Circuit Protector

The *I–V* characteristic of SymFET is shown in Figure 16.7 which indicates the current I_{DS} can exist only for a narrow band of V_{DS}. To prevent the supply voltage-based fault infusion, we implement a current-based circuit protector by using the above property [14]. Only transistor T1 is in direct contact with the power supply V_{DD}, as shown in Figure 16.7. This transistor is the source to initiate a voltage-based fault injection attack.

For all the three SymFETs, V_{BG} is set to 0 V and V_{TG} is set to 0.6 V. The peak current at different power supply ranges can be obtained by varying the gate voltages. The transistors T2 and M3 are in parallel, source to drain voltage V_{DS2} for T2 and V_{DS3} for T3 are equal, which results in the output current I_{OUT} being equal to the input current I_{IN}. The I_{OUT} is the source current for the given circuit in protection. Therefore, the output current only exists for a specified value of the drain-source voltage of SymFET-T3. Some sample power levels provided for the current-based protector are shown in Table 16.2 [14].

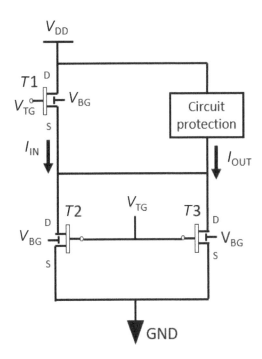

FIGURE 16.7 Schematic diagram of a current-based circuit protector.

TABLE 16.2 List of Sample Power Provided for Current-Based Circuit Protector

Voltage supply /V_{DD} (V)	0.2	0.4	0.6	0.8	1.0
Output current/ I_{out} (µA)	0.02	0.07	0.18	1.2	1.9
Power (µA)	0.009	0.05	0.2	1.93	3.8

Besides IP protection, attention should also be given to attacks on the hardware such as fault injection and analysis of a side-channel signal. Fault injections are also causing a major problem for cryptographic circuits. It is caused by increasing setup time needed for registers, and it also affects the capacitance path. Moreover, it creates the slowest path to the circuit, which makes the hackers attack the hardware. To overcome this problem, the unique properties of graphene-based SymFET have been introduced for the current-based circuit protector [14]. This will be exploited in the side-channel attacks and fault injection. The SymFET devices maintain larger loads to prevent fault injections if the hackers plan to lower the supply voltage of the circuit to trigger a single-bit error of an encryption design. Due to this process, the entire circuit can be shut down by the circuit protector automatically before the triggering of a single-bit error. The circuit protector will act as both a circuit protector and current source [16]. Hence, the graphene SymFET can be utilized for side-channel attack prevention and cryptographic circuits.

16.7 SILICON NANOWIRE FET

16.7.1 Fabrication of the SiNW FET

Two top-down technologies are practiced to fabricate the SiNW field-effect transistors. These top-down nano-lithography methods are oxidation scanning probe lithography (oSPL) and electron beam lithography [17–19]. The two methods are carried out in the same silicon-on-insulator (SOI). The fabrication steps are different for both oSPL and EBL. The lithography process shares some common elements for the fabrication process: SOI substrate, same source and drain metal contacts, and removal of an unmasked region of the top silicon layer by reactive ion etching (RIE), as shown in Figure 16.8. During fabrication, the SOI substrates are cleaned by the sonication. The metal contacts are fabricated by using photolithography. Then the oxidation SPL step is carried out with an atomic force microscope operated in the amplitude modulation mode [20,21]. The relative humidity was controlled by the AFM. Following this step, the SiO_2 masks are developed with electrodes using the photolithography process. The unwanted mask layers are removed by using electron beam lithography. The lithography step is followed for both oSPL and EBL. The bridging nanowire was developed between the source and drain electrodes. The formation of SiNW is to be contingent on the lithography method. The width of the SiNW was developed by the oSPL method, with comparatively less width than the EBL method.

16.7.2 Structure of the Device

Silicon nanowire FETs are designed as shown in Figure 16.9 with a reconfigurable property that is suitable to implement in complementary logic for hardware security. The SiNW FETs can be operated as a p-FET or n-FET by adjusting the suitable external voltage. The structure of SiNW FETs has two gates

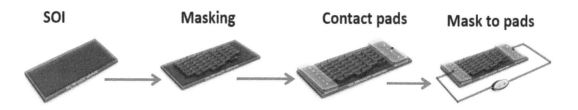

FIGURE 16.8   Fabrication steps of the SiNW FET.

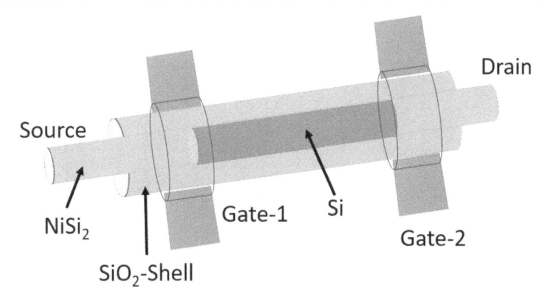

FIGURE 16.9   Reprogrammable silicon nanowire FET.

that are coupled electrostatically by Schottky junction separately from each other. The p-polarity or n-polarity is achieved by one gate, and another gate is used to vary the conductance through the nanowire. The signal transmission across the Schottky junction can be controlled by applying an external voltage source. Device configuration can be addressed by Schottky junctions in Schottky barrier field-effect transistors (SBFETs) separately. One junction is used to block one type of carrier while another is tuned to change its conductance to the other type of carrier (A. Heinzig et.al.) [22]. The choice of the heterostructure materials is $NiSi_2$/intrinsic – $Si/NiSi_2$, as shown in Figure 16.9. The materials are aligned in an axial and embedded in a SiO_2 shell. The thermally grown SiO_2 is used as a gate dielectric in which silicon has a low interface trap density. The presence of two Schottky junctions at each end of the nanowire is the key factor that empowers reconfigurability. Gate 2 is a program gate used to configure the device in p-FET or n-FET, and gate 1 is a control gate used to control the conductance through the nanowire.

16.7.3 Electrical Characteristics and Working Principle of the SiNW FET

The devices can be operated in a bidirectional mode. The similar values are attained by exchanging the signals between V_{G1}/V_{G2} and source/drain, as shown in Figure 16.10. This property of symmetry added the advantage of flexibility and adaptability in the circuit design. The p-FET mode can be programmed by fixing the gate voltage (V_{G2}) program set to −3 V and the voltage (V_D) of the drain-source set to −1 V. At the drain electrode, electron injection is effectively blocked by the applied potential in V_{G2}. The potential voltage (V_{G1}) of the control gate is varied between positive and negative values. The voltage $V_{G1} < 0$ stimulates the hole injection into the active region of the FET at the source site, changing the V_{G1} positive values results in closing off the FET by blocking the injection of holes, as shown in Figure 16.10. The electrical characteristics of the SiNW FET were studied and measured as shown in Figure 16.10 by A. Heinzig et al. [22].

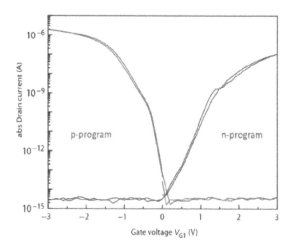

FIGURE 16.10   Transfer characteristic programmed as p-FET and n-FET.

The n-FET mode is initially programmed by setting the gate voltage (V_{G2}) to +3 V and setting the drain voltage (V_D) to +1 V. In this mode, the gate potential voltage (V_{G2}) blocked the injection of holes from the drain electrode into the active region of FET and made the junction transparent for electrons. When V_{G1}< 0, this results in a reduction of electron injection at the source side due to the high-energy barrier. Swapping of gate voltage V_{G1} into positive value reduces the bands at the source electrode, resulting in enhanced electron injection. The p-FET and n-FET configurations are obtained in the same device. This reconfigurable property is used in the complementary design to enable any Boolean logic function. This reconfigurable property supports altering any specific logic functions during operation. In comparison to the CMOS logic, silicon nanowire FETs are used to provide an additional logic function with the same number of transistors. The gate voltage sensitivity is enhanced due to the compact size and geometry of the Schottky contacts. The device used axial metal-semiconductor nanowire heterostructure. It is the most common geometry used in the Schottky junction for coupling of an electric field. The electric field gradient was high at the Schottky junction. In the technology of a silicon CMOS, the holes have lower mobility compared to electrons, which is compensated by using the p-FETs with larger channel width. In the silicon nanowire FETs, the mobility of holes is lower inside the active region and can be compensated by a reduced height of the Schottky barrier for holes compared to electrons. The silicon nanowire FETs are programmed dynamically because this does not require preprogramming. The conductance of the device is turned up to the values of 1 × 10⁹. It is found that at off-state, the source-drain leakage currents are effectively reduced by opposite band bending at source and drain. It does not require any doping to realize n-type and p-type FETs.

16.8 PRACTICAL ASPECTS OF SINW FET TECHNOLOGY

In the IC industry, intellectual property theft and counterfeiting are the most threatening security threads. Camouflaging is the most common solution to prevent attackers from grasping the circuit schematic by reverse engineering [23–25]. This technique depends on layout-level muddling with comparative layouts for dissimilar gates. This method prevents the attackers from recovering the structure of the circuit schematic [26].

16.9 APPLICATIONS

In an existing system, the security developed by the CMOS-based camouflaging gate structure can be easily stolen from the IP protocol, and Trojans also easily occurred. The hackers easily identify the functions and crack the privacy. To improve the security system and prevent the hackers from cracking the function,

SiNW FET-based camouflaging layouts are incorporated in the basic design, which makes the hackers fail to crack the function easily. In SiNW FET-based camouflaging, the functionality has been increased. To crack a single function, the hackers should try up to 4^N times to get the correct design, and it becomes a challenge to them. However, the SiNW FET-based device has more functionality and less area consumption with a higher level of security. The best advantage of the SiNW FETs is IP protection, logic design encryption, and other security applications.

16.9.1 Implementation of Camouflaging of Gates Using SiNW FET

CMOS camouflaging methods of gates can rarely be implemented because it would increase the power consumption along with an area for higher-level protection. To implement XOR, NAND, and NOR in a CMOS camouflaging method requires a minimum of 12 transistors with a huge area of metal connections [26], as shown in Figure 16.11. The PMOS and NMOS have fixed polarities, which require spare transistors in the implementation of CMOS camouflaging of gates. The silicon nanowire FETs have polarity controllable property, which helps to implement camouflaging of gates without using additional FETs. Four silicon nanowire FETs are used to design a NAND or an XOR gate [27]. The four silicon nanowire FETs perform five more functions in addition to that NAND and XOR operation by interconnecting pins with various signals. The polarity controllable feature of silicon nanowire FETs plays a vital role in designing a camouflaging layout in which different gates are shared in a similar structure.

In this chapter, we presented the IP privacy prevention security systems based on the emerging technologies graphene SymFET and SiNW FET and how it can help to prevent IP protocol and protect circuits. The proposed protection schemes are comparatively higher protection than the CMOS logic circuits. The comparative challenges and opportunities are compared with the CMOS logic and mentioned in

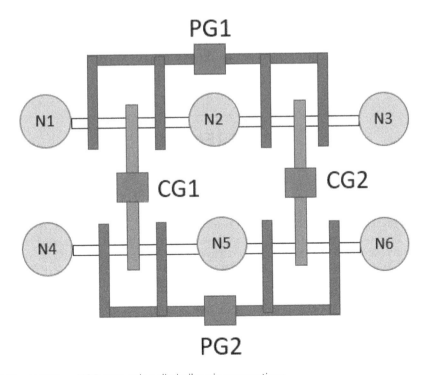

FIGURE 16.11   NAND or XOR gate using dissimilar pin connections.

TABLE 16.3 Comparison of SiNW FET and Graphene SymFET over CMOS in Security Applications

	SINW FETS	GRAPHENE SYMFETS
Benefits over CMOS logic design	• Polarity configurable • Low static power • Fewer transistors for applications	• Low power • Built-in negative differential resistance
Challenges of the devices	• Larger area per transistor • Large dynamic power	• Current-based designs • Non-Boolean computation
Opportunities for the devices	• IP protection • Logic encryption • Other security applications	• Side-channel attack prevention • Cryptographic circuits

Table 16.3 (14). From the comparison, it is clear that the emerging devices can help to guide the researchers for future design in the hardware security area.

16.10 CONCLUSION

In this chapter, the formalism of tunneling effect in the graphene insulator graphene structure for the working of symmetric FET has been discussed for the security applications. Instead of CMOS transistors, two emerging technologies were discussed, including SiNW FETs and graphene SymFETs. The I–V characteristics of the SymFET prove that it is very robust for the temperature changes. The resonant current peak of the SymFET device shows a good potential applicant for the high-speed analog circuit and digital circuit devices. From the resonant current peak, it is understood that the broadening of a current peak can be controlled by chemical doping and applying gate bias. The SiNW was used to design the p- or n-FET for controlling the sensitivity of the charge carrier transport over nanoscale metal-semiconductor junctions. It has the unique advantage of not relying on doping. The device design by using the emerging technologies instead of CMOS provides an extremely efficient and secured circuit that can be accomplished for IP piracy prevention.

REFERENCES

[1] P. Zhao, R. M. Feenstra, G. Gu, and D. Jena, "SymFET: A Proposed Symmetric Graphene Tunneling Field-Effect Transistor", IEEE Transactions on Electron Devices, Vol. 60, No. 3, 2013.

[2] M. Li, D. Esseni, D. Jena, and H. Xing, "Single Particle Transport in Two-Dimensional Heterojunction Interlayer Tunneling Field Effect Transistor", Journal of Applied Physics, Vol. 115, p. 074508, 2014.

[3] M. Li, D. Esseni, D. Jena, and H. Xing, Device Research Conference, Vol. 17, p. 14516299, 2014.

[4] P. R. Wallace, "The Band Theory of Graphite", Physical Review Letters, Vol. 71, No. 9, pp. 622–634, 1947.

[5] T. Ando, T. Nakanishi, and R. Saito, "Berry's Phase and Absence of Back Scattering in Carbon Nanotubes", Journal of the Physical Society of Japan, Vol. 67, No. 8, pp. 2857–2862, 1998.

[6] A. K. Manoharan, S. Chinnathambi, R. Jayavel, and N. Hanagata, "Simplified Detection of the Hybridized DNA Using a Graphene Field Effect Transistor", Science and Technology of Advanced Materials, Vol. 18, pp. 43–50, 2017.

[7] L. Britnell, R. V. Gorbachev, R. Jalil, B. D. Belle, F. Schedin, A. Mishchenko, T. Georgiou, M. I. Katsnelson, L. Eaves, S. V. Morozov, N. M. R. Peres, J. Leist, A. K. Geim, K. S. Novoselov, and L. A. Ponomarenko, "Field-Effect Tunneling Transistor Based on Vertical Graphene Heterostructures", Science, Vol. 335, p. 947, Feb. 2012.

[8] M. Hasan and B. S. Rodriguez, "Effect of the Intra-layer Potential Distributions and Spatial Currents on the Performance of Graphene SymFETs", AIP Advances, Vol. 5, No. 9, p. 097104, 2015.

[9] K. S. Novoselov, A. K. Geim, S. V. Morozov, D. Jiang, Y. Zhang, S. V. Dubonos, I. V. Grigorieva, and A. A. Firsov, "Electric Field Effect in Atomically Thin Carbon Films", Science, Vol. 306, pp. 666–669, 2004.

[10] D. Reddy, L. F. Register, G. D. Carpenter, and S. K. Banerjee, "Graphene Field-Effect Transistors", Journal of Physics D: Applied Physics, Vol. 44, p. 313001, 2011.

[11] H. Liu, Y. Liu, and D. J. Zhu, "Chemical Doping of Graphene", Journal of Materials Chemistry C, Vol. 21, pp. 3335–3345, 2011.

[12] L. S. Panchakarla, K. S. Subrahmanyam, S. K. Saha, A. Govindaraj, H. R. Krishnamurthy, U. V. Waghmare, and C. N. R. Rao, "Synthesis, Structure, and Properties of Boron- and Nitrogen-Doped Graphene", Advanced Materials, Vol. 21, pp. 4726–4730, 2009.

[13] D. Wei, Y. Liu, Y. Wang, H. Zhang, L. Huang, and G. Yu, "Synthesis of N-Doped Graphene by Chemical Vapor Deposition and Its Electrical Properties", Nano Letters, Vol. 9, pp. 1752–1758, 2009.

[14] Y. Bi, P. E. Gaillardon, X. S. Hu, M. Niemier, J. S. Yuan, and Y. Jin, "Leveraging Emerging Technology for Hardware Security-case Study on Silicon Nanowire FETs and Graphene SymFETs", in IEEE Proceeding for Asian Test Symposium, pp. 342–347, 2014.

[15] A. Barenghi, G. M. Bertoni, L. Breveglieri, M. Pellicioli, and G. Pelosi, "Fault Attack on AES with Singlebit Induced Faults", in Proceedings of the 2010 6th International Conference on Information Assurance and Security (IAS'10), pp. 167–172, 2010.

[16] X. Li, W.-Y. Tsai, V. Narayanan, H. Liu, and S. Datta, "A Low-voltage Low-power LC Oscillator Using the Diode-connected SymFET", in Proceedings of the 2014 IEEE Computer Society Annual Symposium on VLSI (ISVLSI'14), pp. 302–307, IEEE, 2014.

[17] R. Garcia, R. V. Martinez, and J. Martinez, "Nanochemistry and Scanning Probe Nanolithographies", Chemical Society Reviews, Vol. 35, pp. 29–38, 2006.

[18] A. W. Knoll, D. Pires, O. Coulembier, P. Dubois, J. L. Hedrick, J. Frommer, and U. Duerig, "Probe-based 3D Nanolithography Using Self-amplified Depolymerization Polymers", Advanced Materials, Vol. 22, pp. 3361–3365, 2010.

[19] A. E. Grigorescu and C. W. Hagen, "Resists for Sub-20-nm Electron Beam Lithography with a Focus on HSQ: State of the Art", Nanotechnology, Vol. 20, p. 292001, 2009.

[20] R. Garcia, M. Calleja, and F. Perez-Murano, "Local Oxidation of Silicon Surfaces by Dynamic Force Microscopy: Nanofabrication and Water Bridge Formation", Applied Physics Letters, Vol. 72, pp. 2295–2297, 1998.

[21] M. Calleja, M. Tello, and R. García, "Size Determination of Field-induced Water Menisci in Noncontact Atomic Force Microscopy", Journal of Applied Physics, Vol. 92, pp. 5539–5542, 2002.

[22] A. Heinzig, S. Slesazeck, F. Kreupl, T. Mikolajick, and W. M. Weber, "Reconfigurable Silicon Nanowire Transistors", ACS Nano Letters, Vol. 12, pp. 119–124, 2012.

[23] L.-W. Chow, J. Baukus, and W. Clark, 2002, "Integrated Circuits Protected Against Reverse Engineering and Method for Fabricating the Same Using an Apparent Metal Contact Line Terminating on Field Oxide", https://patents.google.com/patent/US7294935B2/en

[24] P. Ronald, P. James, and J. Bryan, "Building Block for a Secure Cmos Logic Cell Library", https://patents.google.com/patent/US20100301903A1/en.

[25] Lap Wai Chow, James P. Baukus, Bryan J. Wang, and Ronald P. Cocchi, 2012, "Camouflaging a Standard Cell Based Integrated Circuit", https://patents.google.com/patent/US8151235B2/en

[26] J. Rajendran, M. Sam, O. Sinanoglu, and R. Karri, 2013, "Security Analysis of Integrated Circuit Camouflaging", in Proceedings of the 2013 ACM SIGSAC Conference on Computer & Communications Security, CCS '13, pp. 709–720.

[27] S. Bobba, M. D. Marchi, Y. Leblebici, and G. D. Micheli, 2012, "Physical Synthesis Onto a Sea-of-tiles with Double-gate Silicon Nanowire Transistors", in Proceedings of the 49th Annual Design Automation Conference, DAC'12, pp. 42–47.

[28] Shailendra Singh and Balwinder Raj, "Analytical Modeling and Simulation Analysis of T-shaped III-V Heterojunction Vertical T-FET", Superlattices and Microstructures, Elsevier, Vol. 147, p. 106717, 2020.

[29] Tulika Chawla, Mamta Khosla, and Balwinder Raj, "Optimization of Double-gate Dual Material GeOI-Vertical TFET for VLSI Circuit Design", IEEE VLSI Circuits and Systems Letter, Vol. 6, No. 2, pp. 13–25, Aug. 2020.

[30] Manjit Kaur, Neena Gupta, Sanjeev Kumar, Balwinder Raj, and Arun Kumar Singh, "RF Performance Analysis of Intercalated Graphene Nanoribbon Based Global Level Interconnects", Journal of Computational Electronics, Springer, Vol. 19, pp. 1002–1013, June 2020.

[31] Girish Wadhwa and Balwinder Raj, "An Analytical Modeling of Charge Plasma Based Tunnel Field Effect Transistor with Impacts of Gate underlap Region", Superlattices and Microstructures, Elsevier, Vol. 142, pp. 106512, June 2020.

[32] Shailendra Singh and Balwinder Raj, "Modeling and Simulation analysis of SiGe hetrojunction Double GateVertical t-shaped Tunnel FET", Superlattices and Microstructures, Elsevier, Vol. 142, p. 106496, June 2020.

[33] Shailendra Singh and Balwinder Raj, "A 2-D Analytical Surface Potential and Drain Current Modeling of Double-Gate Vertical t-shaped Tunnel FET", Journal of Computational Electronics, Springer, Vol. 19, pp. 1154–1163, Apr. 2020.

[34] Shradhya Singh, Shashi Bala, Balwant Raj, and Balwinder Raj, "Improved Sensitivity of Dielectric Modulated Junctionless Transistor for Nanoscale Biosensor Design", Sensor Letter, ASP, Vol. 18, pp. 328–333, Apr. 2020.

[35] Vivek Kumar, Santosh Kumar Vishvakarma, and Balwinder Raj, "Design and Performance Analysis of ASIC for IoT Applications", Sensor Letter ASP, Vol. 18, pp. 31–38, Jan. 2020.

[36] Girish Wadhwa and Balwinder Raj, "Design and Performance Analysis of Junctionless TFET Biosensor for High Sensitivity", IEEE Nanotechnology, Vol. 18, pp. 567–574, 2019.

[37] Tanu Wadhera, Deepti Kakkar, Girish Wadhwa, and Balwinder Raj, "Recent Advances and Progress in Development of the Field Effect Transistor Biosensor: A Review", Journal of Electronic Materials, Springer, Vol. 48, No. 12, pp. 7635–7646, Dec. 2019.

[38] Shailendra Singh and Balwinder Raj, "Design and Analysis of Hetrojunction Vertical T-shaped Tunnel Field Effect Transistor", Journal of Electronics Material, Springer, Vol. 48, No. 10, pp. 6253–6260, Oct. 2019.

[39] Candy Goyal, Jagpal Singh Ubhi, and Balwinder Raj, "A Low Leakage CNTFET Based Inexact Full Adder for Low Power Image Processing Applications", International Journal of Circuit Theory and Applications, Wiley, Vol. 47, No. 9, pp. 1446–1458, Sept. 2019.

[40] S. K. Sharma, B. Raj, and M. Khosla, "Enhanced Photosensivity of Highly Spectrum Selective Cylindrical Gate In1-xGaxAs Nanowire MOSFET Photodetector", Modern Physics Letter-B, Vol. 33, No. 12, p. 1950144, 2019.

[41] Jeetendra Singh and Balwinder Raj, "Design and Investigation of 7T2M NVSARM with Enhanced Stability and Temperature Impact on Store/Restore Energy", IEEE Transactions on Very Large Scale Integration Systems, Vol. 27, No. 6, pp. 1322–1328, June 2019.

[42] Anil Kumar Bhardwaj, Sumeet Gupta, Balwinder Raj, and Amandeep Singh, "Impact of Double Gate Geometry on the Performance of Carbon Nanotube Field Effect Transistor Structures for Low Power Digital Design", Computational and Theoretical Nanoscience, ASP, Vol. 16, pp. 1813–1820, 2019.

[43] Candy Goyal, Jagpal Singh Ubhi, and Balwinder Raj, "Low Leakage Zero Ground Noise Nanoscale Full Adder using Source Biasing Technique", Journal of Nanoelectronics and Optoelectronics, American Scientific Publishers, Vol. 14, pp. 360–370, Mar. 2019.

[44] Amandeep Singh, Mamta Khosla, and Balwinder Raj, "Design and Analysis of Dynamically Configurable Electrostatic Doped Carbon Nanotube Tunnel FET", Microelectronics Journal, Elsevier, Vol. 85, pp. 17–24, Mar. 2019.

[45] Candy Goyal, Jagpal Singh Ubhi, and Balwinder Raj, "A Reliable Leakage Reduction Technique for Approximate Full Adder with Reduced Ground Bounce Noise", Journal of Mathematical Problems in Engineering, Hindawi, Vol. 2018, Article ID 3501041, 16 pages, 15 Oct. 2018.

[46] Girish Wadhwa and Balwinder Raj, "Label Free Detection of Biomolecules Using Charge-Plasma-Based Gate Underlap Dielectric Modulated Junctionless TFET", Journal of Electronic Materials (JEMS), Springer, Vol. 47, No. 8, pp. 4683–4693, Aug. 2018.

[47] Girish Wadhwa and Balwinder Raj, "Parametric Variation Analysis of Charge-Plasma-based Dielectric Modulated JLTFET for Biosensor Application", IEEE Sensor Journal, Vol. 18, No. 15, 1 Aug. 2018.

[48] Divya Yadav, Shailesh Singh Chouhan, Santosh Kumar Vishvakarma, and Balwinder Raj, "Application Specific Microcontroller Design for IoT based WSN", Sensor Letter, ASP, Vol. 16, pp. 374–385, May 2018.

[49] Gurmohan Singh, R. K. Sarin, and Balwinder Raj, "Fault-Tolerant Design and Analysis of Quantum-Dot Cellular Automata Based Circuits", IEEE/IET Circuits, Devices & Systems, Vol. 12, pp. 638–664, 2018.

[50] Jeetendra Singh and Balwinder Raj, "Modeling of Mean Barrier Height Levying Various Image Forces of Metal Insulator Metal Structure to Enhance the Performance of Conductive Filament Based Memristor Model", IEEE Nanotechnology, Vol. 17, No. 2, pp. 268–267, Mar. 2018 (SCI).

[51] Aakash Jain, Sanjeev Sharma, and Balwinder Raj, "Analysis of Triple Metal Surrounding Gate (TM-SG) III-V Nanowire MOSFET for Photosensing Application", Opto-electronics Journal, Elsevier, Vol. 26, No. 2, pp. 141–148, May 2018.

[52] Neeraj Jain and Balwinder Raj, "Parasitic Capacitance and Resistance Model Development and Optimization of Raised Source/Drain SOI FinFET Structure for Analog Circuit Applications", Journal of Nanoelectronics and Optoelectronics, ASP, USA, Vol. 13, pp. 531–539, Apr. 2018.

Index